Functional and Impulsive
Differential Equations of Fractional Order

Qualitative Analysis and Applications

Functional and Impulsive Differential Equations of Fractional Order

Qualitative Analysis and Applications

Edited by
Nevijo Zdolec
University of Zagreb, Faculty of Veterinary Medicine
Department of Hygiene
Technology and Food Safety Heinzelova 55
10000 Zagreb, Croatia

Ivanka M. Stamova and Gani Tr. Stamov
The University of Texas at San Antonio
Texas, USA

CRC Press
Taylor & Francis Group
Boca Raton London New York

CRC Press is an imprint of the
Taylor & Francis Group, an **informa** business
A SCIENCE PUBLISHERS BOOK

CRC Press
Taylor & Francis Group
6000 Broken Sound Parkway NW, Suite 300
Boca Raton, FL 33487-2742

First issued in paperback 2020

© 2017 by Taylor & Francis Group, LLC
CRC Press is an imprint of Taylor & Francis Group, an Informa business

No claim to original U.S. Government works

ISBN-13: 978-1-4987-6483-4 (hbk)
ISBN-13: 978-0-367-78272-6 (pbk)

Visit the Taylor & Francis Web site at
http://www.taylorandfrancis.com

and the CRC Press Web site at
http://www.crcpress.com

*This book is dedicated
to Trayan and Alex*

Preface

This book is an exposition of the most recent results related to the qualitative analysis of a variety of fractional-order equations.

The mathematical investigations of derivatives of non-integer order mark their beginning with the correspondence between Leibniz and L'Hospital in 1695 (see [Leibniz 1695], [Podlubny 1999]). Although fractional calculus has a more than three centuries long history, the subject of fractional differential equations has gained considerable popularity and importance during the past few decades, due mainly to its demonstrated applications to many real-world phenomena studies in physics, mechanics, chemistry, engineering, finance, etc.

There are several excellent books devoted to equations of fractional order. The book written by I. Podlubny (see [Podlubny 1999]) played an outstanding role in the development of the theory of fractional ordinary differential equations. It is the most cited and used book which is entirely devoted to a systematic presentation of the basic ideas and methods of fractional calculus in the theory and applications of such equations. The monograph of Kilbas, Srivastava and Trujillo (2006) provides uptodate developments on fractional differential and fractional integro-differential equations involving many different potentially useful operators of fractional calculus. In addition to these, the books of Abbas, Benchohra and N'Guérékata (2012), Baleanu, Diethelm, Scalas and Trujillo (2012), Caponetto, Dongola, Fortuna and Petráš (2010), Das (2011), Diethelm (2010), Hilfer (2000), Kiryakova (1994), Lakshmikantham, Leela and Vasundhara Devi (2009), Magin (2006), Miller and Ross (1993) are good sources for the theory of fractional operators and equations, as well as their numerous applications.

On the other hand, relatively recently functional differential equations of fractional order have started to receive an increasing interest [Baleanu, Sadati, Ghaderi, Ranjbar, Abdeljawad and Jarad 2010], [Banaś and Zając 2011], [Benchohra, Henderson, Ntouyas and Ouahab 2008], [El-Sayed, Gaaraf and

Hamadalla 2010], [Henderson and Ouahab 2009], [Lakshmikantham 2008]. Indeed, fractional operators are a very natural tool to model memory-dependent phenomena. Also, fractional calculus has been incorporated into impulsive differential equations [Ahmad and Nieto 2011], [Ahmad and Sivasundaram 2010], [Anguraj and Maheswari 2012], [Cao and Chen 2012], [Chauhan and Dabas 2011], [Fečkan, Zhou and Wang 2012]. Many interesting results on the fundamental theory of such equations have been reported. However, due to the lack of a book on these topics, many researchers remain unaware of this field.

The questions related to the qualitative theory of different classes of differential equations are the age-old problems of a great importance. The methods used in the qualitative investigation of their solutions, and their wide applications have all advanced to the extent that aspects in each of these areas have demanded individual attention.

The survey published in 2011 by Li and Zhang (see [Li and Zhang 2011]), is a very good overview on the recent stability results of fractional differential equations without impulses and delays, and the analytical methods used. It is seen that, at the time, a few stability results rely on a restrictive modeling of fractional differential systems: the basic hypothesis deals with commensurability, i.e. the fractional derivative orders have to be an integer multiple of minimal fractional order. In the last decades, many researchers have more interests in the stability of linear systems and some methods have emerged in succession. For example, there are the frequency domain methods, Linear Matrix Inequalities methods, and conversion methods [Li and Zhang 2011]. By contrast, the development of stability of nonlinear fractional differential systems even without impulsive perturbations was a bit slow.

During the last few years the authors' research in the area of the qualitative theory of different classes of functional and impulsive fractional-order equations have undergone rapid development. A string of extensive results on the stability, boundedness, asymptotic behavior and almost periodicity for these classes of equations have been obtained. The primary aim of this book is to gather under one cover many of these results which will be of prime importance for researchers on the topic. It fills a void by making available a source book which describes existing literature on the topic, methods and their development. The second motivation comes from the applicable point of view, since the qualitative properties have significant practical applications in the emerging areas such as optimal control, biology, mechanics, medicine, bio-technologies, electronics, economics, etc. For the applied scientists it is important to have an introduction to the qualitative theory of fractional equations, which could help in their initial steps to adopt the results and methods in their research.

The book consists of four chapters. It presents results for different classes of fractional equations, including fractional functional differential equations, fractional impulsive differential equations, fractional impulsive functional differential equations, which have not been covered by other books. It shows the manifestations of different constructive methods by demonstrating how these effective techniques can be applied to investigate qualitative properties of the solutions of fractional systems. Since many applications are also included, the demonstrated techniques and models can be used in training of students in mathematical modeling and as an instigation in the study and development of fractional-order models.

The book is addressed to a wide audience of professionals such as mathematicians, applied researchers and practitioners.

The authors are extremely grateful and very much indebted to Dr. Sandy Norman, Chair of the Department of Mathematics at the University of Texas at San Antonio for ensuring the opportunity for successful work on this book. In addition, the authors have the pleasure to express their sincere gratitude to all their co-authors, the work with whom expanded their experience. They are also thankful to all friends, colleagues and reviewers for their valuable comments and suggestions during the preparation of the manuscript.

I. M. Stamova and G. Tr. Stamov
Corresponding author: ivanka.stamova@utsa.edu

Contents

Chapter 1

Introduction

The present chapter will deal with an introduction of functional and impulsive equations of fractional order and the basic theory necessary for their qualitative investigations.

Section 1.1 will offer the main classes of fractional equations, investigated in the book. Some results from fractional calculus will be also discussed.

Section 1.2 is devoted to definitions of qualitative properties of fractional equations that are used throughout the book.

Section 1.3 will deal with different classes of Lyapunov functions. The definitions of their Caputo and Riemann–Liouville fractional derivatives will be also given.

Finally, in Section 1.4, fractional scalar and vector comparison results will be considered in terms of Lyapunov-like functions. Some auxiliary lemmas for Caputo and Riemann–Liouville fractional derivatives are also presented.

1.1 Preliminary Notes

In this section we shall make a brief description of the main classes of fractional equations that will be used in the book.

Let

$$\Gamma(z) = \int_0^\infty e^{-t} t^{z-1} dt$$

be the *Gamma function* which converges in the right-half of the complex plane $Re(z) > 0$. There are many excellent books (see, for example [Diethelm 2010] and [Podlubny 1999]) in which the definition and main properties of the Gamma function are given, and here we will avoid a repetition.

1

We define the *fractional integral* of order α on the interval $[a,t]$ by

$$_a\mathscr{D}_t^{-\alpha}l(t) = \frac{1}{\Gamma(\alpha)} \int_a^t (t-s)^{\alpha-1}l(s)ds,$$

where $0 < \alpha < 1$ and l is an arbitrary integrable function (see [Gelfand and Shilov 1959], [Podlubny 1999]).

For an arbitrary real number p we denote the *Riemann–Liouville* and *Caputo* fractional derivatives of order p, respectively, as

$$_aD_t^p l(t) = \frac{d^{\lceil p \rceil}}{dt^{\lceil p \rceil}} \left[{_a\mathscr{D}_t^{-(\lceil p \rceil - p)}} l(t) \right]$$

and

$$_a^cD_t^p l(t) = {_a\mathscr{D}_t^{-(\lceil p \rceil - p)}} \left[\frac{d^{\lceil p \rceil}}{dt^{\lceil p \rceil}} l(t) \right],$$

where $\lceil p \rceil$ stands for the smallest integer not less than p; and D and ^cD denote the Riemann–Liouville and Caputo fractional operators, respectively.

For the case $0 < p < 1$, which is investigated in our book, the relation between the two fractional derivatives is given by

$$_a^cD_t^p l(t) = {_aD_t^p} \left[l(t) - l(a) \right].$$

A main difference between both fractional derivatives is that the Caputo derivative of a constant C is zero, i.e.

$$_a^cD_t^p C = 0,$$

while the Riemann–Liouville fractional derivative of a constant C is given by

$$_aD_t^p C = \frac{C}{\Gamma(1-p)} (t-a)^{-p}.$$

Therefore,

$$_a^cD_t^p l(t) = {_aD_t^p} l(t) - \frac{l(a)}{\Gamma(1-p)} (t-a)^{-p}.$$

In particular, if $l(a) = 0$, then

$$_a^cD_t^p l(t) = {_aD_t^p} l(t).$$

1.1.1 Fractional Functional Differential Equations

It is well-known that the delay differential equations of integer order are of great theoretical interest and form an important class as regards to their applications (see [Burton 1985], [Hale 1977], [Kolmanovskii and Myshkis 1999],

[Kolmanovskii and Nosov 1986] and the references therein). The centrality of functional differential equations for theory and applications is witnessed by the current persistency of new contributions in this topic of interest [Bernfeld, Corduneanu and Ignatyev 2003], [Corduneanu and Ignatyev 2005], [Stamova and Stamov G 2014a]. Since 1960 many generalizations and extensions of such equations became a part of the literature ([Stamov 2012], [Stamova 2009]).

It is also known that the functional differential equations of fractional order have several applications. It has been proved that such type of equations are valuable tools in the modeling of many phenomena in various fields of engineering, physics and economics (see [Babakhani, Baleanu and Khanbabaie 2012], [Bhalekar, Daftardar-Gejji, Baleanu and Magin 2011], [El-Sayed, Gaaraf and Hamadalla 2010], [Henderson and Ouahab 2009], [Stamova 2016], [Stamova and Stamov G 2013], [Wang, Huang and Shi 2011], [Wang, Yu and Wen 2014], [Wu, Hei and Chen 2013]). The efficient applications of fractional functional differential equations requires the finding of criteria for qualitative properties of their solutions.

Let $\mathbb{R}^n$ be the n-dimensional Euclidean space with norm $||.||$ and Ω be an open set in $\mathbb{R}^n$ containing the origin, $\mathbb{R}_+ = [0,\infty)$.

For a given $r > 0$ we suppose that $x \in C([t_0 - r, t_0 + A], \mathbb{R}^n)$, $t_0 \in \mathbb{R}$, $A > 0$. For any $t \in [t_0, t_0 + A)$, we denote by x_t an element of $C[[-r,0], \Omega]$ defined by $x_t(\theta) = x(t + \theta)$, $-r \leq \theta \leq 0$.

Consider the system of fractional functional differential equations

$$
{}_{t_0}\mathscr{D}_t^\alpha x(t) = f(t, x_t), \tag{1.1}
$$

where $f : \mathbb{R} \times C[[-r,0], \Omega] \to \mathbb{R}^n$ and ${}_{t_0}\mathscr{D}_t^\alpha$ denotes either Caputo or Riemann–Liouville fractional derivative of order α, $0 < \alpha < 1$. The class of equations (1.1) provides mathematical models for real-world problems in which the fractional rate of change depends on the influence of their hereditary effects.

Instead of an initial point value for an ordinary differential equation, for an initial condition related to (1.1) an initial function is required which is defined over the range of time delimited by the delay.

Let $\varphi_0 \in C[[-r,0], \Omega]$. We shall denote by $x(t) = x(t; t_0, \varphi_0)$ the solution of (1.1) with initial data $(t_0, \varphi_0) \in \mathbb{R} \times C[[-r,0], \Omega]$.

If the fractional derivative ${}_{t_0}\mathscr{D}_t^\alpha = {}_{t_0}^c D_t^\alpha$, then the initial condition is of the same type as for an integer-order equation:

$$
x_{t_0} = \varphi_0. \tag{1.2}
$$

We shall assume that $f(t, \varphi)$ is smooth enough on $\mathbb{R} \times C[[-r,0], \Omega]$ to guarantee the existence and uniqueness of a solution $x(t; t_0, \varphi_0)$ of the initial value problem (IVP) (1.1), (1.2) on $[t_0 - r, \infty)$ for each initial function $\varphi_0 \in C[[-r,0], \Omega]$.

Then, the IVP (1.1), (1.2) is equivalent to the following Volterra fractional integral with memory (see [Lakshmikantham 2008]):

$$x_{t_0} = \varphi_0,$$

$$x(t) = \varphi_0(0) + \frac{1}{\Gamma(\alpha)} \int_{t_0}^{t} (t-s)^{\alpha-1} f(s, x_s) ds, \ t_0 \leq t < \infty.$$

For some basic concept and theorems on the fundamental theory of such equations, we refer readers to [Benchohra, Henderson, Ntouyas and Ouahab 2008], [Henderson and Ouahab 2009], [Lakshmikantham 2008], and [Su and Feng 2012].

It was believed that the case $_{t_0}\mathscr{D}_t^\alpha = {_{t_0}}D_t^\alpha$ leads to initial conditions without physical meaning. This was contradicted by Heymans and Podlubny (2005) who studied several cases and gave physical meaning to the initial conditions of fractional differential equations with Riemann–Liouville derivatives. See also [Lazarević and Spasić 2009], [Li and Zhang 2011] and [Ortigueira and Coito 2010]. It is known that for a zero initial function the Riemann–Liouville and Caputo fractional derivatives coincide.

The system (1.1) is an universal type of fractional functional differential system. In the particular case, it contains a system of fractional differential equations when $r = 0$. Systems of the type (1.1) also include the following systems of fractional functional differential equations:

■ systems of fractional differential-difference equations of the type

$$_{t_0}\mathscr{D}_t^\alpha x(t) = F(t, x(t), x(t-r_1(t)), x(t-r_2(t)), ..., x(t-r_p(t))),$$

where $0 \leq r_j(t) \leq r, \ j = 1, 2, ..., p$;

■ systems of fractional integro-differential equations of the type

$$_{t_0}\mathscr{D}_t^\alpha x(t) = \int_{-r}^{0} g(t, s, x(t+\theta)) d\theta;$$

■ systems of fractional integro-differential equations with infinite delays

$$_{t_0}\mathscr{D}_t^\alpha x(t) = \int_{-\infty}^{t} k(t, \theta) f(t, x(\theta)) d\theta,$$

where $k : \mathbb{R}^2 \to \mathbb{R}_+$ is continuous.

The system (1.1) is *linear*, when $f(t, x_t) = L(t, x_t) + h(t)$, where $L(t, x_t)$ is linear with respect to x_t.

Some systems of more general types are also included in the type (1.1) system.

1.1.2 Fractional Impulsive Differential Equations

The states of many evolutionary processes are often subjects to instantaneous perturbations and experience abrupt changes at certain moments of time. The duration of the changes is very short and negligible in comparison with the duration of the process considered, and can be thought of as "momentarily" changes or as impulses. Systems with short-term perturbations are often naturally described by impulsive differential equations (see for example [Bainov and Simeonov 1989], [Benchohra, Henderson and Ntouyas 2006], [Lakshmikantham, Bainov and Simeonov 1989], [Mil'man and Myshkis 1960], [Pandit and Deo 1982], [Samoilenko and Perestyuk 1995]). Such differential equations have became an active research subject in nonlinear science and have attracted more attention in many fields. For example, important impulsive differential equations have been introduced as mathematical models in population ecology (see [Ahmad and Stamov 2009a], [Ahmad and Stamova 2007a], [Ahmad and Stamova 2012], [Ahmad and Stamova 2013], [Ballinger and Liu 1997], [Dong, Chen and Sun 2006], [Georgescu and Zhang 2010], [Hu 2013], [Jiang and Lu 2007], [Liu and Rohlf 1998], [Stamov 2008], [Stamov 2009a], [Stamov 2012], [Xue, Wang and Jin 2007], [Yan 2003], [Yan 2010]), hematopoiesis (see [Kou, Adimy and Ducrot 2009]), pest control (see [Wang, Chen and Nieto 2010]), drug treatment (see [Liu, Yu and Zhu 2008], [Smith and Wahl 2004]), the chemostat (see [Sun and Chen 2007]) tumor-normal cell interaction (see [Dou, Chen and Li 2004]), plankton allelopathy (see [He, Chen and Li 2010]), vaccination (see [D'Onofrio 2002], [Qiao, Liu and Forys 2013]), radio engineering (see [Joelianto and Sutarto 2009]), communication security (see [Khadra, Liu and Shen 2003]), neural networks (see [Arbib 1987], [Stamov 2004], [Stamov 2009d], [Stamov 2012], [Zhao, Xia and Ding 2008]), etc. Also, in optimal control of economic systems, frequency-modulated signal processing systems, and some flying object motions, many systems are characterized by abrupt changes in their states at certain instants (see [Korn 1999], [Stamova and Stamov 2011], [Stamova and Stamov 2012], [Stamova and Stamov A 2013], [Stamova, Stamov and Simeonova 2013], [Yang 2001]). This type of impulsive phenomenon can also be found in the fields of information science, electronics, automatic control systems, computer networks, artificial intelligence, robotics and telecommunications. Many sudden and sharp changes occur instantaneously in these systems, in the form of impulses which cannot be well described by a pure continuous–time or discrete–time model [Li, Liao, Yang and Huang 2005].

A great progress in studying impulsive functional differential equations and their applications has also been made (see [Ahmad and Stamov 2009b], [Ahmad and Stamova 2007b], [Ahmad and Stamova 2008], [Hui and Chen 2005], [Li and Fan 2007], [Liu and Ballinger 2002], [Liu, Huang and Chen 2012], [Liu and Takeuchi 2007], [Liu, Teo and Hu 2005], [Liu and Wang 2007], [Liu and

Zhao 2012], [Long and Xu 2008], [Luo and Shen 2001], [Stamov 2009b], [Stamov 2009c], [Stamov 2010a], [Stamov 2010b], [Stamov 2012], [Stamov, Alzabut, Atanasov and Stamov 2011], [Stamov and Stamov 2013], [Stamov and Stamov 2001], [Stamov and Stamova 2007], [Stamova 2007], [Stamova 2008], [Stamova 2009], [Stamova 2010], [Stamova 2011a], [Stamova 2011b], [Stamova, Emmenegger and Stamov 2010], [Stamova, Ilarionov and Vaneva 2010], [Stamova and Stamov 2011], [Stamova and Stamov 2012], [Stamova and Stamov A 2013], [Stamova and Stamov T 2014a], [Stamova and Stamov T 2014b], [Stamova, Stamov and Li 2014], [Stamova, Stamov and Simeonova 2013], [Stamova, Stamov and Simeonova 2014], [Teng, Nie and Fang 2011], [Wang, Yu and Niu 2012], [Widjaja and Bottema 2005], [Xia 2007], [Zhou and Wan 2009]). Indeed, impulsive mathematical models with delays are found in almost every domain of applied sciences and they played a very important role in modern mathematical modelling of processes and phenomena studied in physics, population dynamics, chemical technology and economics.

Since, the tools of impulsive fractional differential equations are applicable to various fields of study, the investigation of the theory of such equations has been started quite recently (see [Ahmad and Nieto 2011], [Ahmad and Sivasundaram 2010], [Bai 2011], [Benchohra and Slimani 2009], [Cao and Chen 2012], [Chauhan and Dabas 2011], [Fečkan, Zhou and Wang 2012], [Kosmatov 2013], [Li, Chen and Li 2013], [Liu 2013], [Mahto, Abbas and Favini 2013], [Mophou 2010], [Rehman and Eloe 2013], [Stamov and Stamova 2014a], [Stamov and Stamova 2015b], [Stamova 2014a], [Stamova 2015], [Tariboon, Ntouyas and Agarwal 2015], [Wang, Ahmad and Zhang 2012], [Wang, Fečkan and Zhou 2011], [Wang and Lin 2014], [Wang, Zhou and Fečkan 2012], [Zhang, Zhang and Zhang 2014]). Also, in relation to the mathematical simulation in chaos, fluid dynamics and many physical systems, the investigation of impulsive fractional functional differential equations began (see [Anguraj and Maheswari 2012], [Chang and Nieto 2009], [Chen, Chen and Wang 2009], [Debbouche and Baleanu 2011], [Gao, Yang and Liu 2013], [Guo and Jiang 2012], [Mahto and Abbas, 2013], [Mahto, Abbas and Favini 2013], [Stamov 2015], [Stamov and Stamova 2014b], [Stamov and Stamova 2015a], [Stamova 2014b], [Stamova and Stamov G 2014b], [Wang 2012], [Xie 2014]).

We shall consider the following impulsive systems of fractional order:

I. *Impulsive systems of fractional differential equations.* Let $f : \mathbb{R} \times \Omega \to \mathbb{R}^n$, $\tau_k : \Omega \to \mathbb{R}$, $I_k : \Omega \to \mathbb{R}^n$, $k = \pm 1, \pm 2, ...$ Consider the impulsive fractional-order differential system

$$\begin{cases} {}_{t_0}^c D_t^\alpha x(t) = f(t, x(t)), \, t \neq \tau_k(x(t)), \\ \Delta x(t) = I_k(x(t)), \, t = \tau_k(x(t)), \, k = \pm 1, \pm 2, ..., \end{cases} \tag{1.3}$$

where $\Delta x(t) = x(t^+) - x(t)$.

The second part of the impulsive system (1.3) is called a *jump condition*. The functions I_k, that define the magnitudes of the impulsive perturbations, are the *impulsive functions*.

Let $t_0 \in \mathbb{R}$, $x_0 \in \Omega$. Denote by $x(t) = x(t; t_0, x_0)$ the solution of system (1.3) satisfying the initial condition

$$x(t_0^+; t_0, x_0) = x_0. \tag{1.4}$$

Note that, instead of the initial condition $x(t_0) = x_0$, we have imposed the limiting condition $x(t_0^+) = x_0$ which, in general, is natural for equation (1.3) since (t_0, x_0) may be such that $t_0 = \tau_k(x_0)$ for some k. Whenever $t_0 \neq \tau_k(x_0)$, for all k, we shall understand the initial condition $x(t_0^+) = x_0$ in the usual sense, i.e., $x(t_0) = x_0$.

We suppose that the functions f and I_k, τ_k, $k = \pm 1, \pm 2, ...$, are smooth enough on $\mathbb{R} \times \Omega$ and Ω, respectively, to guarantee existence, uniqueness and continuability of the solution $x(t) = x(t; t_0, x_0)$ of the equation (1.3) on the interval $[t_0, \infty)$ for all suitable initial data $x_0 \in \Omega$ and $t_0 \in \mathbb{R}$. We also assume that the functions $(E + I_k) : \Omega \to \Omega$, $k = \pm 1, \pm 2, ...$, where E is the identity in Ω.

In some qualitative investigations, we shall also consider such solutions of system (1.3) for which the continuability to the left of t_0 is guaranteed. The solutions $x(t; t_0, x_0)$ of system (1.3) are, in general, piecewise continuous functions with points of discontinuity of the first kind at which they are left continuous, i.e., at the moment t_{l_k} when the integral curve of the solution meets the hypersurfaces

$$\sigma_k = \left\{ (t, x) : t = \tau_k(x), \; x \in \Omega \right\}$$

the following relations are satisfied:

$$x(t_{l_k}^-) = x(t_{l_k}) \quad \text{and} \quad x(t_{l_k}^+) = x(t_{l_k}) + I_k(x(t_{l_k})).$$

The points t_{l_k}, at which the impulses occur, are the *moments of impulsive effect*. The impulsive moments for a solution of (1.3) depend on the solution, i.e. different solutions have different points of discontinuity. In general, $k \neq l_k$. In other words, it is possible that the integral curve of the IVP (1.3), (1.4) does not meet the hypersurface σ_k at the moment t_k. This leads to a number of difficulties in the investigation of such systems. We shall assume that for each $x \in \Omega$ and $k = \pm 1, \pm 2, ...,$

$$\tau_k(x) < \tau_{k+1}(x)$$

and

$$\tau_k(x) \to \infty \text{ as } k \to \infty \, (\tau_k(x) \to -\infty \text{ as } k \to -\infty), \text{ uniformly on } x \in \Omega,$$

and the integral curve of each solution of the system (1.3) meets each of the hypersurfaces $\{\sigma_k\}$ at most once.

The above condition means absence of the phenomenon "beating" of the solutions to the system (1.1), i.e. a phenomenon when a given integral curve meets more than once or infinitely many times one and the same hypersurface $\{\sigma_k\}$ (see [Bainov and Dishliev 1997], [Bainov and Simeonov 1989], [Lakshmikantham, Bainov and Simeonov 1989], [Samoilenko and Perestyuk 1995]). A part of the difficulty in the investigations of systems with variable impulsive perturbations is related to the possibilities of "merging" of different integral curves after a given moment, loss of the property of autonomy, etc. For more results about such systems we refer the reader to [Afonso, Bonotto, Federson and Gimenes 2012], [Liu and Ballinger 2002], [Stamov 2012], [Stamova 2009], [Wang and Fu 2007].

Note that, the phenomenon "beating" is not present in the case, when $\tau_k(x) \equiv t_k$, $x \in \Omega$, i.e. when the impulses are realized at fixed moments $t = t_k$, $k = \pm 1, \pm 2, \dots$. Then the system (1.3) reduces to

$$
\begin{cases}
{}^c_{t_0}D^\alpha_t x(t) = f(t, x(t)), \, t \neq t_k, \\
\Delta x(t) = I_k(x(t)), \, t = t_k, \, k = \pm 1, \pm 2, \dots,
\end{cases}
\tag{1.5}
$$

where $t_k < t_{k+1}$, $k = \pm 1, \pm 2, \dots$ and $\lim\limits_{k \to \pm\infty} t_k = \pm\infty$.

It is clear that systems of impulsive differential equations with fixed moments of impulse effect (1.5) can be considered as a particular case of the systems with variable impulsive perturbations (1.3).

II. *Impulsive systems of fractional functional differential equations.* Let $J \subseteq \mathbb{R}$. Define the following class of functions:

$PC[J, \Omega] = \{\sigma : J \to \Omega : \sigma(t) \text{ is a piecewise continuous function with points}$ of discontinuity $\tilde{t} \in J$ at which $\sigma(\tilde{t}^-)$ and $\sigma(\tilde{t}^+)$ exist and $\sigma(\tilde{t}^-) = \sigma(\tilde{t})\}$.

In the present book we shall consider impulsive systems of Caputo-type fractional functional differential equations with fixed moments of impulse effect. A system of this class is written as follows:

$$
\begin{cases}
{}^c_{t_0}D^\alpha_t x(t) = f(t, x_t), \, t \neq t_k, \\
\Delta x(t) = I_k(x(t)), \, t = t_k, \, k = \pm 1, \pm 2, \dots,
\end{cases}
\tag{1.6}
$$

where $f : \mathbb{R} \times PC[[-r, 0], \Omega] \to \mathbb{R}^n$, $I_k : \Omega \to \mathbb{R}^n$, $t_k < t_{k+1}$, $k = \pm 1, \pm 2, \dots$ and $\lim\limits_{k \to \pm\infty} t_k = \pm\infty$.

Let $\varphi_0 \in PC[[-r, 0], \Omega]$. Denote by $x(t) = x(t; t_0, \varphi_0)$ the solution of system (1.6) satisfying the initial conditions

$$
\begin{cases}
x(t) = \varphi_0(t - t_0), \, t_0 - r \leq t \leq t_0, \\
x(t_0^+) = \varphi_0(0),
\end{cases}
\tag{1.7}
$$

and by $J^+(t_0, \varphi_0)$ – the maximal interval of type $[t_0, \gamma)$ in which the solution $x(t; t_0, \varphi_0)$ is defined.

We also suppose that the functions f and I_k, $k = \pm 1, \pm 2, ...$, are smooth enough on their domains to guarantee existence, uniqueness and continuability of the solution $x(t) = x(t; t_0, \varphi_0)$ of the IVP (1.6), (1.7) on the interval $[t_0 - r, \infty)$ for each $\varphi_0 \in PC[[-r, 0], \Omega]$ and $t_0 \in \mathbb{R}$.

Without loss of generality we can assume that ... $t_{-1} < t_0 < t_1 < t_2 < ... < t_k < t_{k+1} < ....$ Then, the solution $x(t) = x(t; t_0, \varphi_0)$ of the initial value problem (1.6), (1.7) is characterized by the following:

1. For $t_0 - r \leq t \leq t_0$ the solution $x(t)$ satisfies the initial conditions (1.7).
2. For $t_0 < t \leq t_1$, $x(t)$ coincides with the solution of the problem

$$
\begin{cases}
{}_{t_0}^{c}D_t^{\alpha} x(t) = f(t, x_t), \, t > t_0, \\
x_{t_0} = \varphi_0(\theta), \, -r \leq \theta \leq 0, \\
x(t_0^+) = \varphi_0(0).
\end{cases}
$$

At the moment $t = t_1$ the mapping point $(t, x(t; t_0, \varphi_0))$ of the extended phase space jumps momentarily from the position $(t_1, x(t_1; t_0, \varphi_0))$ to the position

$$
(t_1, x(t_1; t_0, \varphi_0) + I_1(x(t_1; t_0, \varphi_0))).
$$

3. For $t_1 < t \leq t_2$ the solution $x(t)$ coincides with the solution of

$$
\begin{cases}
{}_{t_0}^{c}D_t^{\alpha} y(t) = f(t, y_t), \, t > t_1, \\
y_{t_1} = \varphi_1, \, \varphi_1 \in PC[[-r, 0], \Omega],
\end{cases}
$$

where

$$
\varphi_1(t - t_1) = \begin{cases}
\varphi_0(t - t_1), \, t \in [t_0 - r, t_0] \cap [t_1 - r, t_1], \\
x(t; t_0, \varphi_0), \, t \in (t_0, t_1) \cap [t_1 - r, t_1], \\
x(t; t_0, \varphi_0) + I_1(x(t; t_0, \varphi_0)), \, t = t_1.
\end{cases}
$$

At the moment $t = t_2$ the mapping point $(t, x(t))$ jumps momentarily, etc.

Thus the solution $x(t; t_0, \varphi_0)$ of problem (1.6), (1.7) is a piecewise continuous function for $t > t_0$ with points of discontinuity of the first kind $t = t_k$, $k = \pm 1, \pm 2, ...$ at which it is continuous from the left, i.e. the following relations are satisfied

$$
x(t_k^-) = x(t_k), \, t_k > t_0, \, k = \pm 1, \pm 2, ...,
$$

$$
x(t_k^+) = x(t_k) + I_k(x(t_k)), \, t_k > t_0, \, k = \pm 1, \pm 2, ...
$$

For equations with Riemann-Liouville fractional derivatives suitable initial data and jump conditions will be used.

Remark 1.1 Efficient existence and uniqueness criteria for impulsive fractional functional systems are given in [Abbas, Benchohra and N'Guérékata 2012], [Anguraj and Maheswari 2012], [Chen, Chen and Wang 2009], [Mahto, Abbas and Favini 2013], [Wang 2012] and we have not included them here. We shall suppose that the functions f and I_k in (1.6) satisfy all conditions that guarantee the existence and uniqueness of the solutions.

1.2   Definitions of Qualitative Properties

This section will offer Lyapunov stability, Mittag–Leffler stability and boundedness definitions, as well as definitions for almost periodic sequences and almost periodic functions that are used throughout the book. In several of the subsequent chapters some extensions of these definitions will be also introduced.

First, we shall introduce Lyapunov stability definitions for system (1.1). For $\varphi \in C[[-r,0],\Omega]$, we equip the space $C[[-r,0],\Omega]$ with the norm $||.||_r$ defined by $||\varphi||_r = \max\limits_{-r \leq \theta \leq 0} ||\varphi(\theta)||$, where $||.||$ is any convenient norm in $\mathbb{R}^n$.

Let $\bar{x}(t) = \bar{x}(t;t_0,\phi_0)$ be a solution of system (1.1) with a suitable initial function $\phi_0 \in C[[-r,0],\Omega]$. We note that a constant solution $\bar{x}^*$ is an equilibrium of system (1.1), if and only if

$$_{t_0}\mathscr{D}_t^\alpha \bar{x}^* = f(t,\bar{x}^*).$$

When zero $\bar{x}(t) \equiv 0$ is an equilibrium of (1.1), we shall use the following stability definition:

Definition 1.1 The zero solution $\bar{x}(t) \equiv 0$ of system (1.1) is said to be:

(a) *stable*, if
$$(\forall t_0 \in \mathbb{R})(\forall \varepsilon > 0)(\exists \delta = \delta(t_0,\varepsilon) > 0)$$
$$(\forall \phi_0 \in C[[-r,0],\Omega] : ||\phi_0||_r < \delta)(\forall t \geq t_0) : ||x(t;t_0,\phi_0)|| < \varepsilon;$$

(b) *uniformly stable*, if the number δ in (a) is independent of $t_0 \in \mathbb{R}$;

(c) *attractive*, if
$$(\forall t_0 \in \mathbb{R})(\exists \lambda = \lambda(t_0) > 0)(\forall \phi_0 \in C[[-r,0],\Omega] : ||\phi_0||_r < \lambda)$$
$$\lim_{t \to \infty} ||x(t;t_0,\phi_0)|| = 0;$$

(d) *equi-attractive*, if
$$(\forall t_0 \in \mathbb{R})(\exists \lambda = \lambda(t_0) > 0)(\forall \varepsilon > 0)(\exists T = T(t_0,\varepsilon) > 0)$$
$$(\forall \phi_0 \in C[[-r,0],\Omega] : ||\phi_0||_r < \lambda)(\forall t \geq t_0 + T) : ||x(t;t_0,\phi_0)|| < \varepsilon;$$

(e) *uniformly attractive*, if the numbers λ and T in (d) are independent of $t_0 \in \mathbb{R}$;

(f) *asymptotically stable*, if it is stable and attractive;

(g) *uniformly asymptotically stable*, if it is uniformly stable and uniformly attractive.

Next, we shall use the Mittag–Leffler function which plays an important role in the theory of non-integer order differential equations. The standard Mittag–Leffler function (see [Li, Chen and Podlubny 2010], [Podlubny 1999]) is given as

$$E_\alpha(z) = \sum_{\kappa=0}^{\infty} \frac{z^\kappa}{\Gamma(\alpha\kappa+1)},$$

where $\alpha > 0$. It is also common to represent the Mittag–Leffler function in two parameters, α and β, such that

$$E_{\alpha,\beta}(z) = \sum_{\kappa=0}^{\infty} \frac{z^\kappa}{\Gamma(\alpha\kappa+\beta)},$$

where $\alpha > 0$ and $\beta > 0$. For $\beta = 1$, we have $E_\alpha(z) = E_{\alpha,1}(z)$ and $E_{1,1}(z) = e^z$. Moreover, the Laplace transform of Mittag–Leffler function in two parameters is

$$\mathscr{L}[t^{\beta-1}E_{\alpha,\beta}(-\gamma t^\alpha)] = \frac{s^{\alpha-\beta}}{s^\alpha+\gamma}, \quad (\mathscr{R}(s) > |\gamma|^{\frac{1}{\alpha}}),$$

where t and s are, respectively, the variables in the time domain and Laplace domain, $\mathscr{R}(s)$ denotes the real part of s, $\gamma \in \mathbb{R}$, and $\mathscr{L}$ stands for the Laplace transform.

We shall introduce the following definition of the Mittag–Leffler stability of the zero solution of (1.1), which extends the definition given in [Li, Chen and Podlubny 2010].

Definition 1.2 The zero solution $\bar{x}(t) \equiv 0$ of system (1.1) is said to be *Mittag–Leffler stable*, if for $t_0 \in \mathbb{R}$, $\varphi_0 \in C[[-r,0],\Omega]$ and $||\varphi_0||_r < \rho$ there exist constants $\mu > 0$ and $d > 0$ such that

$$||x(t;t_0,\varphi_0)|| \leq \{m(\varphi_0)E_\alpha(-\mu(t-t_0)^\alpha)\}^d, \, t \geq t_0,$$

where $\alpha \in (0,1)$, $m(0) = 0$, $m(\varphi) \geq 0$, and $m(\varphi)$ is Lipschitzian with respect to $\varphi \in C[[-r,0],\Omega]$, $||\varphi||_r < \rho$, $\rho > 0$.

If $y(t)$ is a solution of system (1.1), then y is said to be stable, if the zero solution $\bar{x}(t) \equiv 0$ of the system

$$_{t_0}\mathscr{D}_t^\alpha \bar{x}(t) = f(t,\bar{x}_t+y_t) - f(t,y_t)$$

is stable. The other Lyapunov stability concepts and the Mittag–Leffler stability are defined in a similar manner.

We shall use the next Lyapunov stability definitions for the system (1.3).

Let $\bar{x}(t) = \bar{x}(t;t_0,\bar{x}_0)$ be a solution of system (1.3) with a suitable initial value $\bar{x}_0 \in \Omega$.

Definition 1.3 The solution $\bar{x}(t)$ is said to be:

(a) *stable*, if
$$(\forall t_0 \in \mathbb{R})(\forall \varepsilon > 0)(\exists \delta = \delta(t_0,\varepsilon) > 0)$$
$$(\forall x_0 \in \Omega : ||x_0 - \bar{x}_0)|| < \delta)(\forall t \geq t_0) :$$
$$||x(t;t_0,x_0) - \bar{x}(t)|| < \varepsilon;$$

(b) *uniformly stable*, if the number δ in (a) is independent of $t_0 \in \mathbb{R}$;

(c) *attractive*, if
$$(\forall t_0 \in \mathbb{R})(\exists \lambda = \lambda(t_0) > 0)(\forall x_0 \in \Omega : ||x_0 - \bar{x}_0|| < \lambda)$$
$$\lim_{t \to \infty} ||x(t;t_0,x_0) - \bar{x}(t)|| = 0;$$

(d) *equi-attractive*, if
$$(\forall t_0 \in \mathbb{R})(\exists \lambda = \lambda(t_0) > 0)(\forall \varepsilon > 0)(\exists T = T(t_0,\varepsilon) > 0)$$
$$(\forall x_0 \in \Omega : ||x_0 - \bar{x}_0|| < \lambda)(\forall t \geq t_0 + T) : ||x(t;t_0,x_0) - \bar{x}(t)|| < \varepsilon;$$

(e) *uniformly attractive*, if the numbers λ and T in (d) are independent of $t_0 \in \mathbb{R}$;

(f) *asymptotically stable*, if it is stable and attractive;

(g) *uniformly asymptotically stable*, if it is uniformly stable and uniformly attractive;

(h) *exponentially stable*, if
$$(\exists \lambda > 0)(\forall a > 0)(\exists \gamma = \gamma(a) > 0)(\forall t_0 \in \mathbb{R})$$
$$(\forall x_0 \in \Omega : ||x_0 - \bar{x}_0|| < a)(\forall t \geq t_0) :$$
$$||x(t;t_0,x_0) - \bar{x}(t)|| < \gamma(a)||x_0 - \bar{x}_0|| \exp\{-\lambda(t - t_0)\}.$$

In the case, when $\bar{x}(t) \equiv 0$ is a solution of system (1.5), we shall use the following definition.

Definition 1.4 The zero solution $\bar{x}(t) \equiv 0$ of system (1.5) is said to be:

(a) *stable*, if

$$(\forall t_0 \in \mathbb{R})(\forall \varepsilon > 0)(\exists \delta = \delta(t_0, \varepsilon) > 0)$$

$$(\forall x_0 \in \Omega : ||x_0|| < \delta)(\forall t \geq t_0) : ||x(t; t_0, x_0)|| < \varepsilon;$$

(b) *uniformly stable*, if the number δ in (a) is independent of $t_0 \in \mathbb{R}$;

(c) *attractive*, if

$$(\forall t_0 \in \mathbb{R})(\exists \lambda = \lambda(t_0) > 0)(\forall x_0 \in \Omega : ||x_0|| < \lambda) :$$

$$\lim_{t \to \infty} ||x(t; t_0, x_0)|| = 0;$$

(d) *equi-attractive*, if

$$(\forall t_0 \in \mathbb{R})(\exists \lambda = \lambda(t_0) > 0)(\forall \varepsilon > 0)(\exists T = T(t_0, \varepsilon) > 0)$$

$$(\forall x_0 \in \Omega : ||x_0|| < \lambda)(\forall t \geq t_0 + T) : ||x(t; t_0, x_0)|| < \varepsilon;$$

(e) *uniformly attractive*, if the numbers λ and T in (d) are independent of $t_0 \in \mathbb{R}$;

(f) *asymptotically stable*, if it is stable and attractive;

(g) *uniformly asymptotically stable*, if it is uniformly stable and uniformly attractive.

We shall use the following Mittag–Leffler stability definition of the zero solution of system (1.5) which is analogous to the definitions given in [Li, Chen and Podlubny 2010].

Definition 1.5 The zero solution $\bar{x}(t) \equiv 0$ of system (1.5) is said to be *Mittag–Leffler stable*, if given $\delta > 0$ we have $||x_0|| < \delta$ implies

$$||x(t; t_0, x_0)|| \leq \{m(x_0) E_\alpha(-\mu(t - t_0)^\alpha)\}^d, \ t \geq t_0,$$

where $\alpha \in (0, 1)$, $\mu > 0$, $d > 0$, $m(0) = 0$, $m(x) \geq 0$, and $m(x)$ is locally Lipschitz continuous with respect to $x \in \Omega$, $||x|| < \delta$.

When talking about Lyapunov stability and Mittag–Leffler stability of the solutions of system (1.6), the norm $||.||_r$ is defined by $||\varphi||_r = \sup\limits_{-r \leq \theta \leq 0} ||\varphi(\theta)||$. In the case $r = \infty$ we have $||\varphi||_r = ||\varphi||_\infty = \sup\limits_{\theta \in (-\infty, 0]} ||\varphi(\theta)||$.

Let $\bar{x}(t) = \bar{x}(t; t_0, \phi_0)$ be a solution of system (1.6) with a suitable initial function $\phi_0 \in PC[[-r, 0], \Omega]$. When zero is an equilibrium of (1.6), we shall use the following stability definition.

Definition 1.6 The zero solution $\bar{x}(t) \equiv 0$ of system (1.6) is said to be:

(a) *stable*, if
$$(\forall t_0 \in \mathbb{R})(\forall \varepsilon > 0)(\exists \delta = \delta(t_0, \varepsilon) > 0)$$
$$(\forall \varphi_0 \in PC[[-r,0],\Omega] : ||\varphi_0||_r < \delta)(\forall t \geq t_0) : ||x(t;t_0,\varphi_0)|| < \varepsilon;$$

(b) *uniformly stable*, if the number δ in (a) is independent of $t_0 \in \mathbb{R}$;

(c) *attractive*, if
$$(\forall t_0 \in \mathbb{R})(\exists \lambda = \lambda(t_0) > 0)(\forall \varphi_0 \in PC[[-r,0],\Omega] : ||\varphi_0||_r < \lambda)$$
$$\lim_{t \to \infty} ||x(t;t_0,\varphi_0)|| = 0;$$

(d) *equi-attractive*, if
$$(\forall t_0 \in \mathbb{R})(\exists \lambda = \lambda(t_0) > 0)(\forall \varepsilon > 0)(\exists T = T(t_0,\varepsilon) > 0)$$
$$(\forall \varphi_0 \in PC[[-r,0],\Omega] : ||\varphi_0||_r < \lambda)(\forall t \geq t_0 + T) : ||x(t;t_0,\varphi_0)|| < \varepsilon;$$

(e) *uniformly attractive*, if the numbers λ and T in (d) are independent of $t_0 \in \mathbb{R}$;

(f) *asymptotically stable*, if it is stable and attractive;

(g) *uniformly asymptotically stable*, if it is uniformly stable and uniformly attractive.

Definition 1.7 The zero solution $\bar{x}(t) \equiv 0$ of system (1.6) is said to be *Mittag–Leffler stable*, if for $t_0 \in \mathbb{R}$, $\varphi_0 \in PC[[-r,0],\Omega] : ||\varphi_0||_r < \rho$ ($\rho > 0$) there exist constants $\mu > 0$ and $d > 0$ such that

$$||x(t;t_0,\varphi_0)|| \leq \{m(\varphi_0)E_\alpha(-\mu(t-t_0)^\alpha)\}^d, t \geq t_0,$$

where $\alpha \in (0,1)$, $m(0) = 0$, $m(\varphi) \geq 0$, and $m(\varphi)$ is Lipschitzian with respect to $\varphi \in PC[[-r,0],\Omega] : ||\varphi||_r < \rho$.

In this book, we shall apply the second method of Lyapunov for investigating boundedness of the solutions of system (1.1) for $\Omega \equiv \mathbb{R}^n$, i.e. we shall consider the following system

$$_{t_0}\mathscr{D}_t^\alpha x(t) = f(t,x_t), \tag{1.8}$$

where $f : \mathbb{R} \times C[[-r,0],\mathbb{R}^n] \to \mathbb{R}^n$.

Denote by $\mathscr{C}$ the space of all continuous functions mapping $[-r,0]$ into $\mathbb{R}^n$. Let $\varphi_0 \in \mathscr{C}$. Denote again by $x(t) = x(t;t_0,\varphi_0)$ the solution of system (1.8), satisfying a suitable initial condition. Also, for $\rho > 0$ and $\Omega \equiv \mathbb{R}^n$, define $B_\rho(C) = \{\varphi \in \mathscr{C} : ||\varphi||_r < \rho\}$.

We shall use the following boundedness definition.

Definition 1.8 We say that the solutions of system (1.8) are:

(a) *equi-bounded*, if

$$(\forall t_0 \in \mathbb{R})(\forall a > 0)(\exists b = b(t_0, a) > 0)$$

$$(\forall \varphi_0 \in B_a(C))(\forall t \geq t_0): \ ||x(t; t_0, \varphi_0)|| < b;$$

(b) *uniformly bounded*, if the number b in (a) is independent of $t_0 \in \mathbb{R}$;

(c) *quasi-uniformly ultimately bounded*, if

$$(\exists b > 0)(\forall a > 0)(\exists T = T(a) > 0)(\forall t_0 \in \mathbb{R})$$

$$(\forall \varphi_0 \in B_a(C))(\forall t \geq t_0 + T): \ ||x(t; t_0, \varphi_0)|| < b;$$

(d) *uniformly ultimately bounded*, if (b) and (c) hold together.

Similar boundedness definitions will be used for systems of types (1.3), (1.5) and (1.6) in the case when $\Omega \equiv \mathbb{R}^n$ (see [Stamova 2008], [Stamova 2009], [Stamova 2014a], [Yakar, Çiçek and Gücen 2011]. Indeed, boundedness theory has played a significant role in the existence of periodic solutions of differential equations, and it has many applications in areas such as neural network, biological population management, secure communication and chaos control ([Afonso, Bonotto, Federson and Gimenes 2012], [Ballinger and Liu 1997], [Dong, Chen and Sun 2006], [Hu 2013], [Luo and Shen 2001], [Stamova 2016], [Teng, Nie and Fang 2011], [Xia 2007], [Yan 2003]).

Now we shall consider the set $\mathscr{B}$,

$$\mathscr{B} = \left\{ \{t_k\}, \ t_k \in \mathbb{R}, \ t_k < t_{k+1}, \ k = \pm 1, \pm 2, ..., \ \lim_{k \to \pm \infty} t_k = \pm \infty \right\}$$

of all unbounded increasing sequences of real numbers, and let $i(t, t + A)$ be the number of the points t_k in the interval $(t, t + A]$.

We need some essential notations before to introduce the main definitions of almost periodic functions and sequences which are used in this book.

For $\bar{T}, \bar{P} \in \mathscr{B}$, let $seq(\bar{T} \cup \bar{P}): \mathscr{B} \to \mathscr{B}$ be a map such that $seq(\bar{T} \cup \bar{P})$ forms a strictly increasing sequence. For $D \subset \mathbb{R}$, let $\theta_\varepsilon(D) = \{t + \varepsilon, \ t \in D\}$ and $F_\varepsilon(D) = \cap \{\theta_\varepsilon(D)\}$ for $\varepsilon > 0$. By $\Psi = (\psi(t), \bar{T})$, we shall denote an element from the space $PC[\mathbb{R}, \mathbb{R}^n] \times \mathscr{B}$. For every sequence of real numbers $\{s_n\}$, $n = 1, 2, ...,$ $\bar{\theta}_{s_n} \Psi$ means the sets $\{\psi(t + s_n), \bar{T} - s_n\} \subset PC[\mathbb{R}, \mathbb{R}^n] \times \mathscr{B}$ where $\bar{T} - s_n = \{t_k - s_n, \ k = \pm 1, \pm 2, ..., \ n = 1, 2, ...\}$.

Definition 1.9 ([Samoilenko and Perestyuk 1995]) The set of sequences $\{t_k^l\}$, $t_k^l = t_{k+l} - t_k$, $k = \pm 1, \pm 2, ..., l = \pm 1, \pm 2, ...$ is said to be *uniformly almost* *periodic*, if for any $\varepsilon > 0$ there exists a relatively dense set in $\mathbb{R}$ of ε–almost periods common for all sequences $\{t_k^l\}$.

Definition 1.10 ([Stamov 2012]) The sequence $\{\Psi_n\}$,

$$\Psi_n = (\psi_n(t), \bar{T}_n) \in PC[\mathbb{R}, \mathbb{R}^n] \times \mathscr{B},$$

is *convergent to* Ψ, where $\Psi = (\psi(t), \bar{T})$ and $(\psi(t), \bar{T}) \in PC[\mathbb{R}, \mathbb{R}^n] \times \mathscr{B}$, if and only if for any $\varepsilon > 0$ there exists $n_0 > 0$ such that for $n \geq n_0$ the following inequalities

$$\rho(\bar{T}, \bar{T}_n) < \varepsilon, \; \|\psi_n(t) - \psi(t)\| < \varepsilon$$

hold uniformly for $t \in \mathbb{R} \setminus F_\varepsilon(seq(\bar{T}_n \cup \bar{T}))$, where $\rho(.,.)$ is any distance in $\mathscr{B}$.

Definition 1.11 ([Stamov 2012]) The function $\psi \in PC[\mathbb{R}, \mathbb{R}^n]$ is said to be an *almost periodic piecewise continuous function* with points of discontinuity of the first kind t_k, $\{t_k\} \in \mathscr{B}$, if for every sequence of real numbers $\{s'_m\}$ there exists a subsequence $\{s_n\}$, $s_n = s'_{m_n}$, such that $\bar{\theta}_{s_n}\Psi$ is compact in $PC[\mathbb{R}, \mathbb{R}^n] \times \mathscr{B}$.

Consider the impulsive fractional-order system (1.6). Let $\{s'_m\}$ be an arbitrary sequence of real numbers. If there exists a subsequence $\{s_n\}$, $s_n = s'_{m_n}$ such that the system (1.6) moves to the system

$$\begin{cases} {}^c_{t_0}D_t^\alpha x(t) = f^s(t, x_t), \; t \neq t_k^s, \\ \Delta x(t_k^s) = I_k^s(x(t_k^s)), \; k = \pm 1, \pm 2, ..., \end{cases} \qquad (1.9)$$

then the set of systems in the form (1.9) we shall denote by $H(f, I_k, t_k)$.

The system (1.9) is called *Hull* of the system (1.6). In Chapter 3 we shall also consider similar Hull systems related to systems (1.1) and (1.5).

Efficient conditions for the existence of the corresponding subsequences $\{s_n\}$, $s_n = s'_{m_n}$ that move systems (1.1), (1.5) and (1.6) to their Hull systems can be find in [Stamov 2012] and some of the references therein.

1.3 Lyapunov Functions and their Fractional Derivatives

The most useful and general approach for studying stability, asymptotic properties, existence of periodic and almost periodic solutions of a nonlinear system is the Lyapunov direct method (also called the second method of Lyapunov) [Lyapunov 1950]. This method allows us to determine the qualitative properties of a system without explicitly integrating it. After the work of LaSalle (1960) there has been an expansion both in the class of the studied objects and in the mathematical problems investigated by means of the method and many refinements have been developed (see [Dannan and Elaydi

1986], [Hale 1977], [Ikeda, Ohta and Siljak 1991], [Kolmanovskii and Nosov 1986], [Krasovskii 1963], [Lakshmikantham, Leela and Martynyuk 1989], [Lakshmikantham, Leela and Martynyuk 1990], [Lakshmikantham, Matrosov and Sivasundaram 1991], [Razumikhin 1988], [Rouche, Habets and Laloy 1977], [Stamova 2009], [Yoshizawa 1966]).

The objective of this section is to present the main classes of Lyapunov functions that will be used in the qualitative investigation of fractional-order systems considered in the book. The fractional derivatives of these functions will be also defined.

First, we shall define the class of Lyapunov functions that we will be used for the fractional functional systems of the type (1.1). When using the method of Lyapunov functions for functional differential equations, the direct transfer of the Lyapunov theorems leads to significant difficulties when the sign of the derivative of the Lyapunov function with respect to the system has to be determined. For this reason, the application of the Lyapunov second method to systems with delays has been developed in two directions:

(i) The first direction is the use of Lyapunov functions and the Razumikhin technique [Razumikhin 1988]. This technique allows to verify the nonpositivity of the derivative function for some initial data (instead of all initial data) under certain restrictions, and is applied by numerous authors for different classes of functional differential systems. See for example [Hale 1977], [Lakshmikantham, Leela and Martynyuk 1989], [Lakshmikantham, Matrosov and Sivasundaram 1991], [Liu 1995], [Stamova 2009], [Wang and Liu 2007b], [Yan and Shen 1999].

(ii) The second method is the Lyapunov–Krasovskii functional method (replacing the Lyapunov function with a functional) (see [Burton 1985], [Hale 1977], [Kolmanovskii and Myshkis 1999], [Kolmanovskii and Nosov 1986], [Krasovskii 1963], [Wang and Liu 2007a]). For fractional-order systems Lyapunov functionals are used in [Burton 2011].

Since the results obtained by means of the Lyapunov–Razumikhin method are more applicable, it is our aim in this book to contribute to the development of the Lyapunov theory, presenting Razumikhin-type stability theorems.

We shall use the class of continuous Lyapunov-like functions:

$$C_0 = \{V : [t_0, \infty) \times \Omega \to \mathbb{R}_+ : V \in C[[t_0, \infty) \times \Omega, \mathbb{R}_+], V \text{ is locally Lipschitzian}$$
in $x \in \Omega\}$.

For a function $V \in C_0$ we define the following fractional-order Dini derivative.

Definition 1.12 Given a function $V \in C_0$. For $(t, \varphi) \in [t_0, \infty) \times C[[-r, 0], \Omega]$ the *upper right-hand derivative of V in the Caputo sense of order β, $0 < \beta < 1$,* *with respect to system* (1.1) is defined by

$$^cD_+^\beta V(t,\varphi(0)) = \lim_{\chi \to 0^+} \sup \frac{1}{\chi^\beta}[V(t,\varphi(0)) - V(t-\chi,\varphi(0) - \chi^\beta f(t,\varphi))].$$

Using the basic equality between the Caputo and Riemann–Liouville fractional derivatives [Podlubny 1999]

$$^c_{t_0}D_t^\beta l(t) = {}_{t_0}D_t^\beta[l(t) - l(t_0)]$$

for a continuously differentiable function l on $[t_0,b]$, $b > t_0$, we define the upper right-hand derivative of V in the Riemann–Liouville sense $D_+^\beta V(t,\varphi(0))$ of order $\beta, 0 < \beta < 1$, with respect to system (1.1) by the following relation

$$^cD_+^\beta V(t,\varphi(0)) = D_+^\beta[V(t,\varphi(0)) - V(t_0,\varphi_0(0))].$$

In the present book results on qualitative properties for the systems of the type (1.1) will be given where the upper right-hand derivatives of the Lyapunov functions are estimated only by the elements of minimal subsets of the integral curves of the investigated system when the following condition

$$V(t+\theta,\varphi(\theta)) \le V(t,\varphi(0)), \ \theta \in [-r,0] \tag{1.10}$$

holds. The condition (1.10) is called the *Razumikhin condition*, and the corresponding technique is known as the *Razumikhin technique*.

Next, we shall define the class of Lyapunov functions that will be used in the qualitative investigations of fractional impulsive systems. Fractional-order extensions of the Lyapunov direct method for fractional differential systems without impulses have been proposed by several authors in [Aguila-Camacho, Duarte-Mermoud and Gallegos 2014], [Duarte-Mermoud, Aguila-Camacho, Gallegos and Castro-Linares 2015], [Hu, Lu, Zhang and Zhao 2015], [Lakshmikantham, Leela and Sambandham 2008], [Lakshmikantham, Leela and Vasundhara Devi 2009], [Li, Chen and Podlubny 2010], [Liu and Jiang 2014]. Hovewer, the application of continuous Lyapunov functions to the investigation of impulsive systems restricts the possibilities of the Lyapunov second method. The fact that the solutions of impulsive systems are piecewise continuous functions requires introducing some analogous of the classical Lyapunov functions which have discontinuities of the first kind ([Bainov and Simeonov 1989], [Lakshmikantham, Bainov and Simeonov 1989]). By means of such functions it becomes possible to solve basic problems related to the application of Lyapunov second method to impulsive systems.

Let $\tau_0(x) \equiv t_0$ for $x \in \Omega$ and introduce the sets

$$G_k = \Big\{(t,x) \in \mathbb{R} \times \Omega : \ \tau_{k-1}(x) < t < \tau_k(x)\Big\}, \ k = \pm 1, \pm 2, ...,$$

$$G = \bigcup_{k=\pm 1, \pm 2, \dots} G_k.$$

Definition 1.13 A function $V : \mathbb{R} \times \Omega \to \mathbb{R}_+$ *belongs to the class* V_0, if:

1. $V(t,x)$ is continuous in G and locally Lipschitz continuous with respect to its second argument on each of the sets G_k, $k = \pm 1, \pm 2, \dots$.

2. For each $k = \pm 1, \pm 2, \dots$ and $(t_0^*, x_0^*) \in \sigma_k$ there exist the finite limits

$$V(t_0^{*-}, x_0^*) = \lim_{\substack{(t,x) \to (t_0^*, x_0^*) \\ (t,x) \in G_k}} V(t,x), \; V(t_0^{*+}, x_0^*) = \lim_{\substack{(t,x) \to (t_0^*, x_0^*) \\ (t,x) \in G_{k+1}}} V(t,x)$$

and the equality $V(t_0^{*-}, x_0^*) = V(t_0^*, x_0^*)$ holds.

Let the function $V \in V_0$ and $(t,x) \in G$. We define the derivative

$$\dot{V}_{(1.3)}(t,x) = \lim_{\delta \to 0^+} \sup \frac{1}{\delta} \left[V(t+\delta, x+\delta f(t,x)) - V(t,x) \right].$$

Note that if $x = x(t)$ is a solution of system (1.3), then for $t \neq \tau_k(x(t))$, $k = \pm 1, \pm 2, \dots$ we have $\dot{V}_{(1.3)}(t,x) = D_{(1.3)}^+ V(t,x(t))$, where

$$D_{(1.3)}^+ V(t,x(t)) = \lim_{\delta \to 0^+} \sup \frac{1}{\delta} \left[V(t+\delta, x(t+\delta)) - V(t,x(t)) \right]$$

is *the upper right-hand Dini derivative* of $V \in V_0$ (with respect to system (1.3)).

The class of functions V_0 will be also used in the investigation of qualitative properties of the solutions of impulsive systems with fixed moments of impulsive effect (1.5). In this case, $\tau_k(x) \equiv t_k$, $k = \pm 1, \pm 2, \dots$, σ_k are hyperplanes in $\mathbb{R}^{n+1}$, the sets G_k are

$$G_k = \{ (t,x) \in \mathbb{R} \times \Omega : \; t_{k-1} < t < t_k \},$$

and the condition 2 of Definition 1.10 is substituted by the condition:

2'. For each $k = \pm 1, \pm 2, \dots$ and $x \in \Omega$, there exist the finite limits

$$V(t_k^-, x) = \lim_{\substack{t \to t_k \\ t < t_k}} V(t,x), \qquad V(t_k^+, x) = \lim_{\substack{t \to t_k \\ t > t_k}} V(t,x),$$

and the following equalities are valid

$$V(t_k^-, x) = V(t_k, x).$$

For $t \neq t_k$, $k = \pm 1, \pm 2, \dots$ the upper right-hand derivative of a Lyapunov function $V \in V_0$, with respect to system (1.5) is

$$D_{(1.5)}^+ V(t,x(t)) = \lim_{\delta \to 0^+} \sup \frac{1}{\delta} \left[V(t+\delta, x(t+\delta)) - V(t,x) \right].$$

We define, also, the following fractional derivative (Dini-like derivative) in the Caputo sense. Let $t \in [t_{k-1}, t_k)$, $k = \pm 1, \pm 2, \ldots$ and $x \in \Omega$. The upper right-hand derivative of V in the Caputo sense of order β, $0 < \beta < 1$, with respect to system (1.5) is defined by

$$^c D_+^\beta V(t,x) = \lim_{\chi \to 0^+} \sup \frac{1}{\chi^\beta} \left[V(t,x) - V(t - \chi, x - \chi^\beta f(t,x)) \right].$$

By simple calculation (see [Diethelm 2010] and [Wang, Yang, Ma and Sun 2014]), we have that if $x = x(t)$ is a solution of system (1.5), then for $t \neq t_k$, $k = \pm 1, \pm 2, \ldots$,

$$^c D_+^\beta V(t, x(t)) = \frac{1}{\Gamma(1-\beta)} \int_{t_0}^t \frac{D_{(1.5)}^+ V(\tau, x(\tau))}{(t - \tau)^\beta} d\tau.$$

The presence of impulses as well as delays in the impulsive functional differential equations requires using of piecewise continuous Lyapunov functionals or a combination between the method of piecewise continuous Lyapunov functions and the Razumikhin technique. By means of such approaches, many interesting results of the qualitative theory of these equations have been obtained (see [Shen 1999], [Stamov 2012], [Stamova 2009], [Wang and Liu, 2007b], [Yan and Shen 1999]).

We shall employ Lyapunov functions from the class V_0 and develop corresponding stability theory for system (1.6).

Definition 1.14 Given a function $V \in V_0$. For $(t, \varphi) \in \mathbb{R} \times PC[[-r, 0], \Omega]$, $t \neq t_k$, $k = \pm 1, \pm 2, \ldots$ the *upper right-hand derivative of V in the Caputo sense of order β, $0 < \beta < 1$, with respect to system* (1.6) is defined by

$$^c D_+^\beta V(t, \varphi(0)) = \lim_{\chi \to 0^+} \sup \frac{1}{\chi^\beta} [V(t, \varphi(0)) - V(t - \chi, \varphi(0) - \chi^\beta f(t, \varphi))].$$

In the next chapters we shall also use the following classes of piecewise continuous Lyapunov-like functions

$$V_1 = \Big\{ V : \mathbb{R} \times \Omega \times \Omega \to \mathbb{R}_+, \ V \text{ is continuous in } (t_{k-1}, t_k] \times \Omega \times \Omega,$$
$$V(t_k, x, y) = V(t_k^-, x, y) \text{ and } \lim_{t > t_k} V(t, x, y) = V(t_k^+, x, y), \ x, y \in \Omega \Big\}.$$

Definition 1.15 A function $V \in V_1$ *belongs to the class* V_2, if $V(t, x, y)$ is locally Lipschitz continuous with respect to its second and third arguments with a Lipschitz constant $L_V > 0$, i.e. for $x_1, x_2 \in \Omega$, $y_1, y_2 \in \Omega$ and for $t \in \mathbb{R}$ it follows

$$|V(t, x_1, y_1) - V(t, x_2, y_2)| \le L_V \big(||x_1 - x_2|| + ||y_1 - y_2|| \big),$$

$t \neq t_k$, $k = \pm 1, \pm 2, \ldots$.

Since the systems of the type (1.6) are generalizations of types (1.1) and (1.5) systems, next we shall introduce fractional-order derivatives of Lyapunov functions $V \in V_2$ with respect to (1.6).

Definition 1.16 Given a function $V \in V_2$. For $t \in [t_{k-1}, t_k)$, $k = \pm 1, \pm 2, \ldots$ and $\varphi_i \in PC[[-r, 0], \Omega]$, $i = 1, 2$ the *upper right-hand derivative of V in the Caputo sense of order β*, $0 < \beta < 1$, with respect to system (1.6) is defined by

$$^cD^\beta_+ V(t, \varphi_1(0), \varphi_2(0)) = \lim_{\chi \to 0^+} \sup \frac{1}{\chi^\beta} [V(t, \varphi_1(0), \varphi_2(0))$$

$$-V(t - \chi, \varphi_1(0) - \chi^\beta f(t, \varphi_1), \varphi_2(0) - \chi^\beta f(t, \varphi_2))].$$

We shall use the next class of piecewise Lyapunov functions related to system (1.9).

Definition 1.17 A function $W : \mathbb{R} \times \Omega \to \mathbb{R}_+$ *belongs to the class W_0, if:*

1. The function $W(t, x)$ is continuous on $(t, x) \in \mathbb{R} \times \Omega$, $t \neq t^s_k$, $k = \pm 1, \pm 2, \ldots$.

2. The function $W(t, x)$ is locally Lipschitz continuous with respect to its second argument for $t \neq t^s_k$, $k = \pm 1, \pm 2, \ldots$.

3. For each $k = \pm 1, \pm 2, \ldots$ and $x \in \Omega$ there exist the finite limits

$$W(t^{s-}_k, x) = \lim_{\substack{t \to t^s_k \\ t < t^s_k}} W(t, x), \ W(t^{s+}_k, x) = \lim_{\substack{t \to t^s_k \\ t > t^s_k}} W(t, x)$$

and the equality $W(t^{s-}_k, x) = W(t^s_k, x)$ holds.

We define again a fractional order derivative (Dini-like derivative) in the Caputo sense for a function $W \in W_0$.

Definition 1.18 Let $t \in [t^s_{k-1}, t^s_k)$, $k = \pm 1, \pm 2, \ldots$ and $\varphi \in PC[[-r, 0], \Omega]$, then the *upper right-hand derivative of W in the Caputo sense of order β*, $0 < \beta < 1$, with respect to system (1.9) is defined by

$$^cD^\beta_+ W(t, \varphi(0)) = \lim_{\chi \to 0^+} \sup \frac{1}{\chi^\beta} [W(t, \varphi(0)) - W(t - \chi, \varphi(0) - \chi^\beta f^s(t, \varphi))].$$

It is well known that in some situations it is beneficial to use more than one Lyapunov function in obtaining results under weaker conditions. Hence, the corresponding theory, known as the *method of vector Lyapunov functions*, offers a very flexible mechanism. The significance of the method of vector Lyapunov functions is demonstrated by many authors for integer-order systems (see [Lakshmikantham, Leela and Martynyuk 1989], [Lakshmikantham, Leela and

Martynyuk 1990], [Lakshmikantham, Matrosov and Sivasundaram 1991] and the references therein).

Moreover, by means of the method of vector Lyapunov functions we can prove the results in some cases in which using the scalar Lyapunov functions is impossible.

In the present book we shall use vector Lyapunov functions $V : \mathbb{R} \times \Omega \to \mathbb{R}_+^m$, $V = col(V_1, V_2, ..., V_m)$ such that $V_j \in V_0$, $j = 1, 2, ..., m$.

In the presence of delays, we shall use the corresponding modifications and generalizations.

1.4 Fractional Comparison Lemmas

Lyapunov second method and comparison principle are the main techniques in the qualitative investigation of nonlinear systems. In this section we shall present vector and scalar fractional comparison lemmas we use. All comparison results will be given in terms of Caputo fractional derivatives. Applying some relations between Caputo and Riemann–Liouville fractional derivatives, we shall use these lemmas in the case of Riemann–Liouville fractional derivatives.

First, we shall state some scalar comparison results related to system (1.1). Together with system (1.1) we consider the following fractional comparison differential equation

$$_{t_0}\mathscr{D}_t^\beta u = g(t, u), \qquad (1.11)$$

where $0 < \beta < 1$, $g : [t_0, \infty) \times \mathbb{R}_+ \to \mathbb{R}$.

Let $u_0 \in \mathbb{R}_+$. For $_{t_0}\mathscr{D}_t^\beta = {}_{t_0}^c D_t^\beta$ we denote by $u^+(t) = u^+(t; t_0, u_0)$ the maximal solution of equation (1.11), which satisfies the initial condition

$$u^+(t_0; t_0, u_0) = u_0. \qquad (1.12)$$

In the case when the Riemann–Liouville operator is used, i.e. when $_{t_0}\mathscr{D}_t^\beta = {}_{t_0}D_t^\beta$ the initial condition (1.12) is given by

$$_{t_0}D_t^{\beta-1}u^+(t_0; t_0, x_0) = u_0.$$

Note that in this case, the above initial condition can be replaced by the condition

$$u^+(t_0; t_0, u_0) = \frac{u_0}{\Gamma(\beta)}(t - t_0)^{\beta-1}.$$

The following comparison lemma can be proved by the same arguments used for the fractional equations without delays (see Theorem 2.4.3 in [Lakshmikantham, Leela and Vasundhara Devi 2009]) and by the Razumikhin condition (1.10).

Lemma 1.1 Assume that:

1. The function $g : [t_0, \infty) \times \mathbb{R}_+ \to \mathbb{R}$ is continuous in $[t_0, \infty) \times \mathbb{R}_+$, and is nondecreasing in u for each $t \geq t_0$.

2. The maximal solution $u^+(t)$ of the IVP (1.11), (1.12) for $_{t_0}\mathscr{D}_t^\beta = {}^c_{t_0}D_t^\beta$ exists on $[t_0, \infty)$.

3. The function $V \in C_0$ is such that for $t \geq t_0$, $\varphi \in C[[-r,0], \Omega]$ the inequality

$$^cD_+^\beta V(t, \varphi(0)) \leq g(t, V(t, \varphi(0)))$$

is valid whenever $V(t + \theta, \varphi(\theta)) \leq V(t, \varphi(0))$ for $-r \leq \theta \leq 0$.

Then $\max\limits_{-r \leq \theta \leq 0} V(t_0 + \theta, \varphi_0(\theta)) \leq u_0$ implies

$$V(t, x(t; t_0, \varphi_0)) \leq u^+(t; t_0, u_0), \, t \in [t_0, \infty),$$

where $x(t; t_0, \varphi_0)$ is the solution of (1.1) with initial data $t_0 \in \mathbb{R}$ and $\varphi_0 \in C[[-r,0], \Omega]$.

In the case when $g(t, u) = qu$ for $(t, u) \in [t_0, \infty) \times \mathbb{R}_+$, where $q \in \mathbb{R}$ is a constant, we deduce the following corollary from Lemma 1.1.

Corollary 1.1 Assume that the function $V \in C_0$ is such that for $t \geq t_0$ and $\varphi \in C[[-r,0], \Omega]$ the inequality

$$^cD_+^\beta V(t, \varphi(0)) \leq qV(t, \varphi(0))$$

is valid whenever $V(t + \theta, \varphi(\theta)) \leq V(t, \varphi(0))$ for $-r \leq \theta \leq 0$.
Then $\max\limits_{-r \leq \theta \leq 0} V(t_0 + \theta, \varphi_0(\theta)) \leq u_0$ implies

$$V(t, x(t; t_0, \varphi_0)) \leq \max_{-r \leq \theta \leq 0} V(t_0, \varphi_0(\theta)) E_\beta(q(t - t_0)^\beta), \, t \in [t_0, \infty).$$

In the case when $g(t, u) = 0$ for $(t, u) \in [t_0, \infty) \times \mathbb{R}_+$, we deduce the following corollary from Lemma 1.1.

Corollary 1.2 Assume that the function $V \in C_0$ is such that for $t \geq t_0$ and $\varphi \in C[[-r,0], \Omega]$ the inequality

$$^cD_+^\beta V(t, \varphi(0)) \leq 0$$

is valid whenever $V(t + \theta, \varphi(\theta)) \leq V(t, \varphi(0))$ for $-r \leq \theta \leq 0$.
Then

$$V(t, x(t; t_0, \varphi_0)) \leq \max_{-r \leq \theta \leq 0} V(t_0, \varphi_0(\theta)), \, t \in [t_0, \infty).$$

Next, fractional vector comparison lemmas related to systems (1.5) and (1.6) will be given. That is why, together with systems (1.5) and (1.6) we shall consider the following comparison system

$$\begin{cases} {}^{c}_{t_0}D^{\beta}_t u = F(t,u), \, t \neq t_k, \\ \Delta u(t_k) = u(t_k^+) - u(t_k) = J_k(u(t_k)), \, t_k > t_0, \end{cases} \tag{1.13}$$

where $0 < \beta < 1$, $F : \mathbb{R} \times Q \to \mathbb{R}^m$, $J_k : Q \to \mathbb{R}^m$, $k = \pm 1, \pm 2, ...$, Q is an open subset of $\mathbb{R}^m$ containing the origin.

Let $u_0 \in Q$. Denote by $u(t) = u(t;t_0,u_0)$ the solution of system (1.13) satisfying the initial condition $u(t_0^+) = u_0$, and by $J^+(t_0,u_0)$—the maximal interval of type $[t_0,\gamma)$ in which the solution $u(t;t_0,u_0)$ is defined.

Definition 1.19 The solution $u^+ : J^+(t_0,u_0) \to \mathbb{R}^m$ of the system (1.13) for which $u^+(t_0^+;t_0,u_0) = u_0$ is said to be a *maximal solution*, if any other solution $u : [t_0,\tilde{\omega}) \to \mathbb{R}^m$, for which $u(t_0^+) = u_0$ satisfies the inequality $u^+(t) \geq u(t)$ for $t \in J^+(t_0,u_0) \cap [t_0,\tilde{\omega})$.

Analogously, the *minimal solution* $u^-(t;t_0,u_0)$ of system (1.13) is defined.

We introduce into $\mathbb{R}^m$ a partial ordering in the following way: for the vectors $u, v \in \mathbb{R}^m$ we shall say that $u \geq v$ if $u_j \geq v_j$ for each $j = 1,2,...,m$, and $u > v$ if $u_j > v_j$ for each $j = 1,2,...,m$.

Definition 1.20 The function $\psi : Q \to \mathbb{R}^m$ is said to be:

(a) *non-decreasing* in Q if $\psi(u) \geq \psi(v)$ for $u \geq v$, $u,v \in Q$;

(b) *monotone increasing* in Q if $\psi(u) > \psi(v)$ for $u > v$ and $\psi(u) \geq \psi(v)$ for $u \geq v$, $u,v \in Q$.

Definition 1.21 The function $F : \mathbb{R} \times Q \to \mathbb{R}^m$, is said to be *quasi-monotone increasing* in $\mathbb{R} \times Q$ if for each pair of points (t,u) and (t,v) from $\mathbb{R} \times Q$ and for $i \in \{1,2,...,m\}$ the inequality $F_i(t,u) \geq F_i(t,v)$ holds whenever $u_i = v_i$ and $u_j \geq v_j$ for $j = 1,2,...,m$, $i \neq j$, i.e. for any fixed $t \in \mathbb{R}$ and any $i \in \{1,2,...,m\}$ the function $F_i(t,u)$ is non-decreasing with respect to $(u_1,u_2,...,u_{i-1},u_{i+1},...,u_m)$.

In the case when the function $F : \mathbb{R} \times Q \to \mathbb{R}^m$ is continuous and quasi-monotone increasing, all solutions of problem (1.13) starting from the point $(\bar{t}_0,u_0) \in \mathbb{R} \times Q$ lie between two singular solutions - the maximal and the minimal ones.

The next comparison results play a crucial role when a vector Lyapunov function is used. Since the proofs are similar to the proof of Theorem 3.1 in [Stamova and Stamov G 2014b]; working component-wise and using comparison lemmas from [Lakshmikantham, Leela and Vasundhara Devi 2009], we omit them.

Lemma 1.2 Assume that:

1. The function F is quasi-monotonically increasing, continuous in the sets $(t_k, t_{k+1}] \times Q$, $k \in \mathbb{Z}$, and for $k = \pm 1, \pm 2, \ldots$ and $v \in Q$ there exists the finite limit

$$\lim_{\substack{(t,u) \to (t,v) \\ t > t_k}} F(t,u).$$

2. The functions $\psi_k : Q \to \mathbb{R}^m$, $\psi_k(u) = u + J_k(u)$, $k = \pm 1, \pm 2, \ldots$ are non-decreasing in Q.

3. The maximal solution $u^+(t; t_0, u_0)$ of system (1.13) defined on $J^+(t_0, u_0)$, is such that $u^+(t_k^+; t_0, u_0) \in Q$ for all $t_k \in J^+(t_0, u_0)$.

4. The function $V : [t_0, \bar{\omega}) \times \Omega \to Q$, $V = col(V_1, V_2, \ldots, V_m)$, $V_i \in V_0$, $i = 1, 2, \ldots, m$, $[t_0, \bar{\omega}) \subseteq J^+(t_0, u_0)$ is such that

$$V(t^+, x + I_k(x)) \leq \psi_k(V(t,x)), \ x \in \Omega, \ t = t_k, \ t_k \in (t_0, \bar{\omega}),$$

$$^cD_+^\beta V(t,x) \leq F(t, V(t,x)), \ x \in \Omega, \ t \neq t_k, \ t \in [t_0, \bar{\omega}).$$

Then $V(t_0^+, x_0) \leq u_0$ implies

$$V(t, x(t; t_0, x_0)) \leq u^+(t; t_0, u_0), \ t \in [t_0, \bar{\omega}),$$

where $x(t; t_0, x_0)$ is the solution of (1.5) with $x(t_0^+) = x_0 \in \Omega$.

Lemma 1.3 Assume that:

1. Conditions 1–3 of Lemma 1.2 are met.

2. The function $V : [t_0, \bar{\omega}) \times \Omega \to Q$, $V = col(V_1, V_2, \ldots, V_m)$, $V_i \in V_0$, $i = 1, 2, \ldots, m$, $[t_0, \bar{\omega}) \subseteq J^+(t_0, u_0)$ is such that for $t \in [t_0, \bar{\omega})$, $\varphi \in PC[[-r, 0], \Omega]$,

$$V(t^+, \varphi(0) + I_k(\varphi)) \leq \psi_k(V(t, \varphi(0))), \ t = t_k,$$

and the inequality

$$^cD_+^\beta V(t, \varphi(0)) \leq F(t, V(t, \varphi(0))), \ t \neq t_k,$$

is valid whenever $V(t + \theta, \varphi(\theta)) \leq V(t, \varphi(0))$ for $-r \leq \theta \leq 0$.

Then $\sup_{-r \leq \theta \leq 0} V(t_0 + \theta, \varphi_0(\theta)) \leq u_0$ implies

$$V(t, x(t; t_0, \varphi_0)) \leq u^+(t; t_0, u_0), \ t \in [t_0, \bar{\omega}),$$

where $x(t; t_0, \varphi_0)$ is the solution of (1.6) with initial data $t_0 \in \mathbb{R}$ and $\varphi_0 \in PC[[-r, 0], \Omega]$.

Lemma 1.4 Assume that:

1. Conditions 1 and 2 of Lemma 1.2 hold.

2. The minimal solution $u^-(t;t_0,u_0)$ of system (1.13) defined on $J^+(t_0,u_0)$, is such that $u^-(t_k^+;t_0,u_0) \in Q$ for all $t_k \in J^+(t_0,u_0)$.

3. The function $V : [t_0,\bar{\omega}) \times \Omega \to Q$, $V = col(V_1, V_2,...,V_m)$, $V_i \in V_0$, $i = 1,2,...,m$, $[t_0,\bar{\omega}) \subseteq J^+(t_0,u_0)$ is such that for $t \in [t_0,\bar{\omega})$, $\varphi \in PC[[-r,0],\Omega]$,

$$V(t^+,\varphi(0)+I_k(\varphi)) \geq \psi_k(V(t,\varphi(0))), \ t = t_k,$$

and the inequality

$$^cD_+^\beta V(t,\varphi(0)) \geq F(t,V(t,\varphi(0))), \ t \neq t_k,$$

is valid whenever $V(t+\theta,\varphi(\theta)) \leq V(t,\varphi(0))$ *for* $-r \leq \theta \leq 0$.
Then $\sup_{-r \leq \theta \leq 0} V(t_0+\theta,\varphi_0(\theta)) \geq u_0$ implies

$$V(t,x(t;t_0,\varphi_0)) \geq u^-(t;t_0,u_0), t \in [t_0,\bar{\omega}),$$

where $x(t;t_0,\varphi_0)$ is the solution of (1.6) with initial data $t_0 \in \mathbb{R}$ and $\varphi_0 \in PC[[-r,0],\Omega]$.

For the scalar case $m = 1$, we shall consider the comparison equation

$$\begin{cases} {}^c_{t_0}D_t^\beta u = g(t,u), \ t \neq t_k, \\ \Delta u(t_k) = B_k(u(t_k)), \ t_k > t_0, \end{cases} \tag{1.14}$$

where $0 < \beta < 1$, $g : \mathbb{R} \times \mathbb{R}_+ \to \mathbb{R}$, $B_k : \mathbb{R}_+ \to \mathbb{R}_+$, $k = \pm 1, \pm 2, ...$
Let $u_0 \in \mathbb{R}_+$. We denote by $u^+(t) = u^+(t;t_0,u_0)$ the maximal solution of equation (1.14), which satisfies the initial condition

$$u^+(t_0^+;t_0,u_0) = u_0. \tag{1.15}$$

For fractional impulsive differential systems (1.5) the following comparison lemma, whose proof is similar to the proof of the corresponding lemma in [Stamova 2014a], will be used.

Lemma 1.5 Assume that:

1. The function $g : [t_0,\infty) \times \mathbb{R}_+ \to \mathbb{R}$ is continuous in each of the sets $(t_{k-1},t_k] \times \mathbb{R}_+$, $t_k > t_0$.
2. $B_k \in C[\mathbb{R}_+,\mathbb{R}_+]$ and $\tilde{\psi}_k(u) = u + B_k(u) \geq 0$, $k = \pm 1, \pm 2,...$ are non-decreasing with respect to u.

3. The maximal solution $u^+(t;t_0,u_0)$ of the scalar problem (1.14), (1.15) exists on $[t_0,\infty)$.
4. The function $V \in V_0$ is such that

$$V(t^+, x + I_k(x)) \leq \tilde{\psi}_k(V(t,x)), \; x \in \Omega, t = t_k, \; t_k > t_0,$$

$$^cD^{\beta}_+V(t,x) \leq g(t, V(t,x)), \; (t,x) \in G, \; t \in [t_0,\infty).$$

Then $V(t_0^+, x_0) \leq u_0$ implies

$$V(t, x(t;t_0,x_0)) \leq u^+(t;t_0,u_0), \, t \in [t_0,\infty).$$

In the case when $g(t,u) = qu$ for $(t,u) \in \mathbb{R} \times \mathbb{R}_+$, where $q \in \mathbb{R}$ is a constant, and $\tilde{\psi}_k(u) = u$ for $u \in \mathbb{R}_+$, $k = \pm 1, \pm 2, ...$, we deduce the following corollary from Lemma 1.5.

Corollary 1.3 Assume that the function $V \in V_0$ is such that:

$$V(t^+, x + I_k(x)) \leq V(t,x), \; x \in \Omega, t = t_k, \; t_k > t_0,$$

$$^cD^{\beta}_+V(t,x) \leq qV(t,x), \; (t,x) \in G, \; t \in [t_0,\infty).$$

Then

$$V(t, x(t;t_0,x_0)) \leq V(t_0^+, x_0)E_{\beta}(q(t-t_0)^{\beta}), \, t \in [t_0,\infty).$$

In the case when $g(t,u) = 0$ for $(t,u) \in \mathbb{R} \times \mathbb{R}_+$ and $\tilde{\psi}_k(u) = u$ for $u \in \mathbb{R}_+$, $k = \pm 1, \pm 2, ...$, we deduce the following corollary from Lemma 1.5.

Corollary 1.4 Assume that the function $V \in V_0$ is such that:

$$V(t^+, x + I_k(x)) \leq V(t,x), \; x \in \mathbb{R}^n, t = t_k, \; t_k > t_0,$$

$$^cD^{\beta}_+V(t,x) \leq 0, \; (t,x) \in G, \; t \in [t_0,\infty).$$

Then

$$V(t, x(t;t_0,x_0)) \leq V(t_0^+, x_0), \, t \in [t_0,\infty).$$

The next comparison principle for fractional functional differential systems (1.6) is similar to the comparison result in [Stamova and Stamov G 2014b].

Lemma 1.6 Assume that:

1. Conditions 1–3 of Lemma 1.5 are met.
2. The function $V \in V_0$ is such that for $t \in [t_0,\infty)$, $\varphi \in PC[[-r,0],\Omega]$,

$$V(t^+, \varphi(0) + I_k(\varphi)) \leq \tilde{\psi}_k(V(t, \varphi(0))), \; t = t_k, \; t_k > t_0$$

and the inequality

$$^cD^{\beta}_+V(t, \varphi(0)) \leq g(t, V(t, \varphi(0))), t \neq t_k, t \in [t_0,\infty)$$

is valid whenever $V(t+\theta, \varphi(\theta)) \leq V(t, \varphi(0))$ for $-r \leq \theta \leq 0$.

Then $\sup_{-r\leq\theta\leq0} V(t_0+\theta,\varphi_0(\theta)) \leq u_0$ implies

$$V(t,x(t;t_0,\varphi_0)) \leq u^+(t;t_0,u_0), t \in [t_0,\infty).$$

In the case when $g(t,u) = qu$ for $(t,u) \in \mathbb{R} \times \mathbb{R}_+$, where $q \in \mathbb{R}$ is a constant, and $\tilde{\psi}_k(u) = u$ for $u \in \mathbb{R}_+$, $k = \pm1, \pm2, ...$, the following corollary holds.

Corollary 1.5 Assume that the function $V \in V_0$ is such that for $t \in [t_0,\infty)$, $\varphi \in PC[[-r,0],\Omega]$,

$$V(t^+,\varphi(0)+I_k(\varphi)) \leq V(t,\varphi(0)), \ t = t_k, \ t_k > t_0$$

and the inequality

$$^cD_+^\beta V(t,\varphi(0)) \leq qV(t,\varphi(0)), t \neq t_k, t \in [t_0,\infty)$$

is valid whenever $V(t+\theta,\varphi(\theta)) \leq V(t,\varphi(0))$ for $-r \leq \theta \leq 0$.
  Then $\sup_{-r\leq\theta\leq0} V(t_0+\theta,\varphi_0(\theta)) \leq u_0$ implies

$$V(t,x(t;t_0,\varphi_0)) \leq \sup_{-r\leq\theta\leq0} V(t_0^+,\varphi_0(\theta))E_\beta(q(t-t_0)^\beta), t \in [t_0,\infty).$$

In the case when $g(t,u) = 0$ for $(t,u) \in [t_0,\infty) \times \mathbb{R}_+$, and $\tilde{\psi}_k(u) = u$ for $u \in \mathbb{R}_+$, $k = \pm1, \pm2, ...$, we deduce the following corollary from Lemma 1.6.

Corollary 1.6 Assume that the function $V \in V_0$ is such that for $t \in [t_0,\infty)$, $\varphi \in PC[[-r,0],\Omega]$,

$$V(t^+,\varphi(0)+I_k(\varphi)) \leq V(t,\varphi(0)), \ t = t_k, \ t_k > t_0$$

and the inequality

$$^cD_+^\beta V(t,\varphi(0)) \leq 0, t \neq t_k, t \in [t_0,\infty)$$

is valid whenever $V(t+\theta,\varphi(\theta)) \leq V(t,\varphi(0))$ for $-r \leq \theta \leq 0$.
  Then

$$V(t,x(t;t_0,\varphi_0)) \leq \sup_{-r\leq\theta\leq0} V(t_0^+,\varphi_0(\theta)), t \in [t_0,\infty).$$

Remark 1.2 Similar comparison results can be proved in terms of functions from the classes V_2 and W_0 (see [Stamov 2012], [Stamov and Stamov 2013], [Stamova 2009]).

Remark 1.3 Analogous comparison results for impulsive systems (1.5) and (1.6) in which minimal solutions are used (see [Bainov and Simeonov 1992]) can be proved.

When Riemann–Liouville fractional derivatives are used we shall apply the following lemma:

Lemma 1.7 Let $0 < \beta < 1$ and $V \in C_0$. Then for $\varphi \in C[[-r,0],\Omega]$,

$$^cD_+^\beta V(t,\varphi(0)) \le D_+^\beta V(t,\varphi(0)), t \ge t_0.$$

Proof. Using the relation

$$^cD_+^\beta V(t,\varphi(0)) = D_+^\beta [V(t,\varphi(0)) - V(t_0,\varphi_0(0))],$$

we have

$$^cD_+^\beta V(t,\varphi(0)) = D_+^\beta V(t,\varphi(0)) - \frac{V(t_0,\varphi_0(0))}{\Gamma(1-\beta)}(t-t_0)^{-\beta}.$$

Because, $0 < \beta < 1$ and $V \in C_0$, $^cD_+^\beta V(t,\varphi(0)) \le D_+^\beta V(t,\varphi(0)), t \ge t_0$.

Next lemma proved in [Aguila-Camacho, Duarte-Mermoud and Gallegos 2014] presents an important property of Caputo fractional derivatives when $0 < \alpha < 1$, which allows finding a simple Lyapunov candidate function for the corresponding fractional order systems.

Lemma 1.8 ([Aguila-Camacho, Duarte-Mermoud and Gallegos 2014]) Let $x(t) \in \mathbb{R}$ be a continuously differentiable function. Then for any $t \ge t_0$

$$\frac{1}{2}{}^c_{t_0}D_t^\alpha\big(x^2(t)\big) \le x(t){}^c_{t_0}D_t^\alpha x(t)$$

for $0 < \alpha < 1$.

Remark 1.4 Obviously, when $\alpha = 1$, we have

$$\frac{1}{2}\frac{dx^2(t)}{dt} = x(t)\frac{dx(t)}{dt}.$$

The above equality can be considered as a particular case of Lemma 1.8.

In the next lemma a fractional-order generalization of the Gronwall–Bellman inequality is proved.

Lemma 1.9 ([Ye, Gao and Ding 2007]) Suppose $\beta > 0$, $a(t)$ is a non-negative function locally integrable on $0 \le t < T$ (for some $T \le \infty$) and $b(t)$ is a non-negative, non-decreasing continuous function defined on $[0,T)$, $b(t) \le M$ (M = const), and suppose $u(t)$ is non-negative and locally integrable on $0 \le t < T$ with

$$u(t) \le a(t) + b(t)\int_0^t (t-s)^{\beta-1}u(s)ds$$

on this interval. Then

$$u(t) \leq a(t) + \int_0^t \left[\sum_{\kappa=1}^{\infty} \frac{\left(b(t)\Gamma(\beta)\right)^{\kappa}}{\Gamma(\kappa\beta)} (t-s)^{\kappa\beta-1} a(s) \right] ds, \ 0 \leq t < T.$$

Corollary 1.7 ([Ye, Gao and Ding 2007]) Under the hypothesis of Lemma 1.9, let $a(t)$ be a non-decreasing function on $[0, T)$.
   Then

$$u(t) \leq a(t) E_{\beta} \left(b(t)\Gamma(\beta)t^{\beta} \right),$$

where E_{β} is the corresponding Mittag-Leffler function.

1.5   Notes and Comments

The basic notations of fractional derivatives and integrals listed in Section 1.1 are taken from [Li, Chen and Podlubny 2010] and [Podlubny 1999]. The descriptions made for fractional functional differential equations are due to Lakshmikantham (see [Lakshmikantham 2008] and Stamova (see [Stamova 2016]), for fractional impulsive differential equations to Stamova (see [Stamova 2014a]), and for fractional impulsive functional differential equations to Stamova and Stamov (see [Stamova and Stamov G 2014b]). Impulsive and functional differential systems of fractional order were also investigated in [Abbas and Benchohra 2010], [Ahmad and Nieto 2011], [Ahmad and Sivasundaram 2010], [Anguraj and Maheswari 2012], [Benchohra, Henderson, Ntouyas and Ouahab 2008], [Benchohra and Slimani 2009], [Fečkan, Zhou and Wang 2012], [Mahto, Abbas and Favini 2013].

Stability, Mittag–Leffler stability and boundedness definitions in Section 1.2 are introduced by Stamov and Stamova (see [Stamov 2012], [Stamova 2009], [Stamova 2014a], [Stamova 2015], [Stamova 2016], [Stamova and Stamov G 2014b]). The results on almost periodic functions included in Section 1.2 were taken from Samoilenko and Perestuyk (see [Samoilenko and Perestyuk 1995]) and Stamov (see [Stamov 2012], [Stamov and Stamova 2015a], Stamov and Stamova 2015b]).

The Lyapunov–Razumikhin technique was first introduced in [Razumikhin 1988] while, the method of piecewise continuous Lyapunov functions for impulsive systems was introduced by Bainov and Simeonov (see [Bainov and Simeonov 1989]) The definition of the fractional derivative of a Lyapunov function with respect to the system (1.1) is in taken from [Stamova and Stamov G 2013], while the definitions of the fractional derivatives of Lyapunov functions related to systems (1.5) and (1.6) are adapted from [Stamova 2014a] and [Stamova and Stamov G 2014b]. Similar definitions are used by several authors (see [Baleanu, Sadati, Ghaderi, Ranjbar, Abdeljawad and Jarad 2010], [Çiçek, Yaker and Gücen 2014], [Lakshmikantham, Leela and Sambandham

2008], [Lakshmikantham, Leela and Vasundhara Devi 2009], [Sadati, Baleanu, Ranjbar, Ghaderi and Abdeljawad 2010], [Yakar, Çiçek and Gücen 2011]. For more results on the relations between fractional and integer order derivatives of Lyapunov functions see [Wang, Yang, Ma and Sun 2014], and for some main properties of fractional derivatives of functions see [Zhou, Liu, Zhang and Jiang 2013].

The comparison Lemma 1.1 is an extension of the comparison principle in [[Lakshmikantham 2008] in terms of Lyapunov functions. Similar lemmas for fractional-order differential equations without delays are proved in [Lakshmikantham, Leela and Sambandham 2008], [Lakshmikantham, Leela and Vasundhara Devi 2009], [Lakshmikantham and Vatsala 2008] and [Vasundhara Devi, Mc Rae and Drici 2012]. For fractional functional differential systems a special case is considered in [Stamova and Stamov G 2013]. Lemmas 1.2, 1.3 and 1.4 are new. Lemma 1.5 is a result from [Stamova 2014a]. Similar comparison result for impulsive fractional-order differential systems is proved in [Stamova 2015]. Lemma 1.6 is adapted from Stamova and Stamov (see [Stamova and Stamov G 2014b]). Lemma 1.7 is from [Stamova 2016]. The comparison principle is used by several authors for fractional differential systems without impulses (see [Baleanu, Sadati, Ghaderi, Ranjbar, Abdeljawad and Jarad 2010], [Choi and Koo 2011], [Çiçek, Yaker and Gücen 2014], [Lakshmikantham 2008], [Lakshmikantham, Leela and Sambandham 2008], [Lakshmikantham and Vatsala 2008], [Li, Chen and Podlubny 2010], [Sadati, Baleanu, Ranjbar, Ghaderi and Abdeljawad 2010], [Vasundhara Devi, Mc Rae and Drici 2012]. A result similar to Lemma 1.8 is proved in [Yu, Hu, Jiang and Fan 2014]. Lemma 1.9 is from [Ye, Gao and Ding 2007]. Fractional-order impulsive generalizations of the Gronwall-Bellman inequality are given in [Shao 2014].

Chapter 2

Stability and Boundedness

The present chapter will offer stability and boundedness results for fractional order systems using the Lyapunov direct method.

In *Section* 2.1 we shall consider sufficient conditions for Lyapunov stability of solutions for different classes of fractional differential equations.

Section 2.2 will offer boundedness theorems for fractional functional differential equations.

In *Section* 2.3 we shall use scalar Lyapunov functions and investigate global stability of the solutions of impulsive fractional differential systems.

In *Section* 2.4, we shall continue to use the Lyapunov fractional method, and analyze Mittag–Leffler stability properties of functional and impulsive fractional-order differential equations.

Section 2.5 will deal with sufficient conditions for practical stability of nonlinear differential systems of fractional order subject to impulse effects. Scalar and vector Lyapunov-like functions will be used to derive practical stability criteria for the impulsive control fractional differential system. Applications to linear and non-linear systems will be discussed to illustrate the theory.

In *Section* 2.6, we shall study the Lipschitz stability of fractional functional differential systems with Caputo fractional derivatives. Using the fractional Lyapunov method, Lipschitz stability, uniform Lipschitz stability and global uniform Lipschitz stability criteria will be proved.

In *Section* 2.7 results on stability and boundedness of the solutions with respect to sets will be given for fractional impulsive functional differential systems using piecewise continuous vector Lyapunov functions and the fractional vector comparison principle. Examples will be presented to illustrate our theoretical findings.

Finally, *Section* 2.8 is devoted to the existence and stability of integral manifolds for fractional functional differential systems and impulsive fractional functional differential systems.

2.1 Lyapunov Stability

First, we shall investigate the Lyapunov stability of the zero solution of system (1.1). Throughout the rest of the book we shall use the class $K = \{w \in C[\mathbb{R}_+, \mathbb{R}_+] : w(u)$ is strictly increasing and $w(0) = 0\}$ of functions w, which are called wedges.

In the proofs of our main theorems in this section we shall use continuous Lyapunov functions $V : [t_0, \infty) \times \Omega \to \mathbb{R}_+$, $V \in C_0$ for which the following condition is true:

H2.1 $V(t, 0) = 0, t \geq t_0$.

Theorem 2.1 Assume that $f(t, 0) = 0, t \in [t_0, \infty)$, there exists a function $V \in C_0$ such that condition H2.1 holds,

$$w_1(||x||) \leq V(t, x) \leq w_2(||x||), \ w_1, w_2 \in K, \ (t, x) \in [t_0, \infty) \times \Omega, \qquad (2.1)$$

and the inequality

$$^cD_+^\beta V(t, \varphi(0)) \leq 0 \qquad (2.2)$$

is valid whenever $V(t + \theta, \varphi(\theta)) \leq V(t, \varphi(0))$ for $-r \leq \theta \leq 0$, where $t \in [t_0, \infty)$, $\varphi \in C[[-r, 0], \Omega]$, $\beta \in (0, 1)$.

Then the zero solution of system (1.1) is uniformly stable.

Proof. Let $\varepsilon > 0$ be chosen. Choose $\delta = \delta(\varepsilon) > 0$ so that $w_2(\delta) < w_1(\varepsilon)$. Such numbers always exist because of the properties of the wedges w_1 and w_2.

Let $\varphi_0 \in C[[-r, 0], \Omega]: ||\varphi_0||_r < \delta$, and let $x(t) = x(t; t_0, \varphi_0)$ be the solution of the IVP (1.1), (1.2).

It follows from Corollary 1.2 that

$$V(t, x(t; t_0, \varphi_0)) \leq \max_{-r \leq \theta \leq 0} V(t_0, \varphi_0(\theta)), \ t \geq t_0.$$

From the above inequality and (2.1), we get to the inequalities

$$w_1(||x(t; t_0, \varphi_0)||) \leq \max_{-r \leq \theta \leq 0} V(t_0, \varphi_0(\theta)) \leq w_2(||\varphi_0||_r) < w_2(\delta) < w_1(\varepsilon),$$

from which it follows that $||x(t; t_0, \varphi_0)|| < \varepsilon$ for $t \geq t_0$. This proves the uniform stability of the zero solution of system (1.1).

The following corollary is immediate.

Corollary 2.1 If in Theorem 2.1 condition (2.1) is replaced by the condition

$$w_1(||x||) \leq V(t,x), \ w_1 \in K, \ (t,x) \in [t_0,\infty) \times \Omega, \tag{2.3}$$

then the zero solution of system (1.1) is stable.

Theorem 2.2 Assume that $f(t,0) = 0, t \in [t_0,\infty)$, there exists a function $V \in C_0$ such that H2.1 and (2.1) hold, and the inequality

$$^cD_+^\beta V(t,\varphi(0)) \leq -w_3(||\varphi(0)||) \tag{2.4}$$

is valid whenever $V(t+\theta,\varphi(\theta)) \leq p(V(t,\varphi(0)))$ for $-r \leq \theta \leq 0$, where $t \in [t_0,\infty)$, $\varphi \in C[[-r,0],\Omega]$, $\beta \in (0,1)$, $w_3 \in K$, p is continuous and non-decreasing on $\mathbb{R}_+$, and $p(u) > u$ as $u > 0$.

Then the zero solution of system (1.1) is uniformly asymptotically stable.

Proof. It follows from Theorem 2.1 that the zero solution of (1.1) is uniformly stable. Let $H = const > 0 : \{x \in \mathbb{R}^n : ||x|| \leq H\} \subset \Omega$. To complete the proof, suppose $\lambda = const > 0$ be such that $w_2(\lambda) < w_1(H)$. Then, if $t_0 \in \mathbb{R}$ and $\varphi_0 \in C[[-r,0],\Omega]: ||\varphi_0||_r < \lambda$, (2.1) implies

$$\max_{-r \leq \theta \leq 0} V(t_0,\varphi_0(\theta)) \leq w_2(||\varphi_0||_r) < w_2(\lambda) < w_1(H), \ t \geq t_0 - r,$$

and

$$V(t,x(t;t_0,\varphi_0)) < w_2(\lambda), \quad ||x(t;t_0,\varphi_0)|| < H, \ t \geq t_0 - r. \tag{2.5}$$

Let $0 < \varepsilon \leq H$ be chosen. We need to show that there exists $T = T(\varepsilon) > 0$ such that the solution $x(t) = x(t;t_0,\varphi_0)$ of the IVP (1.1), (1.2) satisfies $||x(t;t_0,\varphi_0)|| < \varepsilon$ for $t \geq t_0 + T$. This will be true if we show that $V(t,x(t;t_0,\varphi_0)) \leq a(\varepsilon)$ for $t \geq t_0 + T$.

Let the function $p : \mathbb{R}_+ \to \mathbb{R}_+$ be a continuous and non-decreasing on $\mathbb{R}_+$, and $p(u) > u$ as $u > 0$. We set

$$\eta = \inf\{p(u) - u : \ w_1(\varepsilon) \leq u \leq w_2(\lambda)\}.$$

Then

$$p(u) > u + \eta \text{ as } w_1(\varepsilon) \leq u \leq w_2(\lambda), \tag{2.6}$$

and let ν be the first natural number such that

$$w_1(\varepsilon) + \nu\eta > w_2(\lambda). \tag{2.7}$$

Let us denote

$$\gamma = \inf_{w_2^{-1}(w_1(\varepsilon)) \leq u \leq H} w_3(u),$$

and

$$\xi_k = t_0 + k \left[\frac{\eta \Gamma(1+\beta)}{\gamma} \right]^{1/\beta}, \ k = 0, 1, 2, ..., \nu.$$

We want to prove

$$V(t, x(t)) < w_1(\varepsilon) + (\nu - k)\eta, \ t \geq \xi_k \tag{2.8}$$

for all $k = 0, 1, 2, ..., \nu$.

Indeed, using (2.5) and (2.7) we obtain

$$V(t, x(t; t_0, \varphi_0)) < w_2(\lambda) < w_1(\varepsilon) + \nu\eta, \ t \geq t_0 = \xi_0$$

which means the validity of (2.8) for $k = 0$.

Assume (2.8) to be fulfilled for some integer k, $1 \leq k \leq \nu$, i.e.

$$V(s, x(s)) < w_1(\varepsilon) + (\nu - k)\eta, \ s \geq \xi_k. \tag{2.9}$$

We shall prove that

$$V(t, x(t)) < w_1(\varepsilon) + (\nu - k - 1)\eta, \ t \geq \xi_{k+1}.$$

Suppose now that

$$w_1(\varepsilon) + (\nu - k - 1)\eta \leq V(t, x(t)) < w_1(\varepsilon) + (\nu - k)\eta, \ \xi_k \leq t \leq \xi_{k+1}. \tag{2.10}$$

Then (2.5), (2.6) and (2.10) imply

$$w_1(\varepsilon) \leq w_1(\varepsilon) + (\nu - k - 1)\eta \leq V(t, x(t)) < w_2(\lambda), \ \xi_k \leq t \leq \xi_{k+1}$$

and

$$p(V(t, x(t))) > V(t, x(t)) + \eta \geq w_1(\varepsilon) + (\nu - k)\eta$$
$$> V(t + \theta, x(t + \theta)), \ \xi_k \leq t + \theta \leq t \leq \xi_{k+1}, \ \theta \in [-r, 0].$$

Then by (2.4) for $t \geq \xi_k$ we obtain

$$V(t, x(t)) \leq V(\xi_k, x(\xi_k)) - \frac{1}{\Gamma(\beta)} \int_{\xi_k}^t w_3(||x(\tau)||)(t - \tau)^{\beta - 1} \, d\tau$$

$$< w_2(\lambda) - \frac{\gamma}{\Gamma(\beta + 1)}(t - \xi_k)^\beta \leq 0$$

if $t - \xi_k \geq \left[\frac{\eta \Gamma(1+\beta)}{\gamma} \right]^{1/\beta}$. Consequently, for $t = \xi_{k+1}$

$$V(\xi_{k+1}, x(\xi_{k+1})) < w_1(\varepsilon) + (\nu - k - 1)\eta,$$

and, finally

$$V(t, x(t)) < w_1(\varepsilon) + (\nu - k - 1)\eta$$

for all $t \geq \xi_{k+1}$.

Therefore, (2.8) holds for all $k = 0, 1, 2, ..., \nu$.

Let $T = T(\varepsilon) = \nu \left[\frac{\eta \Gamma(1+\beta)}{\gamma} \right]^{1/\beta}$. Then (2.8) implies

$$V(t, x(t)) < w_1(\varepsilon) \text{ as } t \geq t_0 + T$$

or

$$||x(t)|| < \varepsilon \text{ as } t \geq t_0 + T$$

for any solution $x(t) = x(t; t_0, \varphi_0)$ of (1.1) with $||\varphi_0||_r < \lambda$ and the zero solution of (1.1) is uniformly attractive. Since Theorem 2.1 implies that it is uniformly stable, then the solution $x \equiv 0$ is uniformly asymptotically stable.

Theorem 2.3 If in Theorem 2.2 condition (2.4) is replaced by the condition

$$^cD_+^\beta V(t, \varphi(0)) \leq -wV(t, \varphi(0)), \tag{2.11}$$

whenever $V(t + \theta, \varphi(\theta)) \leq p(V(t, \varphi(0)))$ for $-r \leq \theta \leq 0$, where $w = const > 0$, $t \in [t_0, \infty)$, $\varphi \in C[[-r, 0], \Omega]$, $\beta \in (0, 1)$, p is continuous and non-decreasing on $\mathbb{R}_+$, and $p(u) > u$ as $u > 0$, then the zero solution of system (1.1) is uniformly asymptotically stable.

Proof. Let $H = const > 0 : \{x \in \mathbb{R}^n : ||x|| \leq H\} \subset \Omega$. Choose $\lambda = const > 0$ so that $w_2(\lambda) < w_1(H)$. Let $\varepsilon > 0$ and

$$T = T(\varepsilon) = \left[\frac{-L_\beta \left(\frac{w_1(\varepsilon)}{w_1(H)} \right)}{w} \right]^{1/\beta},$$

where $L_\beta(z)$ is the inverse function of the Mittag–Leffler function $E_\beta(z)$, defined as the solution of the equation $L_\beta(E_\beta(z)) = z$ (see [Hilfer and Seybold 2006]). Then by (2.11) and Corollary 1.1 for $t_0 \in \mathbb{R}$ and $\varphi_0 \in C[[-r, 0], \Omega]: ||\varphi_0||_r < \lambda$, the following inequalities hold

$$V(t, x(t; t_0, \varphi_0)) \leq \max_{-r \leq \theta \leq 0} V(t_0, \varphi_0(\theta)) E_\beta[-w(t-t_0)^\beta] < w_1(\varepsilon),$$

whence, in view of (2.1), we deduce that the solution of system (1.1) is uniformly attractive.

Theorem 2.4 Let the conditions of Theorem 2.2 hold, except replacing the upper right-hand derivative of V in the Caputo sense $^cD_+^\beta V(t, \varphi(0))$ by the upper right-hand derivative of V in the Riemann–Liouville sense of the same order $D_+^\beta V(t, \varphi(0))$.

Then the zero solution of system (1.1) is uniformly asymptotically stable.

Proof. It follows from Lemma 1.7 that the inequality

$$^cD_+^\beta V(t, \varphi(0)) \leq D_+^\beta V(t, \varphi(0)) \leq -w_3(\|\varphi(0)\|)$$

is valid whenever $V(t + \theta, \varphi(\theta)) \leq p(V(t, \varphi(0)))$ for $-r \leq \theta \leq 0$ and $t \geq t_0$. The rest of the proof is analogous to the proof of Theorem 2.2.

Theorem 2.5 Let the conditions of Theorem 2.3 hold, except replacing $^cD_+^\beta V(t, \varphi(0))$ by $D_+^\beta V(t, \varphi(0))$.

Then the zero solution of system (1.1) is uniformly asymptotically stable.

The proof follows from Lemma 1.7.

Next, we shall use the Lyapunov method and the scalar comparison equation (1.11) and we shall give sufficient conditions for stability of the zero solution of system under consideration. We shall consider such solutions $u(t)$ of equation (1.11) for which $u(t) \geq 0$. That is why the following stability definitions of the zero solution of this equation will be used.

Definition 2.1 The zero solution of (1.11) is said to be:

(a) *stable*, if

$$(\forall t_0 \in \mathbb{R})(\forall \varepsilon > 0)(\exists \delta = \delta(t_0, \varepsilon) > 0)$$

$$(\forall u_0 < \delta)(\forall t \geq t_0) : u^+(t; t_0, u_0) < \varepsilon;$$

(b) *uniformly stable*, if the number δ in (a) is independent of $t_0 \in \mathbb{R}$;

(c) *attractive*, if

$$(\forall t_0 \in \mathbb{R})(\exists \lambda = \lambda(t_0) > 0)(\forall u_0 < \lambda)$$

$$\lim_{t \to \infty} u^+(t; t_0, u_0) = 0;$$

(d) *equi-attractive*, if

$$(\forall t_0 \in \mathbb{R})(\exists \lambda = \lambda(t_0) > 0)(\forall \varepsilon > 0)(\exists T = T(t_0, \varepsilon) > 0)$$

$$(\forall u_0 < \lambda)(\forall t \geq t_0 + T) : u^+(t; t_0, u_0) < \varepsilon;$$

(e) *uniformly attractive*, if the numbers λ and T in (d) are independent of $t_0 \in \mathbb{R}$;

(f) *asymptotically stable*, if it is stable and attractive;

(g) *uniformly asymptotically stable*, if it is uniformly stable and uniformly attractive.

Theorem 2.6 Assume that the conditions of Lemma 1.1 and the inequalities (2.1) hold, and $f(t,0) = 0$, $g(t,0) = 0$ for $t \in [t_0, \infty)$.

Then the stability properties of the zero solution of the comparison equation (1.11) for $_{t_0}\mathcal{D}_t^\beta = {}_{t_0}^c D_t^\beta$, $0 < \beta < 1$, imply the corresponding stability properties of the zero solution of system (1.1) for $_{t_0}\mathcal{D}_t^\alpha = {}_{t_0}^c D_t^\alpha$, $0 < \alpha < 1$.

Proof. We shall first prove the stability of the zero solution of (1.1). Suppose that the trivial solution of (1.11) is stable. Then, we have

$$u_0 < w_2(\delta) \text{ implies } u^+(t;t_0,u_0) < w_1(\varepsilon), \ t \geq t_0 \qquad (2.12)$$

for some given $t_0 \in \mathbb{R}$, where $u_0 \in \mathbb{R}_+$ and the maximal solution $u^+(t;t_0,u_0)$ of (1.11) is defined in the interval $[t_0, \infty)$.

Let $\varphi_0 \in C[[-r,0],\Omega]$, and let $x(t) = x(t;t_0,\varphi_0)$ be the solution of the IVP (1.1), (1.2). Setting $u_0 = \max_{-r \leq \theta \leq 0} V(t_0, \varphi_0(\theta))$, we get by Lemma 1.1,

$$V(t,x(t;t_0,\varphi_0)) \leq u^+(t;t_0, \max_{-r \leq \theta \leq 0} V(t_0, \varphi_0(\theta))) \text{ for } t \geq t_0. \qquad (2.13)$$

Let $\|\varphi_0\|_r < \delta$. Then (2.1) imply

$$\max_{-r \leq \theta \leq 0} V(t_0, \varphi_0(\theta)) \leq w_2(\|\varphi_0\|_r) < w_2(\delta), \ t \geq t_0,$$

which due to (2.12) implies

$$u^+(t;t_0, \max_{-r \leq \theta \leq 0} V(t_0, \varphi_0(\theta))) < w_1(\varepsilon), \ t \geq t_0. \qquad (2.14)$$

Consequently, from (2.1), (2.13) and (2.14), we obtain

$$w_1(\|x(t;t_0,\varphi_0)\|) \leq V(t,x(t;t_0,\varphi_0))$$

$$\leq u^+(t;t_0, \max_{-r \leq \theta \leq 0} V(t_0, \varphi_0(\theta))) < w_1(\varepsilon), \ t \geq t_0.$$

Hence, $\|x(t;t_0,\varphi_0)\| < \varepsilon, t \geq t_0$ for the given $t_0 \in \mathbb{R}$, which proves the stability of the zero solution of (1.1).

Suppose now, that the zero solution of (1.11) is uniformly stable. Therefore, we have that

$$u_0 < w_2(\delta) \text{ implies } u^+(t;t_0,u_0) < w_1(\varepsilon), \ t \geq t_0 \qquad (2.15)$$

for every $t_0 \in \mathbb{R}$.

We claim that $\|\varphi_0\|_r < \delta$ implies $\|x(t;t_0,\varphi_0)\| < \varepsilon, t \geq t_0$ for every $t_0 \in \mathbb{R}$. If the claim is not true, there exists $t_0 \in \mathbb{R}$, a corresponding solution $x(t;t_0,\varphi_0)$ of (1.1) with $\|\varphi_0\|_r < \delta$, and $t^* > t_0$ such that,

$$\|x(t^*;t_0,\varphi_0)\| > \varepsilon, \ \|x(t;t_0,\varphi_0)\| \leq \varepsilon, \ t_0 \leq t < t^*.$$

Let $H = const > 0 : \{x \in \mathbb{R}^n : ||x|| \leq H\} \subset \Omega$. To complete the proof, suppose $\delta = const > 0$ be such that $\delta < H$.

Since

$$||\varphi_0||_r < \delta < H,$$

and the function $x(t)$ is continuous on $[t_0, t^*]$, there exists t^0, $t_0 < t^0 \leq t^*$, such that

$$\varepsilon < ||x(t^0)|| < H, \ ||x(t)|| < H, \ t_0 < t \leq t^0. \tag{2.16}$$

Set again $u_0 = \max_{-r \leq \theta \leq 0} V(t_0, \varphi_0(\theta))$, we get by Lemma 1.1

$$V(t, x(t; t_0, \varphi_0)) \leq u^+(t; t_0, \max_{-r \leq \theta \leq 0} V(t_0, \varphi_0(\theta))) \text{ for } t \in [t_0, t^0]. \tag{2.17}$$

From (2.16), (2.1), (2.17) and (2.15) we get

$$w_1(\varepsilon) < w_1(||x(t^0; t_0, \varphi_0)||) \leq V(t^0, x(t^0; t_0, \varphi_0))$$

$$\leq u^+(t^0; t_0, \max_{-r \leq \theta \leq 0} V(t_0, \varphi_0(\theta))) \leq w_1(\varepsilon).$$

The contradiction obtained proves that the zero solution of (1.1) is uniformly stable.

The proofs of the other stability properties are similar.

Remark 2.1 The conclusions of Theorem 2.6 remain true, if in the system (1.1), in the comparison equation (1.11), and in Lemma 1.1 we consider Riemann–Liouville fractional derivatives and suitable initial values.

Remark 2.2 We have assumed, in Theorem 2.6, stronger requirements on V only to unify all the stability results in one theorem. This puts burden on the comparison equation (1.11). However, to obtain only non-uniform stability criteria, we could weaken certain assumptions of Theorem 2.6 as in the next result.

Theorem 2.7 Assume that the conditions of Lemma 1.1 and the inequality (2.3) hold, and $f(t, 0) = 0$, $g(t, 0) = 0$ for $t \in [t_0, \infty)$.

Then the non-uniform stability properties of the zero solution of the comparison equation (1.11) for $_{t_0}\mathscr{D}_t^\beta = {}_{t_0}^c D_t^\beta$, $0 < \beta < 1$, imply the corresponding non-uniform stability properties of the zero solution of system (1.1) for $_{t_0}\mathscr{D}_t^\alpha = {}_{t_0}^c D_t^\alpha$, $0 < \alpha < 1$.

Example 2.1 Let $0 < \alpha < 1$. Consider the following scalar fractional-order equation

$$_0^c D_t^\alpha x(t) = -\alpha_1 x(t) + \alpha_2 x(t - r_0(t)), \quad t \geq 0, \tag{2.18}$$

where $x \in \Omega$, Ω is an open set in $\mathbb{R}$ containing the origin, $\alpha_1 > 0$, $\alpha_2 > 0$ and $\alpha_2 \leq \alpha_1$, $r_0(t)$ is a bounded continuous function on $\mathbb{R}$, $0 \leq r_0(t) \leq r$ for $t \geq 0$.

Let $V(t,x) = |x|$. Then, for $t \geq 0$, we have

$$^c D_+^\alpha V(t, \varphi(0)) = {}^c D_+^\alpha |\varphi(0)|$$

$$\leq -\alpha_1 |\varphi(0)| + \alpha_2 \max_{-r \leq \theta \leq 0} |\varphi(\theta)| \leq -(\alpha_1 - \alpha_2)|\varphi(0)| \leq 0, \qquad (2.19)$$

whenever $\max_{-r \leq \theta \leq 0} |\varphi(\theta)| \leq |\varphi(0)|$, $\varphi \in C[[-r,0], \Omega]$. Since $V(t,x) = |x|$ we have shown that $^c D_+^\alpha V(t, \varphi(0)) \leq 0$ whenever $V(t + \theta, \varphi(\theta)) \leq V(t, \varphi(0))$ for $-r \leq \theta \leq 0$. Theorem 2.1 implies that the zero solution of equation (2.18) is uniformly stable.

If, in addition, there exists a positive constant w such that $\alpha_1 \geq w + \alpha_2$, then

$$^c D_+^\alpha V(t, \varphi(0)) = {}^c D_+^\alpha |\varphi(0)| \leq -w|\varphi(0)| = -wV(t, \varphi(0)),$$

whenever $V(t + \theta, \varphi(\theta)) \leq V(t, \varphi(0))$ for $-r \leq \theta \leq 0$, and according to Theorem 2.3 the zero solution of (2.18) is uniformly asymptotically stable.

Example 2.2 Consider the linear system

$$_0^c D_t^\beta x(t) = Ax(t) + Bx(t - r), \ t \geq t_0, \qquad (2.20)$$

where $0 < \beta < 1$, $t_0 \in \mathbb{R}_+$, $x \in \mathbb{R}^n$, $r > 0$, A and B are diagonal constant $(n \times n)$ matrices.

Note that linear systems of fractional order are used to describe many real-world phenomena involving thresholds in medicine, biology, optimal control models in economics, pharmacokinetics, etc.

Let $V(t,x) = ||x||$, where $||.||$ is any norm in $\mathbb{R}^n$.

If $A = diag(a_1, a_2, ..., a_n)$, $B = diag(b_1, b_2, ..., b_n)$, $a_i \leq 0$, $b_i \geq 0$, $a = \min_i (-a_i)$ and $b = \max_i b_i$, then for $t \geq t_0$ and $\varphi \in C[[-r,0], \Omega]$, we have

$$^c D_+^\beta V(t, \varphi(0)) = {}^c D_+^\beta ||\varphi(0)|| \leq a||\varphi(0)|| + b||\varphi(-r)||.$$

(i) If $a + b \leq 0$, then according to the Example 2.1, the zero solution of the scalar equation (2.18) for $a = -\alpha_1$ and $b = \alpha_2$ is uniformly stable. Therefore, due to Theorem 2.6, the zero solution of the linear system (2.20) is uniformly stable.

(ii) If there exists a constant w such that $a + b \leq -w$, then according to the Example 2.1, the zero solution of the scalar equation (2.18) for $a = -\alpha_1$ and $b = \alpha_2$ is uniformly asymptotically stable. Therefore, due to Theorem 2.6, the zero solution of the linear system (2.20) is uniformly asymptotically stable.

Next, Lyapunov stability results for systems of type (1.5) will be presented. In the proofs of our main theorems we shall use piecewise continuous Lyapunov functions $V : [t_0, \infty) \times \Omega \to \mathbb{R}_+$, $V \in V_0$ for which the condition H2.1 is true. We shall investigate the stability of the zero solution of system (1.5). That is why the following conditions will be used:

H2.2 $f(t,0) = 0, t \geq t_0$.

H2.3 $I_k(0) = 0, k = \pm 1, \pm 2, \ldots$.

Theorem 2.8 Assume that:

1. Conditions H2.2 and H.2.3 are met.

2. There exists a function $V \in V_0$ such that H2.1 holds,

$$w_1(||x||) \leq V(t,x), \ w_1 \in K, \ (t,x) \in [t_0,\infty) \times \Omega, \tag{2.21}$$

$$^cD_+^\beta V(t,x) \leq 0, \ (t,x) \in G_k, \tag{2.22}$$

$$V(t^+, x + I_k(x)) \leq V(t,x), \ x \in \Omega, \ t = t_k, \tag{2.23}$$

where $t \in [t_0,\infty)$, $\beta \in (0,1)$.

Then the zero solution of system (1.5) is stable.

Proof. Let $\varepsilon > 0$. From the properties of the function V, it follows that there exists a constant $\delta = \delta(t_0, \varepsilon) > 0$ such that if $||x|| \leq \delta$, then

$$\sup_{||x|| \leq \delta} V(t_0^+, x) < w_1(\varepsilon). \tag{2.24}$$

Let $x_0 \in \Omega : ||x_0|| < \delta$, and let $x(t) = x(t;t_0,x_0)$ be the solution of the IVP (1.5), (1.4). We shall prove that $||x(t;t_0,x_0)|| < \varepsilon$ for $t \geq t_0$.

Suppose that this is not true. Then there exist a solution $x(t;t_0,x_0)$ of (1.5) for which $||x_0|| < \delta$ and $t^* > t_0$, $t_k < t^* \leq t_{k+1}$, for some fixed integer k such that

$$||x(t^*)|| \geq \varepsilon \ \text{ and } \ ||x(t;t_0,x_0)|| < \varepsilon, \ t \in [t_0,t_k].$$

Then, due to (2.23) and the properties of $E + I_k$, for any k, we can find t^0, $t_k < t^0 \leq t^*$, such that

$$||x(t^0)|| > \varepsilon \ \text{ and } \ x(t^0;t_0,x_0) \in \Omega. \tag{2.25}$$

For $t \in [t_0,t^0]$ it follows from Corollary 1.4 that

$$V(t,x(t;t_0,x_0)) \leq V(t_0^+,x_0), t \in [t_0,t^0]. \tag{2.26}$$

From (2.25), (2.21), (2.26) and (2.24) there follow the inequalities

$$w_1(\varepsilon) < w_1(||x(t^0;t_0,x_0)||) \leq V(t^0,x(t^0;t_0,x_0)) \leq V(t_0^+,x_0) < w_1(\varepsilon).$$

The contradiction obtained shows that

$$||x(t;t_0,x_0)|| < \varepsilon$$

for $||x_0|| < \delta$ and $t \geq t_0$. This implies that the zero solution of system (1.5) is stable.

Theorem 2.9 Let the conditions of Theorem 2.8 hold, and let a function $w_2 \in K$ exists such that

$$V(t,x) \leq w_2(||x||), \ (t,x) \in (t_0,\infty) \times \Omega. \tag{2.27}$$

Then the zero solution of system (1.5) is uniformly stable.

Proof. Let $\varepsilon > 0$ be chosen. Choose $\delta = \delta(\varepsilon) > 0$ so that $w_2(\delta) < w_1(\varepsilon)$.

Let $x_0 \in \Omega : ||x_0|| < \delta$ and $x(t) = x(t;t_0,x_0)$ be the solution of problem (1.5), (1.4).

It follows from Corollary 1.4 that

$$V(t,x(t;t_0,x_0)) \leq V(t_0^+,x_0), \ t \geq t_0.$$

From the above inequalities and (2.27), we get to the inequalities

$$w_1(||x(t;t_0,x_0)||) \leq V(t_0^+,x_0) \leq w_2(||x_0||) < w_2(\delta) < w_1(\varepsilon),$$

from which it follows that $||x(t;t_0,x_0)|| < \varepsilon$ for $t \geq t_0$. This proves the uniform stability of the zero solution of system (1.5).

Theorem 2.10 Assume that:

1. Condition 1 of Theorem 2.8 holds.
2. There exists a function $V \in V_0$ such that H2.1 and (2.23) hold, and

$$w_1(||x||) \leq V(t,x) \leq w_2(||x||), \ w_1,w_2 \in K, \ (t,x) \in [t_0,\infty) \times \Omega, \tag{2.28}$$

$$^cD_+^\beta V(t,x) \leq -w_3(||x||), \ w_3 \in K, \ (t,x) \in G_k, \tag{2.29}$$

where $t \in [t_0,\infty)$, $\beta \in (0,1)$.

Then the zero solution of system (1.5) is uniformly asymptotically stable.

Proof. 1. Let $H = const > 0 : \{x \in \mathbb{R}^n : ||x|| \leq H\} \subset \Omega$.

For any $t \in [t_0,\infty)$ denote

$$V_{t,H}^{-1} = \{x \in \Omega : V(t^+,x) \leq w_1(H)\}.$$

From (2.28), we deduce

$$V_{t,H}^{-1} \subset \{x \in \mathbb{R}^n : ||x|| \leq H\} \subset \Omega.$$

From condition 2 of Theorem 2.10, it follows that for any $t_0 \in \mathbb{R}_+$ and any $x_0 \in V_{t_0,H}^{-1}$ we have $x(t;t_0,x_0) \in V_{t,H}^{-1}, t \geq t_0$.

Let $\varepsilon > 0$ be chosen. Choose $\eta = \eta(\varepsilon)$ so that $w_2(\eta) < w_1(\varepsilon)$, and let $T > \left[\dfrac{w_2(H)\Gamma(1+\beta)}{w_3(\eta)}\right]^{1/\beta}$. If we assume that for each $t \in [t_0,t_0+T]$ the inequality $||x(t;t_0,x_0)|| \geq \eta$ is valid, then from (2.23), (2.29), we get

$$V(t,x(t;t_0,x_0)) \leq V(t_0^+,x_0)$$

$$-\frac{1}{\Gamma(\beta)}\int_{t_0}^{t} w_3(||x(\tau;t_0,x_0)||)(t-\tau)^{\beta-1}d\tau \leq w_2(H) - \frac{w_3(\eta)T^\beta}{\Gamma(\beta+1)} < 0,$$

which contradicts (2.28). The contradiction obtained shows that there exists $t^* \in [t_0,t_0+T]$ such that $||x(t^*;t_0,x_0)|| < \eta$.

Then from (2.23), (2.28) and (2.29) it follows that for $t \geq t^*$ (hence for any $t \geq t_0+T$) the following inequalities hold

$$w_1(||x(t;t_0,x_0)||) \leq V(t;x(t;t_0,x_0)) \leq V(t^*,x(t^*;t_0,x_0))$$

$$\leq w_2(||x(t^*;t_0,x_0)||) < w_2(\eta) < w_1(\varepsilon).$$

Therefore, $||x(t;t_0,x_0|| < \varepsilon$ for $t \geq t_0+T$.

2. Let $\lambda = const > 0$ be such that $w_2(\lambda) < w_1(H)$. Then, if $x_0 \in \Omega : ||x_0|| < \lambda$, (2.18) implies

$$V(t_0^+,x_0) \leq w_2(||x_0||) < w_2(\lambda) < w_1(H),$$

which shows that $x_0 \in V_{t_0,H}^{-1}$. From what we proved in item 1, it follows that the zero solution of system (1.5) is uniformly attractive and since Theorem 2.9 implies that it is uniformly stable, then the solution $x \equiv 0$ is uniformly asymptotically stable.

Theorem 2.11 Let the conditions of Theorem 2.10 hold, except replacing $^cD_+^\beta V$ by $D_+^\beta V$.
Then the zero solution of system (1.5) is uniformly asymptotically stable.

Proof. The proof of Theorem 2.11 is analogous to the proof of Theorem 2.10. It uses the fact that

$$^cD_+^\beta V(t,x) \leq D_+^\beta V(t,x), \ t \neq t_k, \ k = \pm1, \pm2, ...,$$

which implies (2.29).

Example 2.3 Let $0 < \alpha < 1$, $0 < t_1 < t_2 < ...$ and $\lim_{k\to\infty} t_k = \infty$. Consider the following scalar fractional-order equation:

$$\begin{cases} {}_0^cD_t^\alpha x(t) = \dfrac{-\alpha_1 x(t)}{(9+e^t)(1+x(t))}, \ t \neq t_k, \ t \geq 0, \\ x(t_k^+) = \dfrac{k^2}{k^2+1}x(t_k), \ k = 1,2,..., \end{cases} \quad (2.30)$$

where $x \in \mathbb{R}_+$ and $\alpha_1 > 0$.

Let $V(t,x) = |x|$. Then, the inequality

$$^cD_+^\alpha V(t,x) = {}^cD_+^\alpha|x| \leq \frac{-\alpha_1}{10}V(t,x).$$

holds for $t \neq t_k$, $t \geq 0$. Also, for $t = t_k$, $k = 1, 2, \ldots$

$$V(t_k^+, x) = \frac{k^2}{k^2 + 1} |x| \leq V(t_k, x),\ t_k > 0.$$

Hence, the conditions of Theorem 2.10 are satisfied and the zero solution of equation (2.30) is uniformly asymptotically stable.

The proofs of the next Lyapunov stability theorems for system (1.6) are analogous to the proofs of Theorems 2.8–2.11. In this case the Razumikhin condition (1.10) is used, as well as corollaries 1.5 and 1.6. Since the solutions of system (1.6) are piecewise continuous functions, Lyapunov functions $V : [t_0, \infty) \times \Omega \to \mathbb{R}_+$ from the class V_0 are the base of the proofs (see [Stamova and Stamov G 2014b]).

Theorem 2.12 Assume that:

1. Conditions H2.2 and H.2.3 are met for system (1.6).
2. There exists a function $V \in V_0$ such that H2.1 and (2.21) hold,

$$V(t^+, \varphi(0) + I_k(\varphi)) \leq V(t, \varphi(0)),\ t = t_k,\ t_k > t_0, \qquad (2.31)$$

and the inequality

$$^cD_+^\beta V(t, \varphi(0)) \leq 0,\ t \neq t_k,\ t \in [t_0, \infty)$$

is valid whenever $V(t + \theta, \varphi(\theta)) \leq V(t, \varphi(0))$ for $-r \leq \theta \leq 0$, $\varphi \in PC[[-r, 0], \Omega]$, $\beta \in (0, 1)$.
Then the zero solution of system (1.6) is stable.

Theorem 2.13 Let the conditions of Theorem 2.12 and condition (2.27) hold.
Then the zero solution of system (1.6) is uniformly stable.

Theorem 2.14 Assume that:

1. Condition 1 of Theorem 2.12 hold.
2. There exists a function $V \in V_0$ such that H2.1, (2.28) and (2.31) hold, and the inequality

$$^cD_+^\beta V(t, \varphi(0)) \leq -w_3(||\varphi(0)||),\ w_3 \in K,\ t \neq t_k,\ t \in [t_0, \infty)$$

is valid whenever $V(t + \theta, \varphi(\theta)) \leq V(t, \varphi(0))$ for $-r \leq \theta \leq 0$, $\varphi \in PC[[-r, 0], \Omega]$, $\beta \in (0, 1)$.
Then the zero solution of system (1.6) is uniformly asymptotically stable.

Example 2.4 Let $\tau > 0$. Consider the following impulsive fractional single species model of a Lotka-Volterra type

$$\begin{cases} ^cD^{1/2}x(t) = x(t)\big(a(t) - b(t)x(t - \tau)\big),\ t > 0,\ t \neq t_k, \\ \Delta x(t_k) = c_k x(t_k),\ k = 1, 2, \ldots, \end{cases} \qquad (2.32)$$

where $x \in \mathbb{R}_+$, $a, b \in \mathbb{R}_+$, $-1 < c_k \leq 0$, $k = 1, 2, ..., 0 < t_1 < t_2 < ... < t_k < t_{k+1} < ...$ and $\lim_{k \to \infty} t_k = \infty$.

Let $\varphi_0 \in PC[[-\tau, 0], \mathbb{R}_+]$. Denote by $x(t) = x(t; 0, \varphi_0)$ the solution of (2.32) satisfying the initial conditions

$$x(\theta) = \varphi_0(\theta) \geq 0, \quad \theta \in [-\tau, 0); \quad x(0) > 0. \tag{2.33}$$

Let K be the carrying capacity of the environment. Moreover, we assume that solutions of the IVP (2.32), (2.33) are nonnegative, and there exist positive constants a_*, a^*, b_* and b^* such that

$$a_* \leq a(t) \leq a^*, \quad b_* \leq b(t) \leq b^*, \quad t \geq 0. \tag{2.34}$$

Note that these assumptions are natural from the biological point of view.

We shall prove that if

$$\varphi_0(-\tau) \geq \frac{a^*}{b_*}, \tag{2.35}$$

then the zero solution of (2.32) is uniformly stable.

Let $V(t, x) = x^2$. Then, for $t \geq 0$, and $t \neq t_k$, we have

$$^cD_+^{1/2}V(t, \varphi(0)) = \frac{1}{\Gamma(1/2)} \int_0^t \frac{[\varphi^2(0)]'}{(t-s)^{1/2}} ds = \frac{1}{\Gamma(1/2)} \int_0^t \frac{2\varphi(0)\varphi'(0)}{(t-s)^{1/2}} ds.$$

Using (2.34), (2.35), a composition with fractional derivatives, and the linearity of Caputo fractional derivatives (see [Podlubny 1999]), we get

$$
\begin{aligned}
^cD_+^{1/2}V(t, \varphi(0)) \ &\leq \ \frac{1}{\sqrt{\pi}} \int_0^t \frac{2\varphi(0)\,^cD^{1/2}\left[^cD^{1/2}\varphi(0)\right]}{(t-s)^{1/2}} ds \\
&\leq \ \frac{1}{\sqrt{\pi}} \int_0^t \frac{2\varphi(0)\,^cD^{1/2}\left[\varphi(0)(a^* + b^*\varphi(-\tau))\right]}{(t-s)^{1/2}} ds \\
&\leq \ \frac{1}{\sqrt{\pi}} \int_0^t \frac{2\varphi(0)K^cD^{1/2}(a^* + b^*\varphi(0))}{(t-s)^{1/2}} ds \\
&= \ \frac{2Kb^*}{\sqrt{\pi}} \int_0^t \frac{\varphi(0)\,^cD^{1/2}\varphi(0)}{(t-s)^{1/2}} ds \\
&= \ \frac{2Kb^*}{\sqrt{\pi}} \int_0^t \frac{\varphi(0)[\varphi(0))(a(0) - b(0)\varphi(-\tau))}{(t-s)^{1/2}} \\
&\leq \ \frac{2Kb^*V(\varphi(0))}{\sqrt{\pi}} \int_0^t \frac{a^* - b_*\varphi(-\tau)}{(t-s)^{1/2}} ds \\
&= \ \frac{4Kb^*V(\varphi(0))}{\sqrt{\pi}} \left(a^* - b_*\varphi(-\tau)\right)\sqrt{t} \leq 0,
\end{aligned}
$$

whenever $V(t+\theta,\varphi(\theta)) \le V(t,\varphi(0))$, $\theta \in [-\tau,0]$, $\varphi \in PC[[-r,0],\mathbb{R}_+]$, $t \in \mathbb{R}_+$.
Also, for $k = 1,2,...$,

$$V(t_k^+,\varphi(0)+c_k\varphi(0)) = (1+c_k)^2\varphi^2(0) \le V(t_k,\varphi(0)).$$

Thus by Theorem 2.13, the trivial solution of model (2.32) is uniformly stable.

In the above proof, for the case when $t \ne t_k$, we apply the linearity of Caputo fractional derivatives. The proof in that case can be also done by using Lemma 1.8 as follows. For the Caputo fractional derivative ${}^cD_+^{1/2}V(t,\varphi(0))$ with respect to the model (2.32) for $t \ge 0$, and $t \ne t_k$, we have

$$^cD_+^{1/2}V(t,\varphi(0)) \le 2\varphi(0){}^cD_+^{1/2}V(t,\varphi(0))$$

$$= 2\varphi^2(0)\big(a(0)-b(0)\varphi(-\tau)\big).$$

Then by (2.34), (2.35) for $V(t+\theta,\varphi(\theta)) \le V(t,\varphi(0))$, $\theta \in [-\tau,0]$, $\varphi \in PC[[-r,0],\mathbb{R}_+]$, $t \in \mathbb{R}_+$ it follows that

$$^cD_+^{1/2}V(t,\varphi(0)) \le 0, \ t \ge t_0, \ t \ne t_k.$$

2.2   Theorems on Boundedness

In this section boundedness of the solutions of fractional order differential system (1.8) is investigated.

Theorem 2.15 Assume that there exists a function $V \in C_0$ such that H2.1 holds,

$$w_1(||x||) \le V(t,x), \ (t,x) \in [t_0,\infty) \times \mathbb{R}^n, \tag{2.36}$$

where $w_1 \in K$ and $w_1(u) \to \infty$ as $u \to \infty$, and the inequality (2.2) is valid whenever $V(t+\theta,\varphi(\theta)) \le V(t,\varphi(0))$ for $-r \le \theta \le 0$, where $t \in [t_0,\infty)$, $\varphi \in \mathscr{C}$, $\beta \in (0,1)$.
Then the solutions of system (1.8) are equi-bounded.

Proof. Let $a > 0$, $t_0 \in \mathbb{R}$, $\varphi_0 \in \mathscr{C}$. Consider the solution $x(t) = x(t;t_0,\varphi_0)$ of the IVP (1.8), (1.2) for which $||\varphi_0||_r < a$.

From the properties of the function V, it follows that there exists a constant $\bar{\Gamma} = \bar{\Gamma}(t_0,a) > 0$ such that if $x \in \mathbb{R}^n : ||x|| < a$, then $\max\limits_{||x||<a} V(t_0,x) < \bar{\Gamma}(t_0,a)$.

Since for the function $w_1 \in K$ we have $w_1(u) \to \infty$ as $u \to \infty$, then we can choose $b = b(t_0,a) > 0$ so that $b > a$ and $w_1(b) > \bar{\Gamma}(t_0,a)$.

We shall prove that $||x(t;t_0,\varphi_0)|| < b$ for $t \ge t_0$.

Suppose that this is not true. Then, there exists $t^* > t_0$ such that

$$||x(t^*)|| > b, \text{ and } ||x(t;t_0,\varphi_0)|| \le b, t \in [t_0,t^*). \tag{2.37}$$

Since the conditions of Corollary 1.2 are met, then

$$V(t,x(t;t_0,\varphi_0)) \le \max_{-r \le \theta \le 0} V(t_0,\varphi_0(\theta)), t \in [t_0,\infty).$$

From the above inequality, (3.36) and (2.37) we have

$$w_1(||x(t^*;t_0,\varphi_0)||) \le V(t^*,x(t^*;t_0,\varphi_0)) \le \max_{-r \le \theta \le 0} V(t_0,\varphi_0(\theta)) \le \bar{\Gamma}(t_0,a),$$

which contradicts the choice of b.

Therefore, $||x(t;t_0,\varphi_0)|| < b$ for $t \ge t_0$. This implies that the solutions of (1.8) are equi-bounded.

The proofs of the next results are essentially repetitions of the arguments used in the proofs of Theorems 2.1 and 2.2 and we omit the details here.

Theorem 2.16 Assume that there exist a function $V \in C_0$ for which H2.1 holds, and a constant $H > 0$ such that:

1. The inequalities (2.1) hold for $(t,x) \in [t_0,\infty) \times \mathbb{R}^n$, where $w_1, w_2 \in K$ and $w_1(u) \to \infty$ as $u \to \infty$.

2. The inequality (2.2) is valid whenever $||\varphi(0)|| \ge H$, $V(t+\theta,\varphi(\theta)) \le p(V(t,\varphi(0)))$ for $-r \le \theta \le 0$, where $t \in [t_0,\infty)$, $\varphi \in \mathscr{C}$, $\beta \in (0,1)$, p is continuous and non-decreasing on $\mathbb{R}_+$, and $p(u) > u$ as $u > 0$.

Then the solutions of system (1.8) are uniformly bounded.

Theorem 2.17 If in Theorem 2.16 condition (2.2) is replaced by the condition (2.4) whenever $||\varphi(0)|| \ge H$, $V(t+\theta,\varphi(\theta)) \le p(V(t,\varphi(0)))$ for $-r \le \theta \le 0$, where $t \in [t_0,\infty)$, $\varphi \in \mathscr{C}$, $\beta \in (0,1)$, $w_3 \in K$, p is continuous and non-decreasing on $\mathbb{R}_+$, and $p(u) > u$ as $u > 0$, then the solutions of system (1.8) are uniformly ultimately bounded.

Remark 2.3 Theorem 2.17 is related to a result of Hale (see [Hale 1977]) for integer-order functional differential equations. As in that reference, we note that, in the applications one often needs a generalization of this result. More precisely, one may apply inequality (2.4) only for some coordinates of $\varphi(0)$.

Example 2.5 Let $0 < \alpha < 1$ and $x \in \mathbb{R}$. Consider the nonlinear equation

$$_{t_0}^{c}D_t^{\alpha}x(t) = -3\xi x^4(t) + \int_0^t q(t,\theta)x^3(\theta)\,d\theta, \tag{2.38}$$

where $\xi > 0$ and the delay kernel $q \in C[\mathbb{R}_+ \times \mathbb{R}_+, \mathbb{R}_+]$.

Assume that there exists a constant $\mu > 1$ such that

$$\int_0^t q(t,\theta)\,d\theta \leq \frac{\xi}{\mu^3}. \tag{2.39}$$

Then (2.39) is a sufficient condition for uniform ultimate boundedness of the solutions of (2.38). In fact, we can choose $w_1(u) = w_2(u) = u$, $w_3(u) = 2\xi u^4$.
Let $V(t,x) = |x|$, and denote

$$p(u) = \mu u.$$

Thus, using (2.39) we have

$$^cD_+^\alpha V(t,\varphi(0)) = {}^cD_+^\alpha|\varphi(0)| \leq -3\xi|\varphi^4(0)| + \int_0^t q(t,\theta)|\varphi^3(\theta)|\,d\theta$$

$$\leq -3\xi|\varphi(0)|^4 + \mu^3|\varphi(0)|^3 \int_0^t q(t,\theta)\,d\theta$$

$$\leq -3\xi|\varphi(0)|^4 + \mu^3|\varphi(0)|^4 \frac{\xi}{\mu^3} = -2\xi|\varphi(0)|^4,$$

whenever $|\varphi(0)| \geq 1$, $\varphi \in C[[0,t],\mathbb{R}]$ and $p(V(t,\varphi(0))) = \mu|\varphi(0)| \geq |\varphi(\theta)| = V(t+\theta,\varphi(\theta))$ for $0 \leq \theta \leq t$.

Then all conditions of Theorem 2.17 are satisfied. Hence, the solutions of (2.38) are uniformly ultimately bounded.

Example 2.6 As an example for the generalization of Theorem 2.17 discussed in Remark 2.3 consider the following system

$$\begin{cases} {}_0^cD_t^\beta x(t) = y(t), \\ {}_0^cD_t^\beta y(t) = -g(t,y(t)) - f(x(t)) + \int_{-r}^0 h(x(t+\theta))y(t+\theta)d\theta, \end{cases} \tag{2.40}$$

where $0 < \beta < 1$, $r > 0$, $x,y,f,h : \mathbb{R} \to \mathbb{R}$ are continuous, $g : \mathbb{R}^2 \to \mathbb{R}$ is continuous and mapping $\mathbb{R} \times$ (bounded sets of $\mathbb{R}$) into bounded sets of $\mathbb{R}$.
Assume that:

(i) There exist positive constants A and H, such that

$$||g(t,y)|| > A||y|| \text{ for } ||y|| \geq H;$$

(ii)) There exists a positive constant L, such that

$$||h(x)|| \leq L \text{ for all } x \in \mathbb{R},$$

and $Lr < A$.

Let $V(x,y) = ||y|| + V_1(x)$, where $V_1(x) = {}_0\mathscr{D}_x^{-\beta}f(x)$ and $p > 1$, $pLr < A$. Then

$$^cD_+^\beta V(\varphi(0),\phi(0)) \leq -A||\phi(0)||$$

$$+ \int_{-r}^0 h(\varphi(\theta))\phi(\theta)d\theta \leq -(A-pLr)||\phi(0)||,$$

if $||\phi(0)|| \geq H$ and $||\phi(\theta)|| \leq p||\phi(0)||$, $\varphi,\phi \in \mathscr{C}$. By choosing $H_1 \geq H$ appropriately we obtain

$$^cD_+^\beta V(\varphi(0),\phi(0)) \leq -w||\phi(0)||$$

for $w > 0$, $||\phi(0)|| \geq H_1$ and $||\phi(\theta)|| \leq p||\phi(0)||$. Therefore, the Remark 2.3 imply that the second variable of the solutions of (2.40) is uniformly ultimately bounded. The uniform ultimate boundedness of the first variable x, which will implies the uniform ultimate boundedness of the solutions, can be considered as for the integer order case [Hale 1977], only under assumptions $A > 0$ and $f(x)\mathrm{sgn}x \to \infty$ as $||x|| \to \infty$.

2.3 Global Stability

In this section the global stability of systems of impulsive differential equations of fractional order in the Caputo sense with impulsive effect at fixed moments of time will be considered.

Let $f : \mathbb{R} \times \mathbb{R}^n \to \mathbb{R}^n$. Consider the following system of impulsive fractional differential equations

$$\begin{cases} {}_{t_0}^c D_t^\alpha x(t) = f(t,x(t)), t \neq t_k, \\ \Delta x(t_k) = I_k(x(t_k)), k = \pm 1, \pm 2, ..., \end{cases} \qquad (2.41)$$

where $f : \mathbb{R} \times \mathbb{R}^n \to \mathbb{R}^n$, ${}_{t_0}^c D_t^\alpha$ denotes the Caputo fractional derivative of order α, $0 < \alpha < 1$, $\Delta x(t_k) = x(t_k^+) - x(t_k)$, $I_k : \mathbb{R}^n \to \mathbb{R}^n$, $t_k < t_{k+1} < ...$, $k = \pm 1, \pm 2, ...$ and $\lim_{k \to \pm\infty} t_k = \pm\infty$.

For $t_0 \in \mathbb{R}$, we suppose again that the functions f and I_k, $k = \pm 1, \pm 2, ...$, are smooth enough on $[t_0, \infty) \times \mathbb{R}^n$ and $\mathbb{R}^n$, respectively, to guarantee the existence, uniqueness and continuability of the solution $x(t) = x(t; t_0, x_0)$ of the IVP (2.41), (1.4) on the interval $[t_0, \infty)$ for each $x_0 \in \mathbb{R}^n$ and $t \geq t_0$. The solutions $x(t; t_0, x_0)$ are, in general, piecewise continuous functions with points of discontinuity of the first kind at which they are left-continuous (see [Stamova 2014a]), that is, at the moments t_k, $t_k > t_0$, the following relations are satisfied:

$$x(t_k^-) = x(t_k) \text{ and } x(t_k^+) = x(t_k) + I_k(x(t_k)).$$

We shall use the following definitions of global stability of the zero solution of (2.41).

Definition 2.2 The zero solution $\psi(t) \equiv 0$ of system (2.41) is said to be:

(a) *globally equi-attractive*, if

$$(\forall t_0 \in \mathbb{R})(\forall v > 0)(\forall \varepsilon > 0)(\exists \gamma = \gamma(t_0, v, \varepsilon) > 0)$$

$$(\forall x_0 \in \mathbb{R}^n : ||x_0|| < v)(\forall t \geq t_0 + \gamma) : ||x(t; t_0, x_0)|| < \varepsilon;$$

(b) *uniformly globally attractive*, if the number γ in (a) is independent of $t_0 \in \mathbb{R}$;

(c) *globally equi-asymptotically stable*, if it is stable and globally equi-attractive;

(d) *uniformly globally asymptotically stable*, if it is uniformly stable, uniformly globally attractive and the solutions of system (2.41) are uniformly bounded.

In the proofs of the main theorems in this section we shall use piecewise continuous Lyapunov functions $V : \mathbb{R} \times \mathbb{R}^n \to \mathbb{R}_+, V \in V_0$ for which the condition H2.1 is satisfied.

We shall investigate the global stability of the zero solution of system (2.41). That is why the conditions H2.2 and H2.3 for this system will be used.

Theorem 2.18 Assume that:
  1. Conditions H2.2 and H2.3 hold for system (2.41).
  2. There exists a function $V \in V_0$ such that H2.1 holds, and

$$w_1(||x||) \leq V(t, x), \, w_1 \in K, \, (t, x) \in [t_0, \infty) \times \mathbb{R}^n, \tag{2.42}$$

$$V(t^+, x + I_k(x)) \leq V(t, x), \, x \in \mathbb{R}^n, \, t = t_k, \, t_k > t_0, \tag{2.43}$$

$${}^c D_+^\beta V(t, x) \leq -wV(t, x), \, (t, x) \in G, \, t \in [t_0, \infty), \tag{2.44}$$

where $w > 0$, $\beta \in (0, 1)$.

Then the zero solution of system (2.41) is globally equi-asymptotically stable.

Proof. Let $\varepsilon > 0$. From the properties of the function V, it follows that there exists a constant $\delta = \delta(t_0, \varepsilon) > 0$ such that if $x \in \mathbb{R}^n : ||x|| < \delta$, then $\sup_{||x|| < \delta} V(t_0^+, x) < w_1(\varepsilon)$.

Let $x_0 \in \mathbb{R}^n : ||x_0|| < \delta$ and $x(t) = x(t; t_0, x_0)$ be the solution of the IVP (2.41), (1.4).

Since all conditions of Corollary 1.4 are met, then

$$V(t, x(t; t_0, x_0)) \leq V(t_0^+, x_0), \, t \in [t_0, \infty). \tag{2.45}$$

On the other hand, $||x_0|| < \delta$ and hence $V(t_0^+, x_0) < w_1(\varepsilon)$.

From (2.42), (2.45) and the last inequality, there follow the inequalities

$$w_1(||x(t;t_0,x_0)||) \leq V(t,x(t;t_0,x_0)) \leq V(t_0^+,x_0) < w_1(\varepsilon),$$

which imply that $||x(t;t_0,x_0)|| < \varepsilon$ for $t \geq t_0$. This implies that the zero solution of system (2.41) is stable.

Now we shall prove that it is globally equi-attractive.

Let $v = const > 0$ and $x_0 \in \mathbb{R}^n : ||x_0|| < v$.

From conditions (2.43) and (2.44), it follows by Corollary 1.3, that for $t \geq t_0$ the following inequality is valid

$$V(t,x(t;t_0,x_0)) \leq V(t_0^+,x_0)E_\beta[-w(t-t_0)^\beta]. \tag{2.46}$$

Let $N(t_0,v) = \sup\{V(t_0^+,x) : ||x|| < v\}$ and

$$\gamma = \gamma(t_0,v,\varepsilon) > \left(\frac{-1}{w}E_\beta^{-1}\left(\frac{w_1(\varepsilon)}{N(t_0,v)}\right)\right)^{1/\beta}.$$

Then for $t \geq t_0 + \gamma$ from (2.46), it follows that

$$V(t,x(t;t_0,x_0)) < w_1(\varepsilon).$$

From the last inequality and (2.42) we have

$$||x(t;t_0,x_0)|| < \varepsilon,$$

which means that the zero solution of system (2.41) is globally equi-attractive.

Theorem 2.19 Assume that:

1. Condition 1 of Theorem 2.18 holds.

2. There exists a function $V \in V_0$ such that H2.1 and (2.43) hold,

$$w_1(||x||) \leq V(t,x) \leq h(t)w_2(||x||), \ (t,x) \in [t_0,\infty) \times \mathbb{R}^n, \tag{2.47}$$

where $w_1,w_2 \in K$, $h : [t_0,\infty) \to [1,\infty)$, and the inequality

$${}^cD_+^\beta V(t,x) \leq -g(t)w_3(||x||), \ (t,x) \in G, \ t \in [t_0,\infty) \tag{2.48}$$

is valid for $w_3 \in K$, $g : [t_0,\infty) \to (0,\infty)$, $\beta \in (0,1)$.

3. $\dfrac{1}{\Gamma(\beta)}\displaystyle\int_{t_0}^t (t-s)^{\beta-1}g(s)w_3\left[w_2^{-1}\left(\dfrac{\eta}{h(s)}\right)\right]ds = \infty$ for each sufficiently small value of $\eta > 0$ and $t \to \infty$.

Then the zero solution of system (2.41) is globally equi-asymptotically stable.

Proof. We can prove the stability of the zero solution of system (2.41) by the analogous arguments, as in the proof of Theorem 2.18.

Now we shall prove that the zero solution of (2.41) is globally equi-attractive.

Let $v > 0$ be arbitrary, $\varepsilon > 0$ be given and $\eta = \frac{w_1(\varepsilon)}{2}$. Let the number $\gamma = \gamma(t_0, v, \varepsilon) > 0$ be chosen so that

$$\frac{1}{\Gamma(\beta)} \int_{t_0}^{t_0+\gamma} (t_0 + \gamma - s)^{\beta-1} g(s) w_3 \left[w_2^{-1} \left(\frac{\eta}{h(s)} \right) \right] ds > h(t_0) w_2(v). \qquad (2.49)$$

(This is possible in view of condition 3 of Theorem 2.19.)

Let $x_0 \in \mathbb{R}^n$: $||x_0|| < v$ and $x(t) = x(t; t_0, x_0)$ be the solution of the IVP (2.41), (1.4). If we assume that for any $t \in [t_0, t_0 + \gamma]$ the following inequality holds

$$||x(t; t_0, x_0)|| \geq w_2^{-1} \left(\frac{\eta}{h(t)} \right), \qquad (2.50)$$

then by (2.49) and (2.50), it follows that

$$V(t, x(t; t_0, x_0)) \leq V(t_0^+, x_0) - \frac{1}{\Gamma(\beta)} \int_{t_0}^{t} (t - s)^{\beta-1} g(s) w_3 (||x(s; t_0, x_0)||) ds$$

$$\leq V(t_0^+, x_0) - \frac{1}{\Gamma(\beta)} \int_{t_0}^{t} (t - s)^{\beta-1} g(s) w_3 \left[w_2^{-1} \left(\frac{\eta}{h(s)} \right) \right] ds, \ t \in [t_0, t_0 + \gamma].$$

From the above inequalities, (2.47) and (2.49) for $t = t_0 + \gamma$, we obtain

$$V(t_0 + \gamma, x(t_0 + \gamma; t_0, x_0)) \leq h(t_0) w_2(v)$$

$$- \frac{1}{\Gamma(\beta)} \int_{t_0}^{t_0+\gamma} (t_0 + \gamma - s)^{\beta-1} g(s) w_3 \left[w_2^{-1} \left(\frac{\eta}{h(s)} \right) \right] ds < 0,$$

which contradicts (2.47). The contradiction obtained shows that there exists $t^* \in [t_0, t_0 + \gamma]$, such that

$$||x(t^*; t_0, x_0)|| < w_2^{-1} \left(\frac{\eta}{h(t^*)} \right).$$

Then for $t \geq t^*$ (hence for any $t \geq t_0 + \gamma$ as well) the following inequalities are valid

$$w_1 (||x(t; t_0, x_0)||) \leq V(t, x(t; t_0, x_0)) \leq V(t^*, x(t^*; t_0, x_0))$$

$$\leq h(t^*) w_2(||x(t^*; t_0, x_0)||) < \eta < w_1(\varepsilon).$$

Therefore, $||x(t; t_0, x_0)|| < \varepsilon$ for $t \geq t_0 + \gamma$, i.e. the zero solution of (2.1) is globally equi-attractive.

Theorem 2.20 Assume that:

1. Condition 1 of Theorem 2.18 holds.
2. There exists a function $V \in V_0$ such that H2.1 and (2.43) hold,

$$w_1(||x||) \leq V(t,x) \leq w_2(||x||), \ w_1, w_2 \in K, \ (t,x) \in [t_0,\infty) \times \mathbb{R}^n, \qquad (2.51)$$

where $w_1(u) \to \infty$ as $u \to \infty$, and the inequality

$$^cD_+^{\beta}V(t,x) \leq -w_3(||x||), \ (t,x) \in G, \ t \in [t_0,\infty) \qquad (2.52)$$

is valid for $w_3 \in K$, $\beta \in (0,1)$.

Then the zero solution of system (2.41) is uniformly globally asymptotically stable.

Proof. First, we shall show that the zero solution of system (2.41) is uniformly stable.

For an arbitrary $\varepsilon > 0$ choose the positive number $\delta = \delta(\varepsilon)$ so that $w_2(\delta) < w_1(\varepsilon)$.

Let $x_0 \in \mathbb{R}^n : ||x_0|| < \delta$ and $x(t) = x(t;t_0,x_0)$ be the solution of the IVP (2.41), (1.4). Then by (2.51), (2.52) and (2.43) for any $t \geq t_0$, the following inequalities are valid

$$w_1(||x(t;t_0,x_0)||) \leq V(t,x(t;t_0,x_0)) \leq V(t_0^+,x_0)$$
$$\leq w_2(||x_0||) < w_2(\delta) < w_1(\varepsilon).$$

Therefore, $||x(t;t_0,x_0)|| < \varepsilon$ for $t \geq t_0$. Thus, it is proved that the zero solution of system (2.41) is uniformly stable.

Now, we shall prove that the solutions of system (2.41) are uniformly bounded.

Let $a > 0$ and $x_0 \in \mathbb{R}^n : ||x_0|| < a$. Since for the function $w_1 \in K$ we have $w_1(u) \to \infty$ as $u \to \infty$, then we can choose $b = b(a) > 0$ so that $w_1(b) > w_2(a)$.

Since the conditions of Corollary 1.4 are met, then

$$V(t,x(t;t_0,x_0)) \leq V(t_0^+,x_0), t \in [t_0,\infty).$$

From the above inequality, (2.51) and (2.52) we have

$$w_1(||x(t;t_0,x_0)||) \leq V(t,x(t;t_0,x_0)) \leq V(t_0^+,x_0)$$
$$\leq w_2(||x_0||) < w_2(a) < w_1(b)$$

for $t \geq t_0$.

Therefore, $||x(t;t_0,x_0)|| < b$ for $t \geq t_0$. This implies that the solutions of system (2.41) are uniformly bounded.

Finally, we shall prove that the zero solution of system (2.41) is uniformly globally attractive.

Let $v > 0$ be arbitrary, $\varepsilon > 0$ be given. Let the number $\eta = \eta(\varepsilon) > 0$ be chosen so that $w_2(\eta) > w_1(\varepsilon)$ and let $\gamma = \gamma(v, \varepsilon) > 0$ be such that $\gamma > \left(\frac{w_2(v)}{w_3(\eta)} \Gamma(\beta + 1) \right)^{1/\beta}$.

Let $x_0 \in \mathbb{R}^n : ||x_0|| < v$ and $x(t) = x(t; t_0, x_0)$ be the solution of the IVP (2.41), (1.4). If we assume that for any $t \in [t_0, t_0 + \gamma]$ the inequality $||x(t; t_0, x_0)|| \geq \eta$ holds, then by (2.43) and (2.52) it follows that

$$V(t, x(t; t_0, x_0)) \leq V(t_0^+, x_0) - \frac{1}{\Gamma(\beta)} \int_{t_0}^{t} (t - s)^{\beta - 1} w_3(||x(s; t_0, x_0)||) ds$$

$$\leq w_2(v) - \frac{w_3(\eta)}{\Gamma(\beta)} \int_{t_0}^{t} (t - s)^{\beta - 1} ds \leq w_2(v) - \frac{w_3(\eta)}{\Gamma(\beta + 1)} (t - t_0)^{\beta}.$$

For $t = t_0 + \gamma$, from the above we have

$$V(t_0 + \gamma, x(t_0 + \gamma; t_0, x_0)) \leq w_2(v) - \frac{w_2(\eta)}{\Gamma(\beta + 1)} \gamma^{\beta} < 0,$$

which contradicts (2.51).

Thus, in any case, we have $||x(t; t_0, x_0)|| < \varepsilon$ for $t \geq t_0 + \gamma$, whenever $||x_0|| < v$, i.e. the zero solution of (2.41) is uniformly globally attractive.

Corollary 2.2 If in Theorem 2.20 condition (2.52) is replaced by the condition (2.44), then the zero solution of system (2.41) is uniformly globally asymptotically stable.

This follows immediately from Theorem 2.20. However, the proof can be carried out using the estimate (2.46) which follows from (2.43), (2.44) and Corollary 1.3.

Example 2.7 Consider the following impulsive fractional equation:

$$\begin{cases} {}_0^c D_t^\alpha x(t) = -a|x(t)|(1 + x^2(t)), & t > 0, t \neq t_k, \\ \Delta x(t_k) = c_k x(t_k), & k = 1, 2, ..., \end{cases} \tag{2.53}$$

where $x \subset \mathbb{R}$, $0 < \alpha < 1$, $a > 0$, $-2 \leq c_k \leq 0$, $k = 1, 2, ...$, $0 < t_1 < t_2 < ... < t_k < t_{k+1} < ...$ and $\lim\limits_{k \to \infty} t_k = \infty$.

Let $V(t, x) = |x|$. Then, for $t \geq 0$, and $t \neq t_k$, we have

$$^c D_+^\alpha V(t, x) = {}^c D_+^\alpha |x(t)| \leq -a|x(t)| = -aV(t, x).$$

Also, for $k = 1, 2, ...$,

$$V(t_k^+, x + c_k x) = |(1 + c_k)x| \leq V(t_k, x).$$

Thus by Corollary 2.2, the trivial solution of model (2.53) is uniformly globally asymptotically stable.

2.4   Mittag–Leffler Stability

For extending the application of fractional calculus in nonlinear systems, Podlubny and his co-authors propose in [Li, Chen and Podlubny 2010] the Mittag–Leffler stability and the fractional Lyapunov direct method with a view to enrich the knowledge of both system theory and fractional calculus.

In this section, we shall extend first the Lyapunov direct method to the case of functional fractional-order system (1.1), using Mittag–Leffler stability notion. We shall use continuous Lyapunov functions $V : [t_0, \infty) \times \Omega \to \mathbb{R}_+$, $V \in C_0$ for which the condition H2.1 is true.

Theorem 2.21 Assume that $f(t,0) = 0$, $t \in [t_0, \infty)$, there exists a function $V \in C_0$ for which H2.1 holds, for any $H > 0$ there exists $\gamma(H) > 0$ such that

$$||x|| \le V(t,x) \le \gamma(H)||x||, \ (t,x) \in [t_0, \infty) \times \Omega, \tag{2.54}$$

and the inequality (2.11) is valid whenever $V(t + \theta, \varphi(\theta)) \le p(V(t, \varphi(0)))$ for $-r \le \theta \le 0$, where $t \in [t_0, \infty)$, $\varphi \in [[-r, 0], \Omega]$, $\beta \in (0, 1)$, $w = const > 0$, p is continuous and non-decreasing on $\mathbb{R}_+$, and $p(u) > u$ as $u > 0$, then the zero solution of system (1.1) is Mittag–Leffler stable.

Proof.   Let $H = const > 0 : \{x \in \mathbb{R}^n : ||x|| \le H\} \subset \Omega$. Let $\varphi_0 \in C[[-r, 0], \Omega]$: $||\varphi_0||_r < H$ and $x(t) = x(t; t_0, \varphi_0)$ be the solution of the IVP (1.1), (1.2). From (2.11), it follows by Corollary 1.1, that for $t \ge t_0$ the following inequality is valid

$$V(t, x(t; t_0, \varphi_0)) \le \max_{-r \le \theta \le 0} V(t_0, \varphi_0(\theta)) E_\beta[-w(t - t_0)^\beta].$$

From the above inequality and (2.54), we obtain

$$||x(t; t_0, \varphi_0)|| \le \max_{-r \le \theta \le 0} V(t_0, \varphi_0(\theta)) E_\beta[-w(t - t_0)^\beta]$$

$$\le \gamma(H)||\varphi_0||_r E_\beta[-w(t - t_0)^\beta] = m E_\beta[-w(t - t_0)^\beta], \ t \ge t_0,$$

where $m \ge 0$ and $m = 0$ holds only if $\varphi_0(\theta) = 0$ for $\theta \in [-r, 0]$, which implies that the zero solution of system (1.1) is Mittag–Leffler stable.

Remark 2.4   In Theorem 2.21, if $\beta = 1$, it follows that

$$||x(t; t_0, \varphi_0)|| \le \gamma(H)||\varphi_0||_r \exp[-w(t - t_0)], \ t \ge t_0,$$

which means that the zero solution of (1.1) for $\alpha = 1$ is exponentially stable. Therefore, the notion of the Mittag–Leffler stability for fractional-order differential equations is correspondent to the notion of the exponential stability for integer-order differential equations.

Theorem 2.22 If in condition (2.11) of Theorem 2.21 the Caputo upper right-hand fractional derivative ${}^cD_+^\beta V(t, \varphi(0))$ is replaced by the Riemann–Liouville upper right-hand fractional derivative $D_+^\beta V(t, \varphi(0))$, then the zero solution of system (1.1) is Mittag–Leffler stable.

Proof. It follows from Lemma 1.7 and $V(t, x(t)) \geq 0$ that

$$ {}^cD_+^\beta V(t, \varphi(0)) \leq D_+^\beta V(t, \varphi(0)), t \geq t_0, $$

which implies (2.11). Following the same proof as in Theorem 2.21 yields

$$ ||x(t; t_0, \varphi_0)|| \leq \gamma(H) ||\varphi_0||_r E_\beta[-w(t - t_0)^\beta], \ t \geq t_0, $$

i.e., the zero solution of system (1.1) is Mittag–Leffler stable.

Remark 2.5 Mittag–Leffler stability implies asymptotic stability; see [Li, Chen and Podlubny 2010].

Example 2.8 Consider again the scalar fractional equation (2.18). Let $V(t, x) = \gamma(H)|x|$, where $\gamma(H) > 1$. Then, for $t \geq 0$, we have

$$ {}^cD_+^\alpha V(t, \varphi(0)) = \gamma(H) {}^cD_+^\alpha |\varphi(0)| $$

$$ \leq -(\alpha_1 - \alpha_2)\gamma(H)|\varphi(0)| = -(\alpha_1 - \alpha_2)\gamma(H)V(t, \varphi(0)), $$

whenever $V(t + \theta, \varphi(\theta)) \leq V(t, \varphi(0))$ for $-r \leq \theta \leq 0$, $\varphi \in C[[-r, 0], \Omega]$. If there exists a positive constant c_1 such that $\alpha_1 \geq c_1 + \alpha_2$, then

$$ {}^cD_+^\alpha V(t, \varphi(0)) \leq -c_1 \gamma(H)V(t, \varphi(0)), $$

whenever $V(t + \theta, \varphi(\theta)) \leq V(t, \varphi(0))$ for $-r \leq \theta \leq 0$. Thus by Theorem 2.21, the trivial solution of model (2.18) is Mittag–Leffler stable.

Let $t_0 = 0$. Next, Mittag–Leffler stability theorems for the impulsive fractional differential system (1.5) will be presented. In the proof of the main results we shall use the following lemma:

Lemma 2.1 ([Li, Chen and Podlubny 2010]) Let $\alpha \geq 0$, and $f(t, x)$ be continuous on $\mathbb{R} \times \Omega$.
    Then

$$ ||_{t_0}\mathscr{D}_t^{-\alpha} f(t, x)|| \leq {}_{t_0}\mathscr{D}_t^{-\alpha} ||f(t, x)||, (t, x) \in [t_0, \infty) \times \Omega. $$

Theorem 2.23 Assume that:
    1. Conditions H2.2 and H2.3 hold for $t_0 = 0$.
    2. There exists a function $V \in V_0$ such that H2.1 holds, and

$$ \alpha_1 ||x||^a \leq V(t, x) \leq \alpha_2 ||x||^{ab}, (t, x) \in [0, \infty) \times \Omega, \tag{2.55} $$

$$V(t^+, x + I_k(x)) \leq V(t, x), \ x \in \Omega, \ t = t_k, \ t_k > 0, \tag{2.56}$$

$${}_0^c D_t^\beta V(t, x) \leq -\alpha_3 ||x||^{ab}, \ (t, x) \in G_k, \ t \in [0, \infty), \tag{2.57}$$

where $0 < \beta < 1$, α_1, α_2, α_3, a and b are positive constants.

Then the zero solution of system (1.5) is Mittag–Leffler stable. If the assumptions hold globally on $\mathbb{R}^n$, then the zero solution of system (1.5) is globally Mittag–Leffler stable.

Proof. Let $x_0 = x(0) \in \Omega : ||x_0|| < \delta$, and let $x(t) = x(t; 0, x_0)$ be the solution of the IVP (1.5), (1.4) for $t_0 = 0$. It follows from (2.55) and (2.56) that

$$ {}_0^c D_t^\beta V(t, x(t)) \leq -\frac{\alpha_3}{\alpha_2} V(t, x(t)), \ t \neq t_k, \ t > 0. $$

Therefore, for $t \in [0, \infty)$ there exists a nonnegative function $W(t)$ satisfying

$$ {}_0^c D_t^\beta V(t, x(t)) + W(t) = -\alpha_3 \alpha_2^{-1} V(t, x(t)), \ t \neq t_k. \tag{2.58}$$

Taking the Laplace transform of (2.58) for $t \neq t_k$, $t > 0$ gives

$$ s^\beta V(s) - s^{\beta-1} V(0) + W(s) = -\alpha_3 \alpha_2^{-1} V(s), $$

where $V(0) = V(0, x(0))$ and $V(s) = \mathcal{L}[V(t, x(t))]$. From the last equality we obtain

$$ V(s) = \frac{V(0)s^{\beta-1} - W(s)}{s^\beta + \frac{\alpha_3}{\alpha_2}}. $$

It follows from the properties of the function V and from the fractional uniqueness and existence theorem for the continuous case (see [Podlubny 1999]) that the unique solution of (2.58) is

$$ V(t, x(t)) = V(0, x(0))E_\beta\left(-\frac{\alpha_3}{\alpha_2}t^\beta\right) $$

$$ -W(t) * \left[t^{\beta-1}E_{\beta,\beta}\left(-\frac{\alpha_3}{\alpha_2}t^\beta\right)\right], \ t \neq t_k, \ t > 0, $$

where $*$ denotes the convolution operator. Since both $t^{\beta-1}$ and $E_{\beta,\beta}\left(-\frac{\alpha_3}{\alpha_2}t^\beta\right)$ (see [Li, Chen and Podlubny 2010]) are nonnegative for $t \in (t_{k-1}, t_k)$, $k = \pm 1, \pm 2, ...,$ $t > 0$, it follows that for any closed interval contained in $(t_{k-1}, t_k]$

$$ V(t, x(t)) \leq V(0, x(0))E_\beta\left(-\frac{\alpha_3}{\alpha_2}t^\beta\right). $$

Set $R = V(0, x(0))E_\beta\left(-\frac{\alpha_3}{\alpha_2}t^\beta\right)$. From condition (2.56) it follows that, if $V(t_k, x(t_k)) < R$, then

$$ V(t_k^+, x(t_k^+)) = V(t_k, x(t_k) + I_k(x(t_k))) \leq V(t_k, x(t_k)) < R, \ t_k > 0, $$

i.e. $x(t)$ cannot exceed R by jump. Therefore,

$$V(t,x(t)) \leq V(0,x(0))E_\beta(-\frac{\alpha_3}{\alpha_2}t^\beta), \ t \geq 0.$$

From the last inequality and (2.55) we have

$$||x(t)|| \leq \left[\frac{V(0)}{\alpha_1}E_\beta(-\frac{\alpha_3}{\alpha_2}t^\beta)\right]^{\frac{1}{a}}, \ t \geq 0.$$

Let $m = \frac{V(0)}{\alpha_1} = \frac{V(0,x(0))}{\alpha_1} \geq 0$, then we have

$$||x(t)|| \leq \left[mE_\beta(-\frac{\alpha_3}{\alpha_2}t^\beta)\right]^{\frac{1}{a}}, \ t \geq 0,$$

where $m = 0$ holds only if $x(0) = 0$. From the properties of the function V it follows that m is Lipschitz with respect to $x(0)$ and $m(0) = 0$, which imply the Mittag-Leffler stability of the zero solution of system (1.5).

Theorem 2.24 If in condition (2.57) of Theorem 2.23 the Caputo fractional derivative ${}_0^CD_t^\beta$ is replaced by the Riemann–Liouville fractional derivative ${}_0D_t^\beta$, then the zero solution of system (1.5) is Mittag–Leffler stable.

Proof. It follows from the relations between the Caputo and Riemann–Liouville fractional derivatives for the continuous case, $V(t,x(t)) \geq 0$ and (2.56) that

$$ {}_0^CD_t^\beta V(t,x(t)) \leq {}_0D_t^\beta V(t,x(t)), \ t \neq t_k, \ t_k > 0, $$

which implies (2.57). Following the same proof as in Theorem 2.23 yields

$$||x(t)|| \leq \left[\frac{V(0)}{\alpha_1}E_\beta(-\frac{\alpha_3}{\alpha_2}t^\beta)\right]^{\frac{1}{a}}, \ t \geq 0,$$

i.e., the zero solution of system (1.5) is Mittag–Leffler stable.

Now, consider an impulsive system of the type (1.5) with the Riemann-Liouville fractional derivative, i.e. when ${}_{t_0}\mathscr{D}_t^\alpha = {}_{t_0}D_t^\alpha$. The questions about the initial conditions of fractional differential equations with Caputo and Riemann–Liouville derivatives remain quite up-to-date, even in the classic continuous case ([Bagley 2007], [Heymans and Podlubny 2005], [Li, Qian and Chen 2011], [Ortigueira and Coito 2010]).

In the case when the Riemann–Liouville operator is used for the system (1.5) the initial condition (1.4) will be given by

$$ {}_{t_0}D_t^{\alpha-1}x(t_0^+;t_0,x_0) = x_0. $$

Note that in this case, the above initial condition can be replaced by the condition

$$x(t_0^+;t_0,x_0) = \frac{x_0}{\Gamma(\alpha)}(t-t_0)^{\alpha-1}.$$

Also, suitable jump conditions are used (see [Bai 2011], [Kosmatov 2013], [Yukunthorn, Ntouyas and Tariboon 2015]).

Theorem 2.25 Assume that for system (1.5) with $_{t_0}\mathscr{D}_t^\alpha = {}_0D_t^\alpha$:

1. Condition 1 of Theorem 2.23 holds and $_0D_t^{\alpha-1}x(t_k^+) = {}_0D_t^{\alpha-1}x(t_k)$, $t_k > 0$.

2. The function $f(t,x)$ is Lipschitz continuous with respect to $x \in \Omega$ with Lipschitz constant $l > 0$.

3. There exists a function $V \in V_0$ such that H2.1 and (2.56) hold, and

$$\alpha_1\|x\|^a \le V(t,x) \le \alpha_2\|x\|, \ (t,x) \in [0,\infty) \times \Omega, \tag{2.59}$$

$$\dot{V}(t,x) \le -\alpha_3\|x\|, \ (t,x) \in G_k, \ t \in [0,\infty), \tag{2.60}$$

where $\dot{V}(t,x) = \frac{dV(t,x)}{dt}$, a, α_1, α_2, α_3 are positive constants.

Then the zero solution of system (1.5) is Mittag–Leffler stable.

Proof. It follows from (2.59), (2.60) and Lemma 2.1 that for $t \ne t_k$, $t_k > 0$, we have

$$_0^cD_t^{1-\alpha}V(t,x(t)) = {}_0D_t^{-\alpha}\dot{V}(t,x(t)) \le -\alpha_3\left({}_0D_t^{-\alpha}\|x(t)\|\right)$$

$$\le -\alpha_3 l^{-1}{}_0D_t^{-\alpha}\|f(t,x(t))\| \le -\alpha_3 l^{-1}\|{}_0D_t^{-\alpha}f(t,x(t))\| = -\alpha_3 l^{-1}\|x(t)\|,$$

where $_0D_t^{\alpha-1}x(0) = 0$. Following the same proof as in Theorem 2.23 yields

$$\|x(t)\| \le \left[\frac{V(0)}{\alpha_1}E_{1-\alpha}(-\frac{\alpha_3}{\alpha_2}t^{1-\alpha})\right]^{\frac{1}{a}}, \ t \ge 0,$$

i.e., the zero solution of system (1.5) is Mittag–Leffler stable.

Example 2.9 Consider the following system

$$\begin{cases} _0^cD_t^\alpha x(t) = -y(t)\sin(x(t)) - 4x(t) + y(t), \ t \ge 0, t \ne t_k, \\ _0^cD_t^\alpha y(t) = y(t)\sin(x(t)) - 5y(t), \ t \ge 0, t \ne t_k, \\ x(t_k^+) = c_k x(t_k), \ t_k > 0, k = 1,2,..., \\ y(t_k^+) = d_k y(t_k), \ t_k > 0, k = 1,2,..., \end{cases} \tag{2.61}$$

where $0 < \alpha < 1$, $x, y \in \mathbb{R}$, $-1 < c_k \le 0$, $-1 < d_k \le 0$, $k = 1,2,..., 0 < t_1 < t_2 < ... < t_k < t_{k+1} < ...$ and $\lim_{k \to \infty} t_k = \infty$.

Define the function $V(t,x,y) = |x| + |y|$. Then, for $t \ge 0$ and $t \ne t_k$, we have

$$_0^cD_t^\alpha V(t,x(t),y(t)) = sgnx(t) \ _0^cD_t^\alpha x(t) + sgny(t) \ _0^cD_t^\alpha y(t)$$

$$\le -4|x(t)| - 4|y(t)| = -4V(t,x(t),y(t)).$$

Also, for $t = t_k, t > 0$,

$$V(t_k^+, x + c_k x, y + d_k y) = |1 + c_k||x| + |1 + d_k||y| \leq V(t_k, x, y).$$

Thus by Theorem 2.23 for $a = b = \alpha_1 = \alpha_2 = 1$ and $\alpha_3 = 4$, the trivial solution of model (2.61) is globally Mittag–Leffler stable (and, therefore, it is globally asymptotically stable).

2.5 Practical Stability

One of the most important aspects of the stability theory of differential equations is the so-called practical stability. The notion of practical stability for dynamical systems was first discussed by LaSalle and Lefschetz (see [LaSalle and Lefschetz 1961]) in 1960s and since then a great progress in this direction has been made ([Celentano 2012], [Lakshmikantham, Leela and Martynyuk 1990], [Lakshmikantham, Matrosov and Sivasundaram 1991], [Stamova 2007], [Stamova 2009]). The main problem in the theory of practical stability consists of studying the solutions of systems of differential equations, given in advance the domain where the initial conditions change, and the domain where the solutions should remain when the independent variable changes over a fixed interval (finite or infinite). Such a notion is of a significant importance in scientific and engineering problems. For example, it is very useful in estimating the worst-case transient and steady-state responses and in verifying pointwise in time constraints imposed on the state trajectories.

In this section, we shall consider sufficient conditions for practical stability of the impulsive fractional system of differential equations (1.5). First, we shall introduce definitions of practical stability of system (1.5) which are analogous to the definitions given in [Lakshmikantham, Leela and Martynyuk 1990].

Definition 2.3 The system (1.5) is said to be:

(a) *practically stable* with respect to (λ, A), if given (λ, A) with $0 < \lambda < A$, we have $||x_0|| < \lambda$ implies $||x(t; t_0, x_0)|| < A$, $t \geq t_0$ for some $t_0 \in \mathbb{R}$;

(b) *uniformly practically stable* with respect to (λ, A), if (a) holds for every $t_0 \in \mathbb{R}$;

(c) *practically asymptotically stable* with respect to (λ, A), if (a) holds and

$$\lim_{t \to \infty} ||x(t; t_0, x_0)|| = 0.$$

For fractional-order systems (1.5) we shall introduce the new notion of practical Mittag–Leffler stability.

Definition 2.4 The system (1.5) is said to be *practically Mittag–Leffler stable* with respect to (λ, A), if given (λ, A) with $0 < \lambda < A$, we have $||x_0|| < \lambda$ implies

$$||x(t;t_0,x_0)|| \leq \{AE_\beta[-q(t-t_0)^\beta]\}^d, t \geq t_0,$$

where $0 < \beta < 1, q, d > 0$.

Together with system (1.5) we shall consider the scalar comparison equation (1.14). Let $u(t_0^+) = u_0 \geq 0$. We shall consider such solutions $u(t) = u(t;t_0,u_0)$ of equation (1.14) for which $u(t) \geq 0$. That is why the following definitions for practical stability properties of this equation will be used.

Definition 2.5 The equation (1.14) is said to be:

(a) *practically stable* with respect to (λ, A), if given (λ, A) with $0 < \lambda < A$, we have $u_0 < \lambda$ implies $u(t;t_0,u_0) < A, t \geq t_0$ for some $t_0 \in \mathbb{R}$;

(b) *uniformly practically stable* with respect to (λ, A), if (a) holds for every $t_0 \in \mathbb{R}$;

(c) *practically asymptotically stable* with respect to (λ, A), if (a) holds and

$$\lim_{t \to \infty} u(t;t_0,u_0) = 0;$$

(d) *practically Mittag–Leffler stable* with respect to (λ, A), if given (λ, A) with $0 < \lambda < A$, we have $u_0 < \lambda$ implies

$$u(t;t_0,u_0) \leq \{AE_\beta[-q(t-t_0)^\beta]\}^d, t \geq t_0, 0 < \beta < 1, q, d > 0.$$

In the further considerations, we shall use piecewise continuous auxiliary functions $V : \mathbb{R} \times \mathbb{R}^n \to \mathbb{R}_+, V \in V_0$ for which the condition H2.1 is satisfied.

Using the comparison Lemma 1.5 and corollaries 1.3 and 1.4 we shall prove the following result which offers sufficient conditions in a unified way for various practical stability criteria.

Theorem 2.26 Assume that:

1. $0 < \lambda < A$ are given and $B_A \subseteq \Omega$, where $B_A = \{x \in \mathbb{R}^n : ||x|| < A\}$.
2. Conditions H2.1–H2.3 and the conditions of Lemma 1.5 are satisfied for $x \in B_A, g(t,0) = 0$ for $t \geq t_0$, and $B_k(0) = 0$ for $k = \pm1, \pm2, \dots$.
3. There exists $\rho = \rho(A) > 0$ such that $x \in B_A$ implies $x + I_k(x) \in B_\rho$ for all $k = \pm1, \pm2, \dots$, and $B_\rho \subseteq \Omega$.
4. There exist functions $w_1, w_2 \in K$ such that

$$w_1(||x||) \leq V(t,x) \leq w_2(||x||), (t,x) \in [t_0, \infty) \times B_\rho. \tag{2.62}$$

5. $w_2(\lambda) < w_1(A)$ holds.

Then, the practical stability properties of the equation (1.14) with respect to $(w_2(\lambda), w_1(A))$ imply the corresponding practical stability properties of (1.5) with respect to (λ, A).

Proof. We shall first prove the practical stability of (1.5). Let $0 < \lambda < A$ and assume, without loss of generality, that $A < \rho$. Suppose that (1.14) is practically stable with respect to $(w_2(\lambda), w_1(A))$. Then, we have

$$u_0 < w_2(\lambda) \text{ implies } u(t; t_0, u_0) < w_1(A), \ t \geq t_0 \tag{2.63}$$

for some given $t_0 \in \mathbb{R}$, where $u_0 \in \mathbb{R}_+$ and $u(t; t_0, u_0)$ is any solution of (1.14) defined on $[t_0, \infty)$.

Let $u_0 = w_2(||x_0||)$. We claim that if $||x_0|| < \lambda$, then $||x(t; t_0, x_0)|| < A$ for $t \geq t_0$, where $x(t; t_0, x_0)$ is the solution of the initial value problem (1.5), (1.4). If the claim is not true, there exist $t_0 \in \mathbb{R}$, a corresponding solution $x(t; t_0, x_0)$ of (1.5) with $||x_0|| < \lambda$, and $t_0 < t^* < \infty$, such that $t_k < t^* \leq t_{k+1}$ for some k, satisfying

$$||x(t^*)|| \geq A \text{ and } ||x(t; t_0, x_0)|| < A, \ t \in [t_0, t_k).$$

Since $x(t_k) \in B_A$, then condition 3 of Theorem 2.26 shows that $x(t_k^+) = x(t_k) + I_k(x(t_k)) \in B_\rho$. Hence, there exists a t^0 such that $t_k < t^0 \leq t^*$, and

$$A \leq ||x(t^0)|| < \rho. \tag{2.64}$$

Now, using conditions 2 and 3 of Theorem 2.26, we obtain by Lemma 1.5,

$$V(t, x(t)) \leq u^+(t; t_0, w_2(||x_0||)), \ t \in [t_0, t^0], \tag{2.65}$$

where $u^+(t; t_0, u_0)$ is the maximal solution of (1.14) defined on the interval $[t_0, \infty)$.

From (2.64), (2.62), (2.65) and (2.63) we get

$$w_1(A) \leq w_1(||x(t^0)||) \leq V(t^0; t_0, x_0)$$
$$\leq u^+(t^0; t_0, w_2(||x_0||)) < w_1(A),$$

since $u_0 = w_2(||x_0||) < w_2(\lambda)$.

The contradiction obtained shows that

$$||x(t; t_0, x_0)|| \leq A$$

for $||x_0|| < \lambda$ and $t \geq t_0$. It, therefore, follows that system (1.5) is practically stable with respect to (λ, A).

The proof in case $A \geq \rho$ is trivial. In this case by the assumption that $x(t_k) \in B_A$ implies $x(t_k) + I_k(x(t_k)) \in B_\rho$, we get $x(t_k) + I_k(x(t_k)) \in B_A$.

Suppose now, that (1.14) is practically asymptotically stable with respect to $(w_2(\lambda), w_1(A))$. Therefore, we have that (1.14) is practically stable with respect to $(w_2(\lambda), w_1(A))$, and

$$\lim_{t \to \infty} u(t; t_0, u_0) = 0, \tag{2.66}$$

where $u(t;t_0,u_0)$ is any solution of the comparison equation (1.14) with $u_0 \geq 0$.

From what we proved above, the system (1.5) is practically stable with respect to (λ, A). Let $u_0 = V(t_0^+, x_0)$. From Lemma 1.5, we have

$$V(t,x(t;t_0,x_0)) \leq u^+(t;t_0,u_0),\ t \in [t_0,\infty). \tag{2.67}$$

Using condition 4 of Theorem 2.26, together with (2.67), gives

$$w_1(\|x(t)\|) \leq V(t;t_0,x_0)) \leq u^+(t;t_0,u_0),$$

which implies

$$\lim_{t \to \infty} w_1(\|x(t;t_0,x_0)\|) = 0$$

by relation (2.66).

Thus, we have

$$\lim_{t \to \infty} \|x(t;t_0,x_0)\| = 0,$$

which means that system (1.5) is practically asymptotically stable with respect to (λ, A).

To prove the practical Mittag–Leffler stability, we let $w_1 \in K$ to be the identity function. Since the comparison equation (1.14) is practically Mittag–Leffler stable with respect to $(w_2(\lambda), w_1(A)) = (w_2(\lambda), A)$, we have

$$u_0 < w_2(\lambda) \text{ implies } u(t;t_0,u_0) < \{AE_\beta[-q(t-t_0)^\beta]\}^d,\ t \geq t_0, \tag{2.68}$$

where $q,d > 0$ and $u(t;t_0,u_0)$ is any solution of (2.5) defined on $[t_0,\infty)$.

Choosing $u_0 = V(t_0^+, x_0)$, we have from condition 4 of Theorem 2.26, (2.67) and (2.68),

$$\|x(t;t_0,x_0)\| \leq V(t,x(t;t_0,x_0)) \leq u^+(t;t_0,u_0)$$
$$< \{AE_\beta[-q(t-t_0)^\beta]\}^d,\ t \geq t_0,$$

which proves the practical Mittag–Leffler stability of system (1.5).

One can similarly prove other practical stability properties of the system (1.5).

Corollary 2.3 In Theorem 2.26:

(i) $g(t,u) = 0$ for $(t,u) \in [t_0,\infty) \times \mathbb{R}_+$ and $\tilde{\psi}_k(u)$ for $u \in \mathbb{R}_+$, $k = \pm 1, \pm 2, ...,$ are admissible to imply uniform practical stability of (1.5) with respect to (λ, A).

(ii) $g(t,u) = -qu$, $q > 0$, for $(t,u) \in [t_0,\infty) \times \mathbb{R}_+$ and $\tilde{\psi}_k(u)$ for $u \in \mathbb{R}_+$, $k = \pm 1, \pm 2, ...,$ are admissible to imply practical Mittag–Leffler stability of (1.5) with respect to (λ, A).

Proof. The proof of assertions (i) and (ii) of Corollary 2.3 is immediate using Corollary 1.4 and Corollary 1.3, respectively.

For non-uniform practical stability results we shall use functions from the class CK, $CK = \{w \in C[\mathbb{R}_+^2, \mathbb{R}_+] : w(t,u) \in K$ for each $t \in \mathbb{R}_+$ and $w(t,u) \to \infty$ as $u \to \infty\}$.

Theorem 2.27 Assume that:

1. Conditions 1–3 of Theorem 2.26 hold.
2. There exist functions $w_1 \in K$ and $w_2 \in CK$ such that

$$w_1(||x||) \leq V(t,x) \leq w_2(t,||x||), \ (t,x) \in [t_0,\infty) \times B_\rho.$$

3. $w_2(t_0^+, \lambda) < w_1(A)$ holds.

Then, the practical stability properties of the equation (1.14) with respect to $(w_2(t_0^+, \lambda), w_1(A))$ imply the corresponding non-uniform practical stability properties of (1.5) with respect to (λ, A).

In the next, we shall consider criteria for practical asymptotic stability of (1.5). The new notion of practical Mittag-Leffler stability will be also discussed.

Theorem 2.28 Assume that conditions H2.2 and H2.3 are satisfied, and:

1. $0 < \lambda < A$ are such that $B_A \subseteq \Omega$, $g(t,0) = 0$ for $t \geq t_0$, and $B_k(0) = 0$ for $k = \pm 1, \pm 2, \dots$.
2. There exists $\rho = \rho(A) > 0$ such that $x \in B_A$ implies $x + I_k(x) \in B_\rho$ for all $k = \pm 1, \pm 2, \dots$, and $B_\rho \subseteq \Omega$.
3. There exists a function $V \in V_0$ such that condition H2.1 is satisfied and for $w_1, w_3 \in K$, $w_2 \in CK$,

$$w_1(||x||) \leq V(t,x) \leq w_2(t,||x||), \ (t,x) \in [t_0,\infty) \times B_\rho, \tag{2.69}$$

$$V(t^+, x + I_k(x)) \leq V(t,x), \ x \in \mathbb{R}^n, \ t = t_k, \ t_k > t_0, \tag{2.70}$$

$${}^cD_+^\beta V(t,x) \leq -w_3(||x||), \ (t,x) \in G, \ t \in [t_0,\infty), \ 0 < \beta < 1. \tag{2.71}$$

4. $w_2(t_0^+, \lambda) < w_1(A)$ holds.

Then, the system (1.5) is practically asymptotically stable with respect to (λ, A).

Proof. By Theorem 2.27 with $g(t,u) \equiv -w_3(u) \leq 0$ and $\tilde{\psi}_k(u) \equiv u, \ t \geq t_0$, and Corollary 1.4 it follows because of (2.69)–(2.71) and condition 4 of Theorem 2.28, that the system (1.5) is practically stable with respect to (λ, A). Indeed, for a solution $x(t) = x(t;t_0,x_0)$ of (1.5) with $||x_0|| < \lambda$, we have

$$w_1(||x(t;t_0,x_0)||) \leq V(t,x(t;t_0,x_0)) \leq V(t_0^+, x_0)$$

$$\leq w_2(t_0^+, ||x_0||) < w_2(t_0^+, \lambda) < w_1(A), \ t \geq t_0.$$

Therefore $||x(t;t_0,x_0)|| < A, t \geq t_0$ for some $t_0 \in \mathbb{R}$, which proves the practical stability of system (1.5) with respect to (λ, A).

Hence, it is enough to prove that every solution $x(t;t_0,x_0)$ with $||x_0|| < \lambda$ satisfies $\lim_{t\to\infty} ||x(t;t_0,x_0)|| = 0$.

Suppose that this is not true. Then there exist $x_0 \in \Omega$: $||x_0|| < \lambda$, $\delta > 0$, $\gamma > 0$ and a sequence $\{\xi_k\}_{k=1}^\infty \in [t_0, \infty)$ such that for $k = 1,2,...$, the following inequalities are valid

$$\xi_k - \xi_{k-1} \geq \delta, \quad ||x(\xi_k;t_0,x_0)|| \geq \gamma. \tag{2.72}$$

Let $M \in \mathbb{R}_+$ be such that

$$\sup\{^cD_+^\beta ||x(t)|| : t \in G\} \leq M. \tag{2.73}$$

By (2.72) and (2.73) for $t \in [\xi_k - \varepsilon, \xi_k]$, where $0 < \varepsilon < \min\left\{\delta, \left(\gamma \frac{\Gamma(\beta+1)}{2M}\right)^{1/\beta}\right\}$, we have

$$||x(t)|| \geq ||x(\xi_k)|| - \frac{M}{\Gamma(\beta)} \int_t^{\xi_k} (\xi_k - s)^{\beta-1} ds$$

$$\geq \gamma - \frac{M}{\Gamma(\beta+1)} (\xi_k - t)^\beta \geq \gamma - \frac{M}{\Gamma(\beta+1)} \varepsilon^\beta > \frac{\gamma}{2}.$$

From the above estimate using (2.70) and (2.71), for $\xi_R \in \{\xi_k\}_{k=1}^\infty$, we conclude that

$$0 \leq V(\xi_R, x(\xi_R;t_0,x_0))$$

$$\leq V(t_0^+, x_0) - \frac{1}{\Gamma(\beta)} \int_{t_0}^{\xi_R} (\xi_R - s)^{\beta-1} d(||x(s;t_0,x_0)||) ds$$

$$\leq V(t_0^+, x_0) - \frac{1}{\Gamma(q)} \sum_{k=1}^R \int_{\xi_k-\varepsilon}^{\xi_k} (\xi_k - s)^{\beta-1} d(||x(s;t_0,x_0)||) ds$$

$$\leq V(t_0^+, x_0) - R \frac{d(\gamma/2)}{\Gamma(\beta+1)} \varepsilon^\beta,$$

which contradicts (2.69) for large R.

Thus, $\lim_{t\to\infty} ||x(t;t_0,x_0)|| = 0$ which proves the practical asymptotic stability of (1.5) with respect to (λ, A).

Theorem 2.29 Assume that conditions H2.2 and H2.3 are satisfied, and:
1. Conditions 1 and 2 of Theorem 2.28 hold.
2. There exists a function $V \in V_0$ such that condition H2.1 is satisfied,

$$||x|| \leq V(t,x) \leq w_2(t, ||x||), \ (t,x) \in [t_0, \infty) \times B_\rho, \ w_2 \in CK, \tag{2.74}$$

$$V(t^+, x + I_k(x)) \leq V(t,x), \ x \in \mathbb{R}^n, \ t = t_k, \ t_k > t_0, \tag{2.75}$$

$$^cD_+^\beta V(t,x) \leq -qV(t,x), \ (t,x) \in G, \ t \in [t_0, \infty), \ q > 0, \ 0 < \beta < 1. \tag{2.76}$$

3. $w_2(t_0^+, \lambda) < A$ holds.

Then, the system (1.5) is practically Mittag–Leffler stable with respect to (λ, A).

Proof. Let $0 < \lambda < A$, and $B_A \subseteq \Omega$. Let $x_0 \in \Omega : ||x_0|| < \lambda$ and $x(t) = x(t;t_0,x_0)$ be the solution of the IVP (1.5), (1.4). From (2.75) and (2.76), it follows by Corollary 1.3, that for $t \geq t_0$ the following inequality is valid

$$V(t,x(t;t_0,x_0)) \leq V(t_0^+,x_0)E_\beta[-q(t-t_0)^\beta].$$

From the above inequality and (2.74) we obtain

$$||x(t;t_0,x_0)|| \leq V(t_0^+,x_0)E_\beta[-q(t-t_0)^\beta]$$

$$\leq w_2(t_0^+,\lambda)E_\beta[-q(t-t_0)^\beta], \ t \geq t_0.$$

Then condition 3 of Theorem 2.29 gives

$$||x(t;t_0,x_0)|| \leq AE_\beta[-q(t-t_0)^\beta], \ t \geq t_0,$$

which implies that system (1.5) is practically Mittag–Leffler stable with respect to (λ,A).

Example 2.10 Let $0 < \alpha < 1$. Consider the following impulsive scalar fractional-order equation

$$\begin{cases} {}_0^cD_t^\alpha x(t) = -qx(t), t \neq t_k, t \geq 0, \\ \Delta x(t_k) = c_k x(t_k), k = 1,2,..., \end{cases} \tag{2.77}$$

where $x \in \mathbb{R}_+, q \geq 0, -1 < c_k \leq 0, k = 1,2,...,$ are constants, $0 < t_1 < t_2 < ... < t_k < ...,$ and $\lim_{k\to\infty} t_k = \infty$.

Let $x_0 \geq 0$ and $x(t) = x(t;t_0,x_0)$ be the solution of (2.77) such that $x(0) = x_0$. Let $0 < \lambda < A$ and $x_0 < \lambda$.

Let us choose the Lyapunov function $V(t,x) = |x|, x \in \mathbb{R}$. Then, for $t \geq 0$ and $t \neq t_k$, for the upper right-hand derivative of V of order α with respect to (2.77), we have

$$^cD_+^\alpha V(t,x) = {}^cD_+^\alpha|x(t)| \leq -qx(t) = -qV(t,x).$$

Also for $t = t_k, k = 1,2,...,$

$$V(t_k^+,x(t_k) + c_k x(t_k)) = |(1+c_k)x(t_k)| \leq |x(t_k)| = V(t_k,x(t_k)), \ k = 1,2,....$$

(i) If $q > 0$, then all conditions of Theorem 2.28 hold for $w_1(u) = w_2(t,u) = u, w_3(u) = qu$ and, therefore the equation (2.77) is practically asymptotically stable with respect to (λ,A). Also, all conditions of Theorem 2.29 are satisfied and (2.77) is practically Mittag–Leffler stable with respect to (λ,A).

(ii) For the case $q = 0$, using Corollary 2.3, we get that (2.77) is uniformly practically stable with respect to (λ,A).

Example 2.11 As an example consider the linear fractional-order control system

$$\begin{cases} {}^c_0D^\alpha_t x(t) = Cx(t), t \neq t_k, \\ \Delta x(t_k) = C_k x(t_k), k = 1, 2, ..., \end{cases} \tag{2.78}$$

where $x \in \mathbb{R}^n$, C and C_k are constant $(n \times n)$−type matrices, $0 < t_1 < t_2 < ... < t_k < ...$, and $\lim_{k \to \infty} t_k = \infty$.

We define the fractional α-order "logarithmic norm" ${}^cD\mu^\alpha(C)$ of the matrix C by

$$ {}^cD\mu^\alpha(C) = \lim_{h \to 0^+} \frac{1 - ||E - h^\alpha C||}{h^\alpha}, $$

where E denotes the identity $(n \times n)$- matrix and $||C||$ is the matrix norm of C induced by the Euclidean vector norm. For more details on the integer-order logarithmic norm $\mu(C)$ (see [Lakshmikantham, Leela and Martynyuk 1990]).

Choose $V(t, x) = ||x||$. Then

$$ {}^cD^\alpha_+ V(t, x) = {}^cD^\alpha_+ ||x|| \leq {}^cD\mu^\alpha(C)||x||, \ t \neq t_k. $$

Also,

$$ V(t_k^+, x(t_k) + C_k x(t_k)) = ||(E + C_k)x(t_k)||, k = 1, 2, ... $$

If $C_k = diag(c_{1k}, c_{2k}, ..., c_{nk})$ and $-1 < c_{ik} \leq 0, i = 1, 2, ..., n, k = 1, 2, ...,$ then

$$ V(t_k^+, x(t_k) + C_k x(t_k)) \leq ||x(t_k)|| = V(t_k, x(t_k)), k = 1, 2, ... $$

Consider the comparison equation

$$\begin{cases} {}^c_0D^\alpha_t u(t) = {}^cD\mu^\alpha(C)u(t), t \neq t_k, t \geq 0, \\ u(t_k^+) = u(t_k), k = 1, 2, ..., \end{cases} \tag{2.79}$$

where $u \in \mathbb{R}_+$ and $u_0 = ||x_0||$.

(i) If ${}^cD\mu^\alpha(C) \leq 0$, then according to the Example 2.10 (ii), the equation (2.79) is uniformly practically stable with respect to (λ, A). Therefore, due to Theorem 2.26 for $w_1(u) = w_2(u) = u$, the linear system (2.78) is uniformly practically stable with respect to (λ, A).

(ii) If there exists $q > 0$ such that ${}^cD\mu^\alpha(C) \leq -q$, then according to the Example 2.10 (i), the equation (2.79) is practically asymptotically stable (practically Mittag-Leffler stable) with respect to (λ, A). Therefore, due to Theorem 2.26 for $w_1(u) = w_2(u) = u$, the linear system (2.78) is practically asymptotically stable (practically Mittag–Leffler stable) with respect to (λ, A).

Note that the linear impulsive fractional-order system (2.78) is the corresponding closed-loop system to the control system

$$\,^c_0D_t^\alpha x(t) = Cx(t) + \eta(t), \ t > 0,$$

where

$$\eta(t) = \sum_{k=1}^{\infty} C_k x(t)\delta(t - t_k) \tag{2.80}$$

is the control input, $\delta(t)$ is the Dirac impulsive function with discontinuity points

$$0 < t_1 < t_2 < ... < t_k < ....$$

The controller $\eta(t)$ has an effect on suddenly change of the state of (2.78) at the time instants t_k due to which the state $x(t)$ of units change from the position $x(t_k)$ into the position $x(t_k^+)$, C_k are the matrices, which characterize the magnitudes of the impulse effects on $x(t)$ at the moments t_k, i.e., $\eta(t)$ is an impulsive control of the fractional-order linear system

$$\,^c_0D_t^\alpha x(t) = Cx(t), \ t \geq 0. \tag{2.81}$$

Therefore, Theorem 2.26 presents a general design method of impulsive control law (2.80) that practically (asymptotically) stabilizes the linear fractional-order system (2.81). The constant matrices C_k characterize the control gains of synchronizing impulses.

Example 2.12 Consider the non-linear fractional-order system:

$$\begin{cases} \,^c_0D_t^\alpha x(t) = n(t)x(t)y(t) + m(t)x(t), \ t \neq t_k, \ t \geq 0, \\ \,^c_0D_t^\alpha y(t) = -n(t)x(t)y(t) + m(t)y(t), \ t \neq t_k, \ t \geq 0, \\ \Delta x(t_k) = c_k x(t_k), \ \Delta y(t_k) = d_k y(t_k), \ k = 1, 2, ..., \end{cases} \tag{2.82}$$

where $x, y \in \mathbb{R}$, $0 < \alpha < 1$, the functions $n(t)$ and $m(t)$ are continuous in $\mathbb{R}_+$, $-1 < c_k \leq 0, -1 < d_k \leq 0, k = 1, 2, ..., 0 < t_1 < t_2 < ...,$ $\lim_{k\to\infty} t_k = \infty$. Let

$$x(0) = x_0, \ y(0) = y_0, \ x_0, y_0 \in \mathbb{R}.$$

Choose

$$V(t, x, y) = |x| + |y|.$$

If $x(t) = 0$ for $t \geq 0, t \neq t_k$, then $\,^cD_+^\alpha|x(t)| = 0$. If $x(t) > 0$ for $t \geq 0, t \neq t_k$, then

$$\,^c_0D_t^\alpha|x(t)| = \,^c_0D_t^\alpha x(t).$$

If $x(t) < 0$ for $t \geq 0, t \neq t_k$, then

$$\,^c_0D_t^\alpha|x(t)| = -\,^c_0D_t^\alpha x(t).$$

Therefore,

$$_0^cD_t^\alpha|x(t)| = sgn(x(t))_0^cD_t^\alpha x(t), \ t \neq t_k, \ t_k > 0.$$

Analogously,

$$_0^cD_t^\alpha|y(t)| = sgn(y(t))_0^cD_t^\alpha y(t), \ t \neq t_k, \ t_k > 0.$$

Then for $t \geq 0, t \neq t_k$, we have

$$^cD_+^\alpha V(t,x,y) \leq m(t)V(t,x,y).$$

Also,

$$V(t_k^+, x(t_k) + c_k x(t_k), y(t_k) + d_k y(t_k))$$
$$= |(1 + c_k)x(t_k)| + |(1 + d_k)y(t_k)| \leq V(t_k, x(t_k), y(t_k)), \ k = 1, 2, \dots.$$

Consider the comparison equation

$$\begin{cases} _0^cD_t^\alpha u(t) = m(t)u(t), \ t \neq t_k, \ t \geq 0, \\ u(0) = u_0, \\ u(t_k^+) = u(t_k), \ k = 1, 2, \dots, \end{cases} \tag{2.83}$$

where $u \in \mathbb{R}_+$ and $u_0 = |x_0| + |y_0|$. The general solution of the system (2.83) is given by

$$u(t) = u_0 E_\alpha(m(t)t^\alpha). \tag{2.84}$$

Let $A = \lambda$. Suppose that

$$E_\alpha(m(t_1)t_1^\alpha) < \frac{w_1(A)}{w_2(\lambda)}, \ w_1, w_2 \in K.$$

It, therefore, follows from (2.84), that the equation (2.83) is practically stable with respect to $(w_2(\lambda), w_1(A))$.

Hence, we get, by Theorem 2.26, that the system (2.82) is practically stable with respect to (λ, A).

In the successive investigations, we shall use piecewise continuous auxiliary vector functions $V : [t_0, \infty) \times \Omega \to \mathbb{R}_+^m$, $V = col(V_1, \dots, V_m)$ such that $V_i \in V_0$, $i = 1, 2, \dots, m$. The aim of this part of Section 2.5 is to present practical stability criteria for fractional-order impulsive control system (1.5) in terms of vector Lyapunov-like functions.

Naturally, when we employ vector Lyapunov functions, we shall consider the comparison system (1.13) for $Q = \mathbb{R}_+^m$, i.e.,

$$\begin{cases} _{t_0}^cD_t^\beta u(t) = F(t, u(t)), \ t \neq t_k, \\ \Delta u(t_k) = u(t_k^+) - u(t_k) = J_k(u(t_k)), \ t_k > t_0, \end{cases} \tag{2.85}$$

where $F : \mathbb{R} \times \mathbb{R}_+^m \to \mathbb{R}^m$, $J_k : \mathbb{R}_+^m \to \mathbb{R}^m$, $k = \pm 1, \pm 2, \dots$.

Let $u_0 \in \mathbb{R}_+^m$. Denote by $u(t) = u(t;t_0,u_0)$ the solution of system (2.85) satisfying the initial condition $u(t_0^+) = u_0$.

The next theorem gives sufficient conditions, in terms of vector Lyapunov functions, for practical stability properties of the system (1.5).

Theorem 2.30 Assume that:

1. $0 < \lambda < A$ are given and $B_A \subseteq \Omega$.
2. The conditions H2.2, H2.3 and conditions of Lemma 1.2 are satisfied for $x \in B_A$, $F(t,0) = 0$ for $t \geq t_0$, and $J_k(0) = 0$ for $k = \pm 1, \pm 2, \dots$.
3. There exists $\rho = \rho(A) > 0$ such that $x \in B_A$ implies $x + I_k(x) \in B_\rho$ for all $k = \pm 1, \pm 2, \dots$, and $B_\rho \subseteq \Omega$.
4. There exist functions $w_1, w_2 \in K$ such that

$$w_1(||x||) \leq L_0(t,x) \leq w_2(||x||), \ (t,x) \in [t_0,\infty) \times B_\rho, \tag{2.86}$$

where $L_0(t,x) = \displaystyle\sum_{i=1}^{m} V_i(t,x)$ and $V_i(t,0) = 0$ for $t \geq t_0$, $i = 1,2,\dots,m$.

5. $w_2(\lambda) < w_1(A)$.

Then, the practical stability properties of the system (2.85) with respect to $(w_2(\lambda),\ w_1(A))$ imply the corresponding practical stability properties of the system (1.5) with respect to (λ, A).

Proof. The proof of Theorem 2.30 is analogous to the proof of Theorem 2.26. Depending on the vector Lyapunov function used, we need to modify the definitions of practical stability properties of system (2.85). Here, we shall prove the practical stability of (1.5).

Suppose that (2.85) is practically stable with respect to $(w_2(\lambda), w_1(A))$. Then, we have

$$\sum_{i=1}^{m} u_{i0} < w_2(\lambda) \ \text{ implies } \ \sum_{i=1}^{m} u_i^+(t;t_0,u_0) < w_1(A), \ t \geq t_0 \tag{2.87}$$

for some given $t_0 \in \mathbb{R}$, where $u_0 = (u_{10},\dots,u_{m0})^T$ and the maximal solution $u^+(t;t_0,u_0)$ of (2.85) is defined in the interval $[t_0,\infty)$.

Setting $u_0 = V(t_0^+,x_0)$, we get by Lemma 1.2,

$$V(t,x(t;t_0,x_0)) \leq u^+(t;t_0,V(t_0^+,x_0)) \ \text{ for } t \geq t_0. \tag{2.88}$$

Let

$$||x_0|| < \lambda. \tag{2.89}$$

Then, from (2.86) and (2.89), it follows

$$L_0(t_0^+,x_0) \leq w_2(||x_0||) < w_2(\lambda)$$

which due to (2.87) implies

$$\sum_{i=1}^{m} u_i^+ (t;t_0, V(t_0^+, x_0)) < w_1(A), \ t \geq t_0. \tag{2.90}$$

Consequently, from (2.86), (2.88) and (2.90), we obtain

$$w_1(\|x(t;t_0,x_0)\|) \leq L_0(t, x(t;t_0,x_0))$$

$$\leq \sum_{i=1}^{m} u_i^+ (t;t_0, V(t_0^+, x_0)) < w_1(A), \ t \geq t_0.$$

Hence, $\|x(t;t_0,x_0)\| < A$, $t \geq t_0$ for the given $t_0 \in \mathbb{R}$, which proves the practical stability of (1.5).

Other practical stability properties of (1.5) may be proved similarly using corresponding modifications for practical stability properties of (2.85). We omit the details here. Hence the proof is complete.

Remark 2.6 In Theorem 2.30, we have used a vector Lyapunov function $V : [t_0, \infty) \times \Omega \to \mathbb{R}_+^m$, $V = col(V_1, ..., V_m)$ such that $V_i \in V_0$, $i = 1, 2, ..., m$, and the function $L_0(t,x) = \sum_{i=1}^{m} V_i(t,x)$ as a measure with an appropriate modification of the definition of practical stability of the comparison system (2.85). As in the integer-order case (see [Lakshmikantham, Leela and Martynyuk 1990]), we could use other convenient measures such as

$$L_0(t,x) = \max_{1 \leq i \leq m} V_i(t,x),$$

$$L_0(t,x) = \sum_{i=1}^{m} d_i V_i(t,x),$$

where $d_i \in \mathbb{R}_+^m$, $i = 1, 2, ..., m$,

or

$$L_0(t,x) = P(V(t,x)),$$

where $P : \mathbb{R}_+^m \to \mathbb{R}_+$ and $P(u)$ is non-decreasing in u, and appropriate modifications of practical stability definitions are employed for the system (2.85).

In Examples 2.10, 2.11 and 2.12, we have used scalar Lyapunov functions $V \in V_0$, i.e., in all these examples the function $L_0(t,x) = V(t,x)$. To demonstrate the advantage of employing vector Lyapunov functions, let us consider the following example:

Example 2.13 Consider the system

$$\begin{cases} {}_0^c D_t^\alpha x(t) = a(t)x(t) + b(t)y(t), \ t \neq t_k, \\ {}_0^c D_t^\alpha y(t) = b(t)x(t) + a(t)y(t), \ t \neq t_k, \\ \Delta x(t) = a_k x(t) + b_k y(t), \ t = t_k, \ k = 1, 2, ..., \\ \Delta y(t) = b_k x(t) + a_k y(t), \ t = t_k, \ k = 1, 2, ..., \end{cases} \tag{2.91}$$

where $x, y \in \mathbb{R}$, $t \geq 0$, $0 < \alpha < 1$, a, b are continuous functions, $a_k = \frac{1}{2}(c_k + d_k)$, $b_k = \frac{1}{2}(c_k - d_k)$, $-1 < c_k \leq 0$, $-1 < d_k \leq 0$, $k = 1, 2, ...$, $0 < t_1 < t_2 < ...$, and $\lim_{k \to \infty} t_k = \infty$.

Suppose that we choose a single Lyapunov function $V(t, x, y) = |x + y| + |x - y|$. Then for $t \geq 0$, and $t \neq t_k$, for the upper right-hand derivative of V of order α with respect to (2.91), we have

$$^cD_+^\alpha V(t, x, y) = {}^cD_+^\alpha |x + y| + {}^cD_+^\alpha |x - y|$$

$$= sgn(x + y)^c D_+^\alpha (x + y) + sgn(x - y)^c D_+^\alpha (x - y) \leq [a(t) + |b(t)|]V(t, x, y),$$

for $t \geq 0$, $t \neq t_k$.

Also,

$$V(t_k^+, x(t_k) + a_k x(t_k) + b_k y(t_k), y(t_k) + b_k x(t_k) + a_k y(t_k))$$

$$= |(1 + a_k)x(t_k) + b_k y(t_k) + (1 + a_k)y(t_k) + b_k x(t_k)|$$

$$+ |(1 + a_k)x(t_k) + b_k y(t_k) - (1 + a_k)y(t_k) - b_k x(t_k)|$$

$$\leq \beta_k V(t_k, x(t_k), y(t_k)), \quad k = 1, 2, ...,$$

where $\beta_k = 1 + |a_k| + |b_k| > 1$, $k = 1, 2, ...$

It is clear that the fractional scalar equation

$$\begin{cases} {}_0^c D_t^\alpha u(t) = [a(t) + |b(t)|]u(t), \ t \neq t_k, t \geq 0, \\ \Delta u(t_k) = \beta_k u(t_k), \ k = 1, 2, ..., \end{cases}$$

where $u \in \mathbb{R}_+$, is not practically stable and, consequently we cannot deduce any information about the practical stability of the system (2.91) from Theorem 2.26, even in the cases when the system (2.91) is practically stable.

Now, let us take the function $V = (V_1, V_2)$, where the functions V_1 and V_2 are defined by $V_1(t, x, y) = |x + y|$, $V_2(t, x, y) = |x - y|$ so that $L_0(t, x, y) = |x + y| + |x - y|$. Then for $t \geq 0$ the vector inequality

$$^cD_+^\alpha V(t, x, y) \leq F(t, V(t, x, y)), \ t \neq t_k, k = 1, 2, ...,$$

is satisfied with $F = (F_1, F_2)$, where

$$F_1(t, u_1, u_2) = [a(t) + b(t)]u_1,$$

$$F_2(t, u_1, u_2) = [a(t) - b(t)]u_2,$$

and

$$V_1(t_k^+, x(t_k) + \Delta x(t_k), y(t_k) + \Delta y(t_k))$$

$$\leq |1 + c_k|V_1(t_k, x(t_k), y(t_k)) \leq V_1(t_k, x(t_k), y(t_k)), \ k = 1, 2, ...,$$

$$V_2(t_k^+, x(t_k) + \Delta x(t_k), y(t_k) + \Delta y(t_k))$$

$$\leq |1+d_k|V_2(t_k,x(t_k),y(t_k)) \leq V_2(t_k,x(t_k),y(t_k)), \ k=1,2,...,$$

i.e., the functions $\psi_k = u, \ k = 1,2,....$

It is easy to find functions $a(t)$ and $b(t)$ such that the function F satisfies the conditions of Theorem 2.30 and, therefore the comparison system

$$\begin{cases} {}_0^cD_t^{\alpha}u_1(t) = [a(t)+b(t)]u_1(t), \ t \neq t_k, \\ {}_0^cD_t^{\alpha}u_2(t) = [a(t)-b(t)]u_2(t), \ t \neq t_k, \\ u_1(t_k^+) = u_1(t_k), \ u_2(t_k^+) = u_2(t_k), \ k = 1,2,..., \end{cases}$$

to be practically stable with respect to $(w_2(\lambda), w_1(A))$, where $w_1(u) = u$, $w_2(u) = 2u$ for any $0 < \lambda < A$. This is true, for example, if $a(t) = e^{-t}$, $b(t) = \sin t$, and $E_{\alpha}\big([a(t_0) + \sup\limits_{t \geq 0} b(t)](t_0)^{\alpha}\big) < \dfrac{A}{2\lambda}$, $t_0 \in \mathbb{R}$. Hence, Theorem 2.30 implies that the system (2.91) is also practically stable with respect to (λ, A).

2.6 Lipschitz Stability

In this section, the system (1.8) for ${}_{t_0}\mathscr{D}_t^{\alpha} = {}_{t_0}^cD_t^{\alpha}$ will be investigated. Using differential inequalities and Lyapunov-like functions, Lipschitz stability, uniform Lipschitz stability and global uniform Lipschitz stability criteria will be proved.

For nonlinear integer-order dynamic systems, the notion of Lipschitz stability was introduced in [Dannan and Elaydi 1986]. For linear systems, the notions of uniform Lipschitz stability and that of uniform stability are equivalent. However, for nonlinear systems, the two notions are quite distinct (see [Dannan and Elaydi 1986] and [Kulev and Bainov 1991]). In fact, uniform Lipschitz stability lies between uniform stability on one side and the notions of asymptotic stability in variation and uniform stability in variation on the other side. Furthermore, uniform Lipschitz stability neither implies asymptotic stability nor is it implied by it.

The problem of Lipschitz stability of dynamic systems is relevant in various contexts, including many inverse and control problems (see [Bacchelli and Vessella 2006], [Bellassoued and Yamamoto 2007], [Imanuvilov and Yamamoto 2001], [Wang and Zhou 2011]).

Consider the system

$$ {}_{t_0}^cD_t^{\alpha}x(t) = f(t,x_t), \tag{2.92} $$

where $f : \mathbb{R} \times C[[-r,0],\mathbb{R}^n] \to \mathbb{R}^n, 0 < \alpha < 1$.

Let $\varphi_0 \in \mathscr{C}$ and $x(t;t_0,\varphi_0)$ be the solution of the IVP (2.92), (1.2). When the zero solution $x(t) \equiv 0$ is an equilibrium of (2.92), we shall use the following Lipschitz stability definition, which is a generalization of definitions in [Dannan and Elaydi 1986] and [Kulev and Bainov 1991].

Definition 2.6 The zero solution of system (2.92) is said to be:

(a) *Lipschitz stable*, if

$$(\forall t_0 \in \mathbb{R})(\exists M > 0)(\exists \delta = \delta(t_0) > 0)(\forall \varphi_0 \in B_\delta(C))$$

$$(\forall t \geq t_0) : ||x(t;t_0,\varphi_0)|| \leq M||\varphi_0||_r;$$

(b) *uniformly Lipschitz stable*, if the number δ in (a) is independent of $t_0 \in \mathbb{R}$;

(c) *globally uniformly Lipschitz stable*, if

$$(\exists M > 0)(\forall \varphi_0 \in \mathscr{C} : ||\varphi_0||_r < \infty)$$

$$(\forall t \geq t_0) : ||x(t;t_0,\varphi_0)|| \leq M||\varphi_0||_r.$$

Together with system (2.92) we consider the scalar fractional differential equation (1.11) with ${}_{t_0}\mathscr{D}_t^\beta = {}_{t_0}^c D_t^\beta$ and a nonnegative right-hand side, i.e., we consider the following comparison equation

$$ {}_{t_0}^c D_t^\beta u = g(t,u), \tag{2.93}$$

where $0 < \beta < 1$, $g : [t_0,\infty) \times \mathbb{R}_+ \to \mathbb{R}_+$.

Let $u_0 \in \mathbb{R}_+$. We denote by $\eta(t) = \eta(t;t_0,u_0)$ the maximal solution of equation (2.93), which satisfies the initial condition

$$\eta(t_0;t_0,u_0) = u_0. \tag{2.94}$$

We shall consider such solutions $u(t)$ of equation (2.93) for which $u(t) \geq 0$. That is why the following definitions for Lipschitz stability properties of the zero solution of this equation will be used.

Definition 2.7 The zero solution of equation (2.93) is said to be:

(a) *Lipschitz stable*, if

$$(\forall t_0 \in \mathbb{R})(\exists M > 0)(\exists \delta = \delta(t_0) > 0)$$

$$(\forall u_0 \in \mathbb{R}_+ : u_0 < \delta)(\forall t \geq t_0) : \eta(t;t_0,u_0) \leq Mu_0;$$

(b) *uniformly Lipschitz stable*, if the number δ in (a) is independent of $t_0 \in \mathbb{R}$;

(c) *globally uniformly Lipschitz stable*, if

$$(\exists M > 0)(\forall u_0 \in \mathbb{R}_+ : u_0 < \infty)(\forall t \geq t_0) : \eta(t;t_0,u_0) \leq Mu_0;$$

(d) *uniformly stable*, if

$$(\forall \varepsilon > 0)(\exists \delta = \delta(\varepsilon) > 0)(\forall t_0 \in \mathbb{R})$$

$$(\forall u_0 \in \mathbb{R}_+ : u_0 < \delta)(\forall t \geq t_0) : \eta(t;t_0,u_0) \leq \varepsilon.$$

Following Dannan and Elaydi (see [Dannan and Elaydi 1986]), we shall first show the equivalence of various notions of stability in linear differential systems of fractional order without delays.

We shall consider the linear system

$$
{}_{t_0}^c D_t^\alpha x(t) = Ax(t), \ t \geq t_0, \tag{2.95}
$$

where $x \in \mathbb{R}^n$ and A is an $n \times n$ matrix.

Let $x_0 \in \mathbb{R}^n$ and $x(t) = x(t;t_0,x_0)$ be the solution of (2.95) with the initial condition

$$
x(t_0) = x_0. \tag{2.96}
$$

We shall note that, if A is a sectorial operator, then the unique solution of the IVP (2.95), (2.96) is (see [Chauhan and Dabas 2011] and [Podlubny 1999])

$$
x(t) = T_\alpha(t - t_0)x_0, \tag{2.97}
$$

where

$$
T_\alpha(t) = E_{\alpha,1}(At^\alpha)
$$

is the solution operator, generated by A.

Theorem 2.31 For system (2.95) the following assertions are equivalent:

(i) The zero solution of (2.95) is uniformly Lipschitz stable in variations.

(ii) The zero solution of (2.95) is globally uniformly Lipschitz stable.

(iii) The zero solution of (2.95) is uniformly Lipschitz stable.

(iv) The zero solution of (2.95) is uniformly stable.

(v) The zero solution of (2.95) is globally uniformly Lipschitz stable in variations.

Proof. (i)$\Rightarrow$(ii) Let the zero solution of (2.95) be uniformly Lipschitz stable in variations. Then there exist constants $M > 0$ and $\delta > 0$ such that

$$
||T_\alpha(t - t_0)|| \leq M \text{ for } t \geq t_0, \ ||x_0|| < \delta.
$$

Since the solution operator $T_\alpha(t)$ of (2.95) does not depend on x_0, then (2.97) holds for any $x_0 \in \mathbb{R}^n$. Consequently,

$$
||x(t;t_0,x_0)|| = ||T_q(t - t_0)x_0|| \leq ||T_\alpha(t - t_0)|| ||x_0|| \leq M||x_0||,
$$

for $t \geq t_0$ and $x_0 \in \mathbb{R}^n$, i.e. (ii) holds.

(ii)$\Rightarrow$(iii) This follows immediately from Definitions 2.6(b) and 2.6(c).

(iii)$\Rightarrow$(iv) Let (iii) holds. Then there exist constants $M > 0$ and $\delta_1 > 0$ such that $||x(t;t_0,x_0)|| \leq M||x_0||$ whenever $||x_0|| < \delta_1$ and $t \geq t_0$.

Let $\varepsilon > 0$ be given and let $\delta = \delta(\varepsilon) = \min(\delta_1, \varepsilon/M)$. Then for $||x_0|| < \delta$ and $t \geq t_0$, we have $||x(t;t_0,x_0)|| \leq M||x_0|| \leq M\delta < \varepsilon$. It follows that the zero solution of (2.95) is uniformly stable.

(iv)⇒(v) Let the zero solution of (2.95) is uniformly stable. Then there exists a constant $M > 0$ such that $||T_\alpha(t-t_0)|| \leq M$, where $T_\alpha(t)$ is the solution operator, generated by A (see [Chauhan and Dabas 2011]). Consequently, (v) is obtained.

(v)⇒(i) This follows from the corresponding definitions.

Remark 2.7 As in the integer-order cases (see [Dannan and Elaydi 1986] and [Kulev and Bainov 1991]), we proved that for linear fractional-order systems, the notions of uniform Lipschitz stability and that of uniform stability are equivalent. But, it is easy to see, that for functional differential systems of fractional order the two notions are different, even for the linear systems

$$\,_{t_0}^{c}D_t^\alpha x(t) = Ax(t) + Bx_t, \, t \geq t_0,$$

whose solutions are given by

$$x(t) = T_\alpha(t-t_0)x_0 + \int_{t_0}^{t} BS_\alpha(t-\theta)x_\theta d\theta,$$

where $S_\alpha(t) = t^{\alpha-1}E_{\alpha,\alpha}(At^\alpha)$ is the $\alpha-$ resolvent family, and B is an $n \times n$ matrix.

The efficient applications of fractional functional equations requires the finding of criteria for various types of Lipschitz stability of such systems, and an investigation of the relations between these notions.

Theorem 2.32 Let $f(t,0) = 0$, $t \in [t_0,\infty)$, and let the zero solution of (2.92) is uniformly Lipschitz stable. Then the zero solution of (2.92) is uniformly stable.

Proof. Let the zero solution of (2.92) is uniformly Lipschitz stable. Then there exist constants $M > 0$ and $\delta_1 > 0$ such that $||x(t;t_0,\varphi_0)|| \leq M||\varphi_0||_r$ whenever $||\varphi_0||_r < \delta_1$, $\varphi_0 \in \mathscr{C}$ and $t \geq t_0$.

Let $\varepsilon > 0$ be given and let $\delta = \delta(\varepsilon) = \min(\delta_1, \varepsilon/M)$. Then for $\varphi_0 \in \mathscr{C}$, $||\varphi_0||_r < \delta$ and $t \geq t_0$, the inequalities $||x(t;t_0,\varphi_0)|| \leq M||\varphi_0||_r \leq M\delta < \varepsilon$ are valid. Hence, the zero solution of (2.92) is uniformly stable.

In the proof of the next result we shall use the following lemma:

Lemma 2.2 ([Lakshmikantham, Leela and Vasundhara Devi 2009]) Let $m \in C[[t_0 - r,\infty),\mathbb{R}]$ and satisfy the inequality

$$\,_{t_0}^{c}D_t^\beta m(t) \leq g(t,||m_t||_r), \, t > t_0,$$

where $g \in C[[t_0,\infty) \times \mathbb{R}_+,\mathbb{R}_+]$. Assume that the maximal solution $\eta(t)$ of the IVP (2.93), (2.94) exists on $[t_0,\infty)$. Then, if $||m_{t_0}||_r \leq u_0$, we have

$$m(t) \leq \eta(t), \, t \in [t_0,\infty).$$

Theorem 2.33 Assume that:

1. $f(t,0) = 0$ and $g(t,0) = 0$, $t \in [t_0, \infty)$.
2. $\alpha = \beta$ and there exists $\rho > 0$ such that for $(t,\varphi) \in [t_0, \infty) \times B_\rho(C)$,

$$||f(t,\varphi)|| \leq g(t, ||\varphi||_r), \tag{2.98}$$

   where $g \in C[[t_0, \infty) \times \mathbb{R}_+, \mathbb{R}_+]$.
3. The maximal solution $\eta(t)$ of the IVP (2.93), (2.94) exists on $[t_0, \infty)$.

Then the Lipschitz stability properties of the zero solution of (2.93) imply the corresponding Lipschitz stability properties of the zero solution of system (2.92).

Proof. We shall first prove Lipschitz stability of the zero solution of (2.92). Suppose that the zero solution of (2.93) is Lipschitz stable. Then, there exist $M > 0$ and $\delta = \delta(t_0) > 0$, $(M > 1, M\delta < \rho)$, such that

$$0 \leq u_0 < \delta \quad \text{implies} \quad \eta(t;t_0,u_0) \leq Mu_0, \ t \geq t_0 \tag{2.99}$$

for some given $t_0 \in \mathbb{R}$, where the maximal solution $\eta(t;t_0,u_0)$ of (2.93) is defined in the interval $[t_0, \infty)$, and

$$\eta(t;t_0,u_0) \leq Mu_0 < M\delta < \rho, \ t \geq t_0.$$

Let $\varphi_0 \in \mathscr{C}$ be such that

$$||\varphi_0||_r < \delta \tag{2.100}$$

and $x(t;t_0,\varphi_0)$ be the solution of the IVP (2.92), (1.2).

Then

$$||\varphi_0||_r < \delta < M\delta < \rho.$$

Setting $m(t) = ||x(t;t_0,\varphi_0)||$ and $||m_{t_0}||_r = ||\varphi_0||_r = u_0$, we get by Lemma 2.2,

$$||x(t;t_0,\varphi_0)|| \leq \eta(t;t_0,||\varphi_0||_r) \text{ for } t \geq t_0. \tag{2.101}$$

From (2.99) and (2.100), it follows

$$\eta(t;t_0,||\varphi_0||_r) \leq M||\varphi_0||_r,$$

which due to (2.101) implies

$$||x(t;t_0,\varphi_0)|| \leq \eta(t;t_0,||\varphi_0||_r) \leq M||\varphi_0||_r, \ t \geq t_0.$$

Hence, $||x(t;t_0,\varphi_0)|| \leq M||\varphi_0||_r$, $t \geq t_0$ for the given $t_0 \in \mathbb{R}$, which proves the Lipschitz stability of the zero solution of (2.92).

Suppose now, that the zero solution of (2.93) is uniformly Lipschitz stable. Therefore, there exist constants $M > 0$ and $\delta > 0$ $(M > 1, M\delta < \rho)$ such that $0 \leq u_0 < \delta$ implies

$$\eta(t;t_0,u_0) \leq Mu_0, \ t \geq t_0 \tag{2.102}$$

for every $t_0 \in \mathbb{R}$.

We claim that $\varphi_0 \in \mathscr{C}$, $||\varphi_0||_r < \delta$ implies $||x(t;t_0,\varphi_0)|| \leq M||\varphi_0||_r$, $t \geq t_0$ for every $t_0 \in \mathbb{R}$. If the claim is not true, there exists $t_0 \in \mathbb{R}$, a corresponding solution $x(t;t_0,\varphi_0)$ of (2.92) with $||\varphi_0||_r < \delta$, and $t^* > t_0$, $t^* < \infty$, such that

$$||x(t^*)|| > M||\varphi_0||_r \text{ and } ||x(t;t_0,\varphi_0)|| \leq M||\varphi_0||_r, \ t \in [t_0,t^*).$$

Since

$$M||\varphi_0||_r < M\delta < \rho,$$

and the function $x(t)$ is continuous on $[t_0,t^*]$, there exists t^0, $t_0 < t^0 \leq t^*$, such that

$$M||\varphi_0||_r < ||x(t^0)|| < \rho, \ ||x(t)|| < \rho, \ t_0 < t \leq t^0. \tag{2.103}$$

Define for $t \in [t_0 - r, t^0]$ the notation $m(t) = ||x(t;t_0,\varphi_0)||$. Using the assumption (2.98), it is easy to obtain for $t \in [t_0, t^0]$, the inequality

$${}_{t_0}^c D_t^\alpha m(t) \leq g(t, ||m_t||_r).$$

Choosing $||m_{t_0}||_r = ||\varphi_0||_r = u_0$, we obtain by Lemma 2.2,

$$||x(t;t_0,\varphi_0)|| \leq \eta(t;t_0,u_0), \ t \in [t_0,t^0]. \tag{2.104}$$

From (2.103), (2.104), and (2.102) we get

$$M||\varphi_0||_r < ||x(t^0)|| \leq \eta(t^0) \leq Mu_0 = M||\varphi_0||_r.$$

The contradiction obtained shows that

$$||x(t;t_0,\varphi_0)|| \leq M||\varphi_0||_r$$

for $||\varphi_0||_r < \delta$ and $t \geq t_0$. It, therefore, follows that the zero solution of system (2.92) is uniformly Lipschitz stable.

In the next, we shall apply the basic comparison Lemma 1.1 in terms of Lyapunov functions from the class C_0 and the Razumikhin technique, and we shall obtain sufficient conditions for uniform Lipschitz stability (global uniform Lipschitz stability) of the zero solution of (2.92).

Theorem 2.34 Assume that:

1. Condition 1 of Theorem 2.33 and conditions of Lemma 1.1 hold.
2. There exist positive constants A and B, such that

$$A||x|| \leq V(t,x) \leq B||x||, \ (t,x) \in [t_0,\infty) \times B_\rho, \tag{2.105}$$

   where $B_\rho = \{x \in \mathbb{R}^n : |x| < \rho\}$ and the function $V \in C_0$ satisfies H2.1.
3. The zero solution of equation (2.93) is uniformly Lipschitz stable (globally uniformly Lipschitz stable).

Then, the zero solution of system (2.92) is uniformly Lipschitz stable (globally uniformly Lipschitz stable).

Proof. Let the zero solution of (2.93) be uniformly Lipschitz stable. Therefore, there exist constants $M > 0$ and $\delta_1 > 0$ such that $0 \leq u_0 < \delta_1$ implies (2.102) for every $t_0 \in \mathbb{R}$.

Choose $M_1 > 1$ and $\delta_2 > 0$ such that

$$M_1 > \frac{B}{A}, \quad M_1 > \frac{MB}{A}, \quad M_1 \delta_2 < \rho. \tag{2.106}$$

Let $\delta = \min\{\frac{\delta_1}{B}, \delta_1, \delta_2\}$, and $\varphi_0 \in \mathscr{C}$ be such that inequality (2.100) holds. Then, using (2.105), we get

$$||\varphi_0||_r < \frac{B}{A}\delta < M_1 \delta_2 < \rho,$$

i.e., $\varphi_0 \in B_\rho(C)$.

We shall prove that, if inequality (2.100) is satisfied, then

$$||x(t;t_0,\varphi_0)|| \leq M||\varphi_0||_r, \ t \geq t_0, \tag{2.107}$$

for every $t_0 \in \mathbb{R}$, where $x(t;t_0,\varphi_0)$ is the solution of the IVP (2.92), (1.2) with the chosen above initial function $\varphi_0 \in B_\rho(C)$.

Suppose that (2.107) is not true. Then, there exists a solution $x(t;t_0,\varphi_0)$ of (2.92) for which $||\varphi_0||_r < \delta$ and $t^* > t_0, t^* < \infty$, such that

$$||x(t^*)|| > M||\varphi_0||_r \ \text{and} \ ||x(t;t_0,\varphi_0)|| \leq M||\varphi_0||_r, \ t \in [t_0,t^*).$$

Then, due to the choice of δ, and the fact that the function $x(t)$ is continuous on $[t_0,t^*]$, we can find $t^0, t_0 < t^0 \leq t^*$, such that

$$M_1||\varphi_0||_r < ||x(t^0)|| < \rho, \ ||x(t)|| < \rho, \ t_0 < t \leq t^0. \tag{2.108}$$

Set $\max\limits_{-r \leq \theta \leq 0} V(t_0 + \theta, \varphi_0(\theta)) = u_0$. From the choice of δ, we have

$$V(t_0 + \theta, \varphi_0(\theta)) \leq B||\varphi_0||_r < B\delta \leq \delta_1, \ \theta \in [-r,0],$$

i.e., $u_0 < \delta_1$.

Since for $t \in [t_0,t^0]$ all the conditions of Lemma 1.1 are met, then

$$V(t,x(t;t_0,\varphi_0)) \leq \eta(t;t_0,u_0), \ t \in [t_0,t^0]. \tag{2.109}$$

From (2.108), (2.105), (2.109) and (2.106) there follow the inequalities

$$M_1||\varphi_0||_r < ||x(t^0)|| \leq \frac{1}{A}V(t^0,x(t^0;t_0,\varphi_0)) \leq \frac{1}{A}\eta(t^0;t_0,u_0) \leq \frac{M}{A}u_0$$

$$= \frac{M}{A}\max\limits_{-r \leq s \leq 0} V(t_0 + \theta, \varphi_0(\theta)) \leq \frac{MB}{A}||\varphi_0||_r < M_1||\varphi_0||_r.$$

The obtained contradiction proves the validity of inequality (2.107) and the claim of Theorem 2.34.

Remark 2.8 As we can see, again, from Theorem 2.34, when we apply Lyapunov-like functions in the stability analysis of a fractional-order system, the Lyapunov functions fractional differential operator may not be of the same fractional order as that associated with the system, as it is for integer-order systems. Studying the relations between the two fractional orders is important in elaborating a research methodology.

The next theorem is a particular case of Theorem 2.34.

Theorem 2.35 Assume that $\alpha = \beta$, and the function $g(t,u)$ is nondecreasing in u for each $t \geq t_0$. If in Theorem 2.33 condition (2.98) is replaced by the condition

$$||\varphi(0)|| \leq ||\varphi(0) - \chi^\alpha f(t,\varphi)|| + \chi^\alpha g(t,||\varphi(0)||) + \varepsilon(\chi^\alpha), \ t \geq t_0, \quad (2.110)$$

where $\chi > 0$ is sufficiently small, $\frac{\varepsilon(\chi^\alpha)}{\chi^\alpha} \to 0$ as $\chi \to 0^+$, and $\varphi \in B_\rho(C)$ is such that $||\varphi(t + \theta)|| \leq ||\varphi(t)||$, $\theta \in [-r,0]$, then the uniform Lipschitz stability (global uniform Lipschitz stability) of the zero solution of (2.93) imply the uniform Lipschitz stability (global uniform Lipschitz stability) of the zero solution of system (2.92).

Proof. Let $V(t,x(t)) = ||x(t)||$. Note that $V \in C_0$ and $V(t,0) = 0$ for $t \geq t_0$.

According to (2.110), for $\varphi \in B_\rho(C)$ such that $||\varphi(t + \theta)|| \leq ||\varphi(t)||$, $\theta \in [-r,0]$, we have

$$^cD_+^\alpha V(t,\varphi(0))$$

$$= \lim_{\chi \to 0^+} \sup \frac{1}{\chi^\alpha} [V(t,\varphi(0)) - V(t - \chi, \varphi(0) - \chi^\alpha f(t,\varphi))]$$

$$= \lim_{\chi \to 0^+} \sup \frac{1}{\chi^\alpha} [||\varphi(0)|| - ||\varphi(0) - \chi^\alpha f(t,\varphi)||]$$

$$\leq g(t,||\varphi(0)||) + \lim_{\chi \to 0^+} \sup \frac{\varepsilon(\chi^\alpha)}{\chi^\alpha}$$

$$= g(t,||\varphi(0)||) = g(t,V(t,\varphi(0))).$$

Later the proof of Theorem 2.35 is completed as the proof of Theorem 2.33.

Theorem 2.36 If in Theorem 2.35 condition (2.110) is replaced by the condition

$$[\varphi(0), f(t,\varphi)]_+ \leq g(t,||\varphi(0)||),$$

where $\varphi \in B_\rho(C)$ is such that $||\varphi(t + \theta)|| \leq ||\varphi(t)||$, $\theta \in [-r,0]$, and $[x,y]_+ = \lim_{\chi \to 0^+} \sup \frac{1}{\chi^\alpha} [||x|| - ||x - \chi^\alpha y||]$, $x,y \in \mathbb{R}^n$, then the uniform Lipschitz stability (global uniform Lipschitz stability) of the zero solution of (2.93) imply the uniform Lipschitz stability (global uniform Lipschitz stability) of the zero solution of system (2.92).

The proof of Theorem 2.36 is analogous to the proof of Theorem 2.35.

Theorem 2.37 Assume that $f(t,0) = 0$, $t \in [t_0, \infty)$, $f \in C[[t_0, \infty) \times B_\rho(C), \mathbb{R}^n]$ and for $(t, \varphi) \in [t_0, \infty) \times B_\rho(C)$,

$$||f(t, \varphi)|| \leq m(t)p(||\varphi(t)||),$$

where $p(u)$ is non-decreasing positive submultiplicative continuous function on $(0, \infty)$, and

$$G^{-1}\left(G(1) + \frac{p(||\varphi_0||_r)}{||\varphi_0||_r} \frac{1}{\Gamma(\alpha)} \int_\theta^t (t-s)^{\alpha-1} m(s)\,ds\right) \leq M,$$

$0 < M = const$, for all $\theta \geq t_0 \geq 0$, and $G(u) = \int_{u_0}^u \frac{ds}{p(s)}$, $u \geq u_0 > 0$.

Then the zero solution of (2.92) is globally uniformly Lipschitz stable.

Proof. Since the solution $x(t) = x(t; t_0, \varphi_0)$ satisfies

$$x_{t_0} = \varphi_0,$$

$$x(t) = \varphi_0(0) + \frac{1}{\Gamma(\alpha)} \int_{t_0}^t (t-s)^{\alpha-1} f(s, x_s)\,ds, \quad t_0 \leq t < \infty,$$

it follows that

$$||x(t; t_0, \varphi_0)|| \leq ||\varphi_0(0)|| + \frac{1}{\Gamma(\alpha)} \int_{t_0}^t (t-s)^{\alpha-1} ||f(s, x_s)||\,ds$$

$$\leq ||\varphi_0||_r + \frac{1}{\Gamma(\alpha)} \int_{t_0}^t (t-s)^{\alpha-1} m(s)p(||x(s)||)\,ds,$$

from which, we obtain the estimates

$$\frac{||x(t; t_0, \varphi_0)||}{||\varphi_0||_r} \leq 1 + \frac{1}{\Gamma(\alpha)} \int_{t_0}^t (t-s)^{\alpha-1} \frac{m(s)}{||\varphi_0||_r} p\left(||\varphi_0||_r \frac{||x(s; t_0, \varphi_0)||}{||\varphi_0||_r}\right) ds$$

$$\leq 1 + \frac{1}{\Gamma(\alpha)} \int_{t_0}^t (t-s)^{\alpha-1} \frac{p(||\varphi_0||_r)}{||\varphi_0||_r} m(s)p\left(\frac{||x(s; t_0, \varphi_0)||}{||\varphi_0||_r}\right) ds, \quad t \geq t_0.$$

To the last inequality, we apply a generalization of Bihari's inequality (see [Pachpatte 2005]), and we are led to the inequality

$$||x(t; t_0, \varphi_0)|| \leq ||\varphi_0||_r G^{-1}\left(G(1) + \frac{p(||\varphi_0||_r)}{||\varphi_0||_r} \frac{1}{\Gamma(\alpha)} \int_\theta^t (t-s)^{\alpha-1} m(s)\,ds\right).$$

Hence, $||x(t;t_0,\varphi_0)|| \leq M||\varphi_0||_r$ for all $\varphi_0 \in \mathscr{C}$, $||\varphi_0||_0 < \infty$, and $t \geq t_0$, and the conclusion of the theorem follows.

Example 2.14 Let $x \in \mathbb{R}^n$, $t_0 \in \mathbb{R}_+$ and $\tau = const > 0$. Consider the following fractional linear functional differential system

$$^c_{t_0}D^\alpha_t x(t) = Ax(t) + Bx(t-\tau), t \in \mathbb{R}_+, \qquad (2.111)$$

where $0 < \alpha < 1$, A and B are constant matrices of type $(n \times n)$.

Let $\mu(A+B)$ be the logarithmic norm of Lozinskii (see [Dannan and Elaydi 1986] and [Lakshmikantham, Leela and Martynyuk 1990])

$$\mu(A+B) = \lim_{h \to 0^+} \frac{||E+h(A+B)|| - 1}{h},$$

where E denotes the identity $(n \times n)-$matrix.

We have that for $t \geq t_0$,

$$||Ax(t) + Bx(t-\tau)|| \leq \mu(A+B) \max_{\theta \in [t-\tau,t]} ||x(\theta)||.$$

Together with system (2.111) consider the equation

$$^c_{t_0}D^\beta_t u = \mu(A+B)u, \qquad (2.112)$$

where $0 < \beta < 1$, $u \in \mathbb{R}_+$.

If there exists a continuous function $q(t)$ on $[t_0,\infty)$ such that

(i) $\mu(A+B) \leq q(t)$ for $t \geq t_0 \geq 0$, $u_0 \leq \delta$;

(ii) $\int_\theta^\infty q(s)ds < \infty$ for $\theta \geq t_0 \geq 0$,

then the zero solution of the equation (2.112) is uniformly stable (see [El-Sayed, Gaaraf and Hamadalla 2010]). Therefore, according to Theorem 2.31, the zero solution of (2.112) is globally uniformly Lipschitz stable, and by Theorem 2.34 we conclude that the zero solution of (2.111) is globally uniformly Lipschitz stable.

Remark 2.9 Note that in the above example α may not be equal to β which leads to great flexibility in the choice of the comparison equation.

Example 2.15 Consider the following fractional single–species model exhibiting the so-called Allee effect (see [Gopalsamy 1992], [Stamova 2009]) in which the per-capita growth rate is a quadratic function of the density and subject to delays:

$$^c_0D^\alpha_t N(t) = N(t)[a + bN(t-\tau(t)) - cN^2(t-\tau(t))], t \geq 0, \qquad (2.113)$$

where $0 < \alpha < 1$, $a,c \in (0,\infty)$, $b \in \mathbb{R}$, and $0 < \tau(t) < \tau$.

Let $\varphi_0 \in C[[-\tau,0],\mathbb{R}_+]$. Denote by $N(t) = N(t;0,\varphi_0)$ the solution of (2.113) satisfying the initial conditions

$$N(s) = \varphi_0(\theta) \geq 0, \quad \theta \in [-\tau,0); \qquad N(0) > 0.$$

Define the function $V(t,N) = |N|$. Let there exist functions $d \in C[\mathbb{R}_+,\mathbb{R}]$ and $w \in K$, such that for $t \geq 0$,

$$[N(0),N(t)(a+bN(t-\tau(t)) - cN^2(t-\tau(t)))]_+ \leq d(t)w(N(t)),$$

whenever $|N(\theta)| \leq |N(t)|$, $t - \tau \leq \theta \leq t$, and $|N(t)| < \rho$, $\rho = const > 0$.

Hence, by Theorem 2.36, if the zero solution of the equation

$$_0^c D_t^\alpha N(t) = d(t)w(N(t)), t \geq 0,$$

is uniformly Lipschitz stable, then the zero solution of (2.113) is uniformly Lipschitz stable.

2.7 Stability of Sets

In this section, we shall present the techniques of using vector Lyapunov functions for finding stability and boundedness criteria for the solutions of impulsive functional differential equations of fractional order (1.6) for $\Omega = \mathbb{R}^n$ with respect to sets.

We shall consider the system

$$\begin{cases} _{t_0}^c D_t^\alpha x(t) = f(t,x_t), t \neq t_k, \\ \Delta x(t) = I_k(x(t)), t = t_k, k = \pm 1, \pm 2, ..., \end{cases} \tag{2.114}$$

where $f : [t_0,\infty) \times PC[[-r,0],\mathbb{R}^n] \to \mathbb{R}^n$, $I_k : \mathbb{R}^n \to \mathbb{R}^n$, $t_k < t_{k+1}$, $k = \pm 1, \pm 2, ...$ and $\lim_{k \to \pm\infty} t_k = \pm\infty$.

Let $\mathscr{PC} = PC[[-r,0],\mathbb{R}^n]$ and $\varphi_0 \in \mathscr{PC}$. We denote by $x(t) = x(t;t_0,\varphi_0)$ the solution of system (2.114), satisfying the initial conditions (1.7) and by $J^+(t_0,\varphi_0)$—the maximal interval of the type $[t_0,\gamma)$ in which the solution $x(t;t_0,\varphi_0)$ is defined.

Let $h : [t_0 - r,\infty) \times \mathbb{R}^n \to \mathbb{R}^k$ ($k \leq n$) be a function in $[t_0 - r,\infty) \times \mathbb{R}^n$. We introduce the sets

$$M_t(n-k) = \{x \in \mathbb{R}^n : h(t,x) = 0, t \in [t_0,\infty)\},$$

$$M_{t,r}(n-k) = \{x \in \mathbb{R}^n : h(t,x) = 0, t \in [t_0 - r,t_0]\},$$

$$M_t(n-k)(\varepsilon) = \{x \in \mathbb{R}^n : ||h(t,x)|| < \varepsilon, \ t \in [t_0,\infty)\}, \ \varepsilon > 0,$$

$$M_t(n-k)(\bar{\varepsilon}) = \{x \in \mathbb{R}^n : ||h(t,x)|| \leq \varepsilon, \ t \in [t_0,\infty)\},$$

$$M_{t,r}(n-k)(\varepsilon) = \{\phi \in \mathscr{PC} : \sup_{-r \leq \theta \leq 0} ||h(t,\phi(\theta))|| < \varepsilon, \ t \in [t_0-r,t_0]\},$$

$$M_{t,r}(n-k)(\bar{\varepsilon}) = \{\phi \in \mathscr{PC} : \sup_{-r \leq \theta \leq 0} ||h(t,\phi(\theta))|| \leq \varepsilon, \ t \in [t_0-r,t_0]\}.$$

Introduce the following assumptions:

H2.4 The function h is continuous on $[t_0-r,\infty) \times \mathbb{R}^n$ and the sets $M_t(n-k)$, $M_{t,r}(n-k)$ are $(n-k)-$dimensional manifolds in $\mathbb{R}^n$.

H2.5 H2.5. Each solution $x(t;t_0,\varphi_0)$ of the IVP (2.114), (1.7) satisfying

$$||h(t,x(t;t_0,\varphi_0))|| \leq L < \infty \quad \text{for} \quad t \in J^+(t_0,\varphi_0)$$

is defined on the interval $[t_0,\infty)$.

We shall introduce the following definitions of stability of the zero solution of (2.114) with respect to the function h.

Definition 2.8 The zero solution $x(t) \equiv 0$ of system (2.114) is said to be:

(a) *stable with respect to the function h*, if

$$(\forall t_0 \in \mathbb{R})(\forall \varepsilon > 0)(\exists \delta = \delta(t_0,\varepsilon) > 0)(\forall \varphi_0 \in M_{t,r}(n-k)(\bar{\delta}))$$

$$(\forall t \in J^+(t_0,\varphi_0)) : x(t;t_0,\varphi_0) \in M_t(n-k)(\varepsilon);$$

(b) *uniformly stable with respect to the function h*, if the number δ in (a) depends only on ε;

(c) *globally equi-attractive with respect to the function h*, if

$$(\forall t_0 \in \mathbb{R})(\forall \eta > 0)(\forall \varepsilon > 0)(\exists T = T(t_0,\eta,\varepsilon) > 0, \ t_0+T \in J^+(t_0,\varphi_0))$$

$$(\forall \varphi_0 \in M_{t,r}(n-k)(\bar{\eta}))(\forall t \geq t_0+T, \ t \in J^+(t_0,\varphi_0)) : x(t;t_0,\varphi_0) \in M_t(n-k)(\varepsilon);$$

(d) *uniformly globally attractive with respect to the function h*, if the number T in (c) does not depend on $t_0 \in \mathbb{R}$;

(e) *globally equi-asymptotically stable with respect to the function h*, if it is stable and globally equi-attractive with respect to the function h;

(f) *uniformly globally asymptotically stable with respect to the function h*, if it is uniformly stable and uniformly globally attractive with respect to the function h;

(g) *unstable with respect to the function h*, if

$$(\exists t_0 \in \mathbb{R})(\exists \varepsilon > 0)(\forall \delta > 0)(\exists \varphi_0 \in M_{t,r}(n-k)(\bar{\delta}))$$

$$(\exists t^* \in J^+(t_0,\varphi_0)) : ||h(t^*,x(t^*;t_0,\varphi_0))|| \geq \varepsilon.$$

Remark 2.10 The stability of the zero solution of (2.114) with respect to the function h guarantee that the set $\{(t,x) : t \in [t_0,\infty),\, x \in M_t(n-k)\}$ is a positively invariant set of the system (2.114).

We shall introduce a definition of the Mittag–Leffler stability of the zero solution of (2.114) with respect to the function h, which generalizes definitions given in [Li, Chen and Podlubny 2010].

Definition 2.9 The zero solution of (2.114) is said to be *globally Mittag–Leffler stable with respect to the function h*, if for $t_0 \in \mathbb{R}$ and $\varphi_0 \in M_{t,r}(n-k)(\bar{\delta})$ there exist constants $\mu > 0$ and $d > 0$ such that

$$x(t;t_0,\varphi_0) \in M_t(n-k)(\{m(\varphi_0)E_\beta(-\mu(t-t_0)^\beta)\}^d),\, t \in J^+(t_0,\varphi_0),$$

where E_β is the corresponding Mittag–Leffler function, $0 < \beta < 1$, $m(0) = 0$, $m(\phi) \geq 0$, and $m(\phi)$ is Lipschitz with respect to $\phi \in M_{t,r}(n-k)(\bar{\delta})$.

We shall give definitions of boundedness of the solutions of system (2.114) with respect to the function h.

Definition 2.10 We say that the solutions of system (2.114) are:

(a) *equi-bounded with respect to the function h*, if

$$(\forall t_0 \in \mathbb{R})(\forall A > 0)(\exists B = B(t_0,A) > 0)(\forall \varphi_0 \in M_{t,r}(n-k)(\bar{A}))$$

$$(\forall t \in J^+(t_0,\varphi_0)) : x(t;t_0,\varphi_0) \in M_t(n-k)(B);$$

(b) *uniformly bounded with respect to the function h*, if the number B in (a) is independent of $t_0 \in \mathbb{R}$;

(c) *ultimately bounded with respect to the function h* for bound N, if

$$(\exists N > 0)(\forall t_0 \in \mathbb{R})(\forall A > 0)(\exists T = T(t_0,A) > 0)(\forall \varphi_0 \in M_{t,r}(n-k)(\bar{A}))$$

$$(\forall t \geq t_0 + T,\, t \in J^+(t_0,\varphi_0)) : x(t;t_0,\varphi_0) \in M_t(n-k)(N);$$

(d) *uniformly ultimately bounded with respect to the function h* for bound N, if the number T from (c) does not depend on $t_0 \in \mathbb{R}$.

Remark 2.11 In the case $n = k$ and $h(t,x) = x$, Definition 2.8 is reduced to the definition of the Lyapunov-like stability of the zero solution of system (2.114) (see [Stamova and Stamov G 2014b]), Definition 2.9 can be reduced to the definition of the Mittag–Leffler stability of the zero solution of system (2.114) (see [Stamova 2014a]), and Definition 2.10 is a generalization of the boundedness definitions for the solutions of (2.114).

In this section we shall use the fractional vector comparison principle. That is why, together with system (2.114) we shall consider the comparison system (1.13).

Let $e \in \mathbb{R}^m$ be the vector $(1,1,...,1)$. Introduce the following assumption:

H2.6 If $\{u \in \mathbb{R}^m : 0 \le u \le Le\} \subset Q$ and the maximal solution $u^+(t;t_0,u_0)$ of (1.13) satisfies the estimate

$$u^+(t;t_0,u_0) \le Le \text{ for } t \in J^+(t_0,u_0),$$

then this solution is defined on the interval $[t_0,\infty)$.

Let $\{u \in \mathbb{R}^m : 0 \le u \le e\} \subset Q$. Further we shall consider only such solutions $u(t)$ of system (1.13) for which $u(t) \ge 0$. That is why the following stability and boundedness definitions for system (1.13) are appropriate.

Definition 2.11 The zero solution $u(t) \equiv 0$ of system (1.13) is said to be:

(a) *stable*, if

$$(\forall t_0 \in \mathbb{R})(\forall \varepsilon > 0)(\exists \delta = \delta(t_0,\varepsilon) > 0)(\forall u_0 \in Q : 0 \le u_0 \le \delta e)$$

$$(\forall t \in J^+(t_0,u_0)) : u^+(t;t_0,u_0) < \varepsilon e;$$

(b) *uniformly stable*, if the number δ in (a) depends only on ε;

(c) *globally equi-attractive*, if

$$(\forall t_0 \in \mathbb{R})(\forall \eta > 0)(\forall \varepsilon > 0)(\exists T = T(t_0,\eta,\varepsilon) > 0, t_0 + T \in J^+(t_0,u_0))$$

$$(\forall u_0 \in Q : 0 \le u_0 \le \eta e)(\forall t \ge t_0 + T, t \in J^+(t_0,u_0)) : u^+(t;t_0,u_0) < \varepsilon e;$$

(d) *uniformly globally attractive*, if the number T in (c) does not depend on $t_0 \in \mathbb{R}$;

(e) *globally equi-asymptotically stable*, if is stable and globally equi-attractive;

(f) *uniformly globally asymptotically stable*, if is uniformly stable and uniformly globally attractive;

(g) *globally Mittag–Leffler stable*, if for $t_0 \in \mathbb{R}$ and $u_0 \in Q$, $0 \le u_0 \le \delta e$,

$$u^+(t;t_0,u_0) < \{m(u_0)E_\beta(-\mu(t-t_0)^\beta)\}^d e, \ t \in J^+(t_0,u_0),$$

where $0 < \beta < 1$, $\mu > 0$, $d > 0$, $m(0) = 0$, $m(u) \ge 0$, and $m(u)$ is locally Lipschitz with respect to $u \in Q : 0 \le u \le \delta e$;

(h) *unstable*, if

$$(\exists t_0 \in \mathbb{R})(\exists \varepsilon > 0)(\forall \delta > 0)(\exists u_0 \in Q : 0 \le u_0 \le \delta e)(\exists t^* \in J^+(t_0,u_0)) :$$

the inequality $u^-(t^*;t_0,u_0) < \varepsilon e$ does not hold.

(Note that the fact that the inequality "<" does not hold is not equivalent to the fact that the inequality "≥" holds, and means that there exists $j \in \{1,2,...,m\}$ such that $u_j^-(t;t_0,u_0) \geq \varepsilon e$).

Definition 2.12 We say that the solutions of system (1.13) are:

(a) *equi-bounded*, if

$$(\forall t_0 \in \mathbb{R})(\forall A > 0)(\exists B = B(t_0,A) > 0)(\forall u_0 \in Q : 0 \leq u_0 \leq Ae)$$

$$(\forall t \in J^+(t_0,u_0)) : u^+(t;t_0,u_0) < Be;$$

(b) *uniformly bounded*, if the number B in (a) is independent of $t_0 \in \mathbb{R}$;

(c) *ultimately bounded* for bound N, if

$$(\exists N > 0)(\forall t_0 \in \mathbb{R})(\forall A > 0)(\exists T = T(t_0,A) > 0)(\forall u_0 \in Q : 0 \leq u_0 \leq Ae)$$

$$(\forall t \geq t_0 + T, \ t \in J^+(t_0,u_0)) : u^+(t;t_0,u_0) < Ne;$$

(d) *uniformly ultimately bounded* for bound N, if the number T from (c) does not depend on $t_0 \in \mathbb{R}$.

In the further considerations, we shall use piecewise continuous auxiliary functions $V : [t_0,\infty) \times \mathbb{R}^n \to \mathbb{R}^m$, $V_i \in V_0$, $i = 1,2,...,m$, for which the condition H2.1 is satisfied.

Theorem 2.38 Assume that:

1. Conditions H2.2–H2.6 hold.
2. Conditions of Lemma 1.3 hold, and $F(t,0) = 0$ for $t \in [t_0,\infty)$, $J_k(0) = 0$, $k = \pm 1, \pm 2, ...$
3. The function $V : [t_0,\infty) \times \mathbb{R}^n \to Q$, $V = col(V_1, V_2,...,V_m)$, $V_i \in V_0$, $i = 1,2,...,m$, is such that H2.1 holds, $\sup_{[t_0,\infty) \times \mathbb{R}^n} \|V(t,x)\| < P \leq \infty$, $Q = \{u \in \mathbb{R}^m : \|u\| < P\}$, and the inequalities

$$w_1(\|h(t,x)\|)e \leq V(t,x) \leq \gamma(t)w_2(\|h(t,x)\|)e, \ (t,x) \in [t_0,\infty) \times \mathbb{R}^n \tag{2.115}$$

hold, where $w_1, w_2 \in K$ and the function $\gamma(t) \geq 1$ is defined and continuous for $t \in [t_0,\infty)$.

Then:

(a) If the zero solution of system (1.13) is stable, then the zero solution of system (2.114) is stable with respect to the function h.

(b) If the zero solution of system (1.13) is globally equi-asymptotically stable, then the zero solution of system (2.114) is globally equi-asymptotically stable with respect to the function h.

Proof. (a) Let $t_0 \in \mathbb{R}$ and $\varepsilon > 0$ $(w_1(\varepsilon) < P)$ be chosen. From the stability of the zero solution of system (1.13) it follows that there exists $\delta^* = \delta^*(t_0, \varepsilon)$ such that $u_0 \in Q$ and $0 \le u_0 \le \delta^* e$ imply $u^+(t; t_0, u_0) < w_1(\varepsilon)e$ for $t \in J^+(t_0, u_0)$. Then from H2.6 it follows that $J^+(t_0, u_0) = [t_0, \infty)$. Set

$$\delta = \delta(t_0, \varepsilon) = w_2^{-1}\left(\frac{\delta^*(t_0, \varepsilon)}{\gamma(t_0^+)}\right).$$

Let $\varphi_0 \in M_{t,r}(n-k)(\bar{\delta})$. This means that

$$\gamma(t_0^+)w_2\left(\sup_{-r \le \theta \le 0} ||h(t_0^+, \varphi_0(\theta))||\right) \le \delta^*.$$

Then from (2.115), we have

$$\sup_{-r \le \theta \le 0} V(t_0^+, \varphi_0(\theta)) \le \gamma(t_0^+)w_2\left(\sup_{-r \le \theta \le 0} ||h(t_0^+, \varphi_0(\theta))||\right)e \le \delta^* e.$$

Hence

$$u^+(t; t_0, \sup_{-r \le \theta \le 0} V(t_0^+, \varphi_0(\theta))) < w_1(\varepsilon)e \qquad (2.116)$$

for $t \ge t_0$.

Let $x(t) = x(t; t_0, \varphi_0)$ be the solution of the IVP (2.114), (1.7). Since the conditions of Lemma 1.3 are met, then

$$V(t, x(t; t_0, \varphi_0)) \le u^+(t; t_0, \sup_{-r \le \theta \le 0} V(t_0^+, \varphi_0(\theta))), \ t \in J^+(t_0, \varphi_0). \qquad (2.117)$$

From (2.115), (2.117) and (2.116), there follow the inequalities

$$w_1(||h(t, x(t; t_0, \varphi_0))||)e \le V(t, x(t; t_0, \varphi_0))$$

$$\le u^+(t; t_0, \sup_{-r \le \theta \le 0} V(t_0^+, \varphi_0(\theta))) < w_1(\varepsilon)e, \ J^+(t_0, \varphi_0).$$

Hence, $||h(t, x(t; t_0, \varphi_0))|| < \varepsilon$ for $t \ge t_0$, i.e. the zero solution of (2.114) is stable with respect to the function h.

(b) From (a) it follows that the zero solution of system (2.114) is stable with respect to the function h. Hence, it is enough to prove that it is globally equi-attractive with respect to the function h. Since the zero solution of system (1.13) is globally equi-attractive, then for any $t_0 \in \mathbb{R}$, $\eta > 0$ and $\varepsilon > 0$ $(w_1(\varepsilon) < P$, $\eta < w_2^{-1}(P/\gamma(t_0^+))$) there exists a $T = T(t_0, \eta, \varepsilon) > 0$, $t_0 + T \in J^+(t_0, u_0)$ such that $u_0 \in Q$ and $0 \le u_0 \le \gamma(t_0^+)w_2(\eta)e = \eta^* e$ imply $u^+(t; t_0, u_0) < w_1(\varepsilon)e$ for $t \ge t_0 + T$, $t \in J^+(t_0, u_0)$. Hence $J^+(t_0, u_0) = [t_0, \infty)$.

Let $\varphi_0 \in M_{t,r}(n-k)(\bar{\eta})$. This means that

$$\gamma(t_0^+)w_2\left(\sup_{-r \le \theta \le 0} ||h(t_0^+, \varphi_0(\theta))||\right) \le \eta^*.$$

Then from (2.115) we have

$$\sup_{-r \le \theta \le 0} V(t_0^+, \varphi_0(\theta)) \le \gamma(t_0^+) w_2(\sup_{-r \le \theta \le 0} ||h(t_0^+, \varphi_0(\theta))||) e \le \eta^* e.$$

Hence

$$u^+(t; t_0, \sup_{-r \le \theta \le 0} V(t_0^+, \varphi_0(\theta))) < w_1(\varepsilon) e \qquad (2.118)$$

for $t \ge t_0 + T$.

Let $x(t) = x(t; t_0, \varphi_0)$ be the solution of the IVP (2.114), (1.7). From (2.115), (2.117) and (2.118), there follow the inequalities

$$w_1(||h(t, x(t; t_0, \varphi_0))||) e \le V(t, x(t; t_0, \varphi_0))$$

$$\le u^+(t; t_0, \sup_{-r \le \theta \le 0} V(t_0^+, \varphi_0(\theta))) < w_1(\varepsilon) e, \ t \in J^+(t_0, \varphi_0).$$

Hence, $||h(t, x(t; t_0, \varphi_0))|| < \varepsilon$ for $t \ge t_0 + T$, i.e. the zero solution of (2.114) is globally equi-attractive with respect to the function h. From what we proved in (a) it follows that the zero solution of (2.114) is globally equi-asymptotically stable with respect to the function h.

The next theorem gives sufficient conditions, in terms of vector Lyapunov functions, for uniform stability properties of the system (2.114) with respect to the function h.

Theorem 2.39 If in Theorem 2.38 condition (2.115) is replaced by the condition

$$w_1(||h(t, x)||) e \le V(t, x) \le w_2(||h(t, x)||) e, \ w_1, w_2 \in K, \ (t, x) \in [t_0, \infty) \times \mathbb{R}^n,$$

then:

(a) If the zero solution of system (1.13) is uniformly stable, then the zero solution of system (2.114) is uniformly stable with respect to the function h.

(b) If the zero solution of system (1.13) is uniformly globally asymptotically stable, then the zero solution of system (2.114) is uniformly globally asymptotically stable with respect to the function h.

The proof of Theorem 2.39 is analogous to the proof of Theorem 2.38. In this case we can choose δ^* (as well as δ) and T independent of t_0.

Theorem 2.40 If in Theorem 2.38 condition (2.115) is replaced by the condition

$$||h(t, x)|| e \le V(t, x) \le \lambda(L) ||h(t, x)|| e, \ (t, x) \in [t_0, \infty) \times \mathbb{R}^n, \qquad (2.119)$$

where $\lambda(L) \ge 1$ exists for any $0 < L \le \infty$, then the global Mittag–Leffler stability of the zero solution of system (1.13) implies the global Mittag-Leffler stability of the zero solution of system (2.114) with respect to the function h.

Proof. Let $t_0 \in \mathbb{R}$ and $u_0 \in Q$, $0 \le u_0 \le \delta e$, $\delta > 0$. Since the zero solution of system (1.13) is globally Mittag-Leffler stable, then

$$u^+(t;t_0,u_0) < \{m(u_0)E_\beta(-\mu(t-t_0)^\beta)\}^d e,$$

where $\mu > 0$, $d > 0$, $m(0) = 0$, $m(u) \ge 0$, and $m(u)$ is locally Lipschitz with respect to $u \in \mathbb{R}_+^m$, $0 \le u \le \delta e$. Hence, $J^+(t_0,u_0) = [t_0,\infty)$.

Let $\varphi_0 \in M_{t,r}(n-k)(\bar{\delta})$ and $x(t) = x(t;t_0,\varphi_0)$ be the solution of the IVP (2.114), (1.7). From Lemma 1.3 for $u_0 = \sup_{-r \le \theta \le 0} V(t_0^+,\varphi_0(\theta))$, we get (2.117).

Then, from (2.119) and (2.117), we have

$$||h(t,x(t;t_0,\varphi_0))||e \le V(t,x(t;t_0,\varphi_0))$$

$$\le u^+(t;t_0, \sup_{-r \le \theta \le 0} V(t_0^+,\varphi_0(\theta)))$$

$$< \{m(\sup_{-r \le \theta \le 0} V(t_0^+,\varphi_0(\theta)))E_\beta(-\mu(t-t_0)^\beta)\}^d e$$

$$\le \{m(\lambda(L) \sup_{-r \le \theta \le 0} ||h(t_0^+,\varphi_0(\theta))||e)E_\beta(-\mu(t-t_0)^\beta)\}^d e, \; t \in J^+(t_0,\varphi_0).$$

Hence,

$$x(t;t_0,\varphi_0) \in M_t(n-k)(\{m(\lambda(L) \sup_{-r \le \theta \le 0} ||h(t_0^+,\varphi_0(\theta))||e)E_\beta(-\mu(t-t_0)^\beta)\}^d)$$

for $t \ge t_0$, $0 < \beta < 1$, i.e., the zero solution of (2.114) is globally Mittag–Leffler stable with respect to the function h.

Corollary 2.4 Let $Q = \{u \in \mathbb{R}^m : ||u|| < P\}$, there exists a function $V : [t_0,\infty) \times \mathbb{R}^n \to Q$, $V = col(V_1, V_2, ..., V_m)$, $V_i \in V_0$, $i = 1,2,...,m$, is such that H2.1 holds, $\sup_{[t_0,\infty) \times \mathbb{R}^n} ||V(t,x)|| < P \le \infty$, the condition (2.119) hold,

$$V(t^+,\varphi(0) + I_k(\varphi)) \le V(t,\varphi(0)), \; t = t_k, \; t_k > t_0,$$

and the inequality

$$^cD_+^\beta V(t,\varphi(0)) \le -\Lambda V(t,\varphi(0))$$

is valid whenever $V(t+\theta,\varphi(\theta)) \le V(t,\varphi(0))$ for $-r \le \theta \le 0$, where Λ is a positively definite $m \times m$ matrix, $0 < \beta < 1$.

Then the zero solution of system (2.114) is globally Mittag–Leffler stable with respect to the function h.

Proof. Let $t_0 \in \mathbb{R}$. For the comparison system (1.13) in case $F(t,u) = -\Lambda u$ and $J_k(u) = 0$ for $t \in [t_0,\infty)$, $u \in \mathbb{R}^m$, $k = \pm 1, \pm 2, ...$, we deduce from Lemma 1.3

$$V(t,x(t;t_0,\varphi_0)) \le \sup_{-r \le \theta \le 0} V(t_0^+,\varphi_0(\theta))E_\beta(-\Lambda(t-t_0)^\beta) \qquad (2.120)$$

for $t \in J^+(t_0,\varphi_0)$.

From (2.119) and (2.120), we have

$$||h(t,x(t;t_0,\varphi_0))||e \leq V(t,x(t;t_0,\varphi_0)) \leq \sup_{-r \leq \theta \leq 0} V(t_0^+,\varphi_0(\theta))E_\beta(-\Lambda(t-t_0)^\beta)$$

$$\leq \lambda(L) \sup_{-r \leq \theta \leq 0} ||h(t_0^+,\varphi_0(\theta))||eE_\beta(-\Lambda(t-t_0)^\beta), \ t \in J^+(t_0,\varphi_0).$$

Let $m = m(\varphi_0) > \lambda(L) \sup_{-r \leq \theta \leq 0} ||h(t_0^+,\varphi_0(\theta))||$. Then

$$||h(t,x(t;t_0,\varphi_0))|| < mE_\beta(-\Lambda(t-t_0)^\beta), \ t \geq t_0,$$

where $m \geq 0$ and $m = 0$ holds only if $h(t_0^+,\varphi_0(\theta)) = 0$ for $\theta \in [-r,0]$.

From the last estimate and the fact that the matrix Λ is positively definite it follows that the zero solution of (2.114) is globally Mittag–Leffler stable with respect to the function h.

The boundedness results will be state in the next two theorems.

Theorem 2.41 Assume that conditions of Theorem 2.38 hold and $w_1(u) \to \infty$ as $u \to \infty$. Then:

(a) If the solutions of system (1.13) are equi-bounded, then the solutions of system (2.114) are equi-bounded with respect to the function h.

(b) If the solutions of system (1.13) are ultimately bounded for a bound N, then the zero solution of system (2.114) are ultimately bounded for the bound $w_1^{-1}(N)$, with respect to the function h.

Proof. (a) Let $t_0 \in \mathbb{R}$ and $A > 0$, $A < w_2^{-1}\left(\frac{P}{\gamma(t_0^+)}\right)$ be given. Set $A^* = \gamma(t_0^+)w_2(A)$. Then, $w_1(u) \to \infty$ as $u \to \infty$, implies $A \to \infty$ as $A^* \to \infty$.

From the equi-boundedness of the solutions of system (1.13) it follows that there exists $B_1 = B_1(t_0,A)$ such that $u_0 \in Q$ and $0 \leq u_0 \leq A^*e$ imply

$$u^+(t;t_0,u_0) < B_1e, \ t \in J^+(t_0,u_0).$$

Then from H2.6 it follows that $J^+(t_0,u_0) = [t_0,\infty)$. Set

$$B = B(t_0,A) = w_1^{-1}(B_1(t_0,A)).$$

Let $\varphi_0 \in M_{t,r}(n-k)(\bar{A})$. This means that $\gamma(t_0^+)w_2(\sup_{-r \leq \theta \leq 0} ||h(t_0^+,\varphi_0(\theta))||) \leq A^*$ and since

$$\sup_{-r \leq \theta \leq 0} V(t_0^+,\varphi_0(\theta)) \leq \gamma(t_0^+)w_2(\sup_{-r \leq \theta \leq 0} ||h(t_0^+,\varphi_0(\theta))||)e,$$

then

$$\sup_{-r \leq \theta \leq 0} V(t_0^+,\varphi_0(\theta)) \leq A^*e.$$

Hence

$$u^+(t;t_0, \sup_{-r\leq\theta\leq 0} V(t_0^+,\varphi_0(\theta))) < B_1 e, \ t \geq t_0. \tag{2.121}$$

Let $x(t) = x(t;t_0,\varphi_0)$ be the solution of the IVP (2.114), (1.7). From (2.115), (2.117) and (2.121), there follow the inequalities

$$w_1(||h(t,x(t;t_0,\varphi_0))||)e \leq V(t,x(t;t_0,\varphi_0))$$

$$\leq u^+(t;t_0, \sup_{-r\leq\theta\leq 0} V(t_0^+,\varphi_0(\theta))) < B_1 e, \ J^+(t_0,\varphi_0).$$

Hence, $||h(t,x(t;t_0,\varphi_0))|| < w_1^{-1}(B_1) = B$ for $t \geq t_0$, i.e. the solutions of (2.114) are equi-bounded with respect to the function h.

(b) Let $t_0 \in \mathbb{R}$, $N > 0$ and $A > 0$, $A < w_2^{-1}(\frac{P}{\gamma(t_0^+)})$ be given. Set again $A^* = \gamma(t_0^+)w_2(A)$. From the ultimate boundedness of the solutions of system (1.13) for a bound N it follows that there exists $T = T(t_0,A) > 0$ such that $u_0 \in Q$ and $0 \leq u_0 \leq A^*e$ imply

$$u^+(t;t_0,u_0) < Ne, \ t \geq t_0 + T, \ t \in J^+(t_0,u_0).$$

Hence $J^+(t_0,u_0) = [t_0,\infty)$. Let $\varphi_0 \in M_{t,r}(n-k)(\bar{A})$. This means that

$$\gamma(t_0^+)w_2(\sup_{\theta\in[-r,0]} ||h(t_0^+,\varphi_0(\theta))||) \leq A^*$$

and since

$$\sup_{-r\leq\theta\leq 0} V(t_0^+,\varphi_0(\theta)) \leq \gamma(t_0^+)w_2(\sup_{-r\leq\theta\leq 0} ||h(t_0^+,\varphi_0(\theta))||)e,$$

then

$$\sup_{-r\leq\theta\leq 0} V(t_0^+,\varphi_0(\theta)) \leq A^*e.$$

Hence

$$u^+(t;t_0, \sup_{-r\leq\theta\leq 0} V(t_0^+,\varphi_0(\theta))) < Ne, \ t \geq t_0 + T. \tag{2.122}$$

Let $x(t) = x(t;t_0,\varphi_0)$ be the solution of the IVP (2.114), (1.7). From (2.115), (2.117) and (2.122), there follow the inequalities

$$w_1(||h(t,x(t;t_0,\varphi_0))||)e \leq V(t,x(t;t_0,\varphi_0))$$

$$\leq u^+(t;t_0, \sup_{-r\leq\theta\leq 0} V(t_0^+,\varphi_0(\theta))) < Ne, \ J^+(t_0,\varphi_0).$$

Hence, $||h(t,x(t;t_0,\varphi_0))|| < w_1^{-1}(N)$ for $t \geq t_0 + T$, i.e. the solutions of (2.114) are ultimately bounded with respect to the function h for the bound $w_1^{-1}(N)$.

Theorem 2.42 Assume that conditions of Theorem 2.39 hold and $w_1(u) \to \infty$ as $u \to \infty$. Then:

(a) If the solutions of system (1.13) are uniformly bounded, then the solutions of system (2.114) are uniformly bounded with respect to the function h.

(b) If the solutions of system (1.13) are uniformly ultimately bounded for a bound N, then the zero solution of system (2.114) are uniformly ultimately bounded for the bound $w_1^{-1}(N)$, with respect to the function h.

The proof of Theorem 2.42 is similar to that of Theorem 2.41. In this case B and T can be chosen independent of t_0.

Theorem 2.43 Assume that:

1. Conditions H2.2–H2.6 hold.
2. Conditions of Lemma 1.4 hold, and $F(t,0) = 0$ for $t \in [t_0,\infty)$, $J_k(0) = 0$, $k = \pm 1, \pm 2, ...$
3. The function $V : [t_0,\infty) \times \mathbb{R}^n \to Q$, $V = col(V_1, V_2, ..., V_m)$, $V_i \in V_0$, $i = 1, 2, ..., m$, is such that H2.1 holds, $\sup_{[t_0,\infty) \times \mathbb{R}^n} ||V(t,x)|| < P \leq \infty$, $Q = \{u \in \mathbb{R}^m : ||u|| < P\}$, and the inequality

$$V(t,x) \leq w_1(||h(t,x)||)e, \ (t,x) \in [t_0,\infty) \times \mathbb{R}^n \tag{2.123}$$

holds, where $w_1 \in K$.

4. For any $\delta > 0$ and $t_0 \in \mathbb{R}$ there exists $\varphi_0 \in M_{t,r}(n-k)(\delta)$ such that $V(t_0, \varphi_0(0)) > 0$.

Then, if the zero solution of system (1.13) is unstable, then the zero solution of system (2.114) is unstable with respect to the function h.

Proof. From the instability of the zero solution of system (1.13) it follows that there exist $\varepsilon^* > 0$ and $t_0 \in \mathbb{R}$ such that for any $\delta^* > 0$ there exists $u_0 \in Q : 0 \leq u_0 \leq \delta^* e$ and $t^* \geq t_0$, $t^* \in J^+(t_0, u_0)$, for which the inequality

$$u^-(t^*; t_0, u_0) < \varepsilon^* e \tag{2.124}$$

does not hold.

Choose the number $\varepsilon > 0$ so that $w_1(\varepsilon) < \varepsilon^*$.

Case 1 Let $t_0 \neq t_k$, $k = \pm 1, \pm 2, ...$ In this case

$$\sup_{-\tau \leq \theta \leq 0} V(t_0, \varphi_0(\theta)) = \sup_{-\tau \leq \theta \leq 0} V(t_0^+, \varphi_0(\theta)).$$

Let $\delta > 0$ be given. From condition 4 of Theorem 2.43 it follows that we can choose $\varphi_0 \in M_{t,r}(n-k)(\delta)$ such that $\sup_{-r \leq \theta \leq 0} V(t_0^+, \varphi_0(\theta)) \geq V(t_0, \varphi_0(0)) > 0$. Let $\delta^* > 0$ be such that $0 < \delta^* e \leq \sup_{-r \leq \theta \leq 0} V(t_0^+, \varphi_0(\theta))$. Let $x(t; t_0, \varphi_0)$ be the solution of the IVP (2.114), (1.7). From Lemma 1.4 it follows that for $t \in J^+(t_0, u_0) \cap J^+(t_0, \varphi_0)$ the following inequality holds

$$V(t, x(t; t_0, \varphi_0)) \geq u^-(t; t_0, u_0). \tag{2.125}$$

If we assume that for any $t \in J^+(t_0, \varphi_0)$, we have

$$||h(t, x(t; t_0, \varphi_0))|| < \varepsilon, \tag{2.126}$$

then from H2.5 it follows that $J^+(t_0, \varphi_0) = [t_0, \infty)$. From (2.123), (2.125) and (2.126) we obtain

$$\varepsilon^* e > w_1(\varepsilon)e > w_1(||h(t^*, x(t^*; t_0, \varphi_0))||)e$$

$$\geq V(t^*, x(t^*; t_0, \varphi_0)) \geq u^-(t^*; t_0, u_0),$$

which contradicts the fact that (2.124) does not hold.

Hence, there exists a $t \in J^+(t_0, \varphi_0)$ for which the inequality (2.126) does not hold.

Case 2 Let $t_0 = t_k$ for some $k = \pm 1, \pm 2, \ldots$. We shall prove that in this case for any $\delta > 0$ there exists $\varphi_0 \in M_{t,r}(n-k)(\delta)$ such that $V(t_0^+, \varphi_0(0)) > 0$.

Suppose that this is not true, i.e. that there exists $\delta > 0$ such that for any $\varphi_0 \in M_{t,r}(n-k)(\delta)$ and for some $i \in \{1, 2, \ldots, m\}$ the following inequality holds

$$V_i(t_0^+, \varphi_0(0)) \leq 0. \tag{2.127}$$

From the properties of the function $h(t_0, x)$ at $x = 0$ and H2.4 it follows that there exists a number δ_1 $(0 < \delta_1 < \delta)$ such that $\varphi \in \mathscr{PC}$ and $||\varphi(0) + I_k(\varphi)|| < \delta_1$ imply

$$||h(t_0, \varphi(0) + I_k(\varphi))|| < \delta. \tag{2.128}$$

On the other hand, from the continuity of the functions $I_k(x)$ at $x = 0$ and H2.4 it follows that there exists a number δ_2 $(0 < \delta_2 < \delta_1)$ such that $\varphi \in M_{t,r}(n-k)(\delta_2)$ imply

$$||\varphi(0) + I_k(\varphi)|| < \delta_1. \tag{2.129}$$

Let $\varphi_0 \in M_{t,r}(n-k)(\delta_2)$ be such that $V(t_0, \varphi_0(0)) > 0$. Then from (2.129), (2.128) and (2.127) and Lemma 1.4 we obtain the contradiction

$$0 \geq V_i(t_0^+, \varphi(0) + I_k(\varphi)) \geq \psi_k(V_i(t_0, \varphi_0(0))) > 0.$$

Then $\sup_{-r \leq \theta \leq 0} V(t_0^+, \varphi_0(\theta)) \geq V(t_0, \varphi_0(0)) > 0$. Further on we can carry out the proof as in the Case 1.

This completes the proof of Theorem 2.43.

Example 2.16 Let $0 < r_i \leq r$, $i = 1, 2, 3$. Consider the following fractional chaotic system

$$\begin{cases} {}_0^c D_t^\alpha x_1(t) = a(t)x_3(t - r_3) - b(t)x_1(t), \ t \neq t_k, \\ {}_0^c D_t^\alpha x_2(t) = a(t)x_1(t - r_1) - b(t)x_2(t), \ t \neq t_k, \\ {}_0^c D_t^\alpha x_3(t) = a(t)x_2(t - r_2) - b(t)x_3(t), \ t \neq t_k, \\ \Delta x_i(t) = a_i x_i(t), \ t = t_k, \ k = 1, 2, \ldots, \end{cases} \tag{2.130}$$

where $x_i \in \mathbb{R}$, $i = 1, 2, 3$, $t \geq 0$, $0 < \alpha < 1$, $a, b \in C[\mathbb{R}_+, \mathbb{R}_+]$, $-1 < a_i \leq 0$, $i = 1, 2, 3$, $0 < t_1 < t_2 < ...$, and $\lim_{k \to \infty} t_k = \infty$.

We shall use the Lyapunov function

$$V(t, x_1, x_2, x_3) = |x_1 + x_2 + x_3|.$$

Then for $t \geq 0$, $\varphi_i \in PC[[-r, 0], \mathbb{R}]$, $i = 1, 2, 3$ and $t \neq t_k$, for the upper right-hand derivative of V of order α with respect to (1.130), we have

$$^cD_+^\alpha V(t, \varphi_1(0), \varphi_2(0), \varphi_3(0))$$

$$\leq (a(t) - b(t))V(t, \varphi_1(0), \varphi_2(0), \varphi_3(0)), \tag{2.131}$$

whenever $V(t + \theta, \varphi_1(\theta), \varphi_2(\theta), \varphi_3(\theta)) \leq V(t, \varphi_1(0), \varphi_2(0), \varphi_3(0))$ for $-r \leq \theta \leq 0$.

Also,

$$V(t_k^+, \varphi_1(0) + a_1 \varphi_1(0), \varphi_2(0) + a_2 \varphi_2(0), \varphi_3(0) + a_3 \varphi_3(0))$$

$$= |(1 + a_1)\varphi_1(0) + (1 + a_2)\varphi_2(0) + (1 + a_3)\varphi_3(0)|$$

$$\leq V(t_k, \varphi_1(0), \varphi_2(0), \varphi_3(0)). \tag{2.132}$$

(i) If $b(t) \geq a(t)$ for $t \in \mathbb{R}_+$, then from (2.131) we get

$$^cD_+^\alpha V(t, \varphi_1(0), \varphi_2(0), \varphi_3(0)) \leq 0, \ t \neq t_k,$$

whenever $V(t + \theta, \varphi_1(\theta), \varphi_2(\theta), \varphi_3(\theta)) \leq V(t, \varphi_1(0), \varphi_2(0), \varphi_3(0))$ for $-r \leq \theta \leq 0$.

Since the zero solution of the fractional scalar equation

$$\begin{cases} {}^c_0D_t^\alpha u(t) = 0, \ t \neq t_k, t \geq 0, \\ \Delta u(t_k) = 0, \ t_k > 0 \end{cases}$$

is stable (see [Podlubny 1999]), then by Theorem 2.38 (a) it follows that the zero solution of the chaotic system (2.130) is stable with respect to the function $h(t, x_1, x_2, x_3) = x_1 + x_2 + x_3$.

(ii) If there exists a constant $w > 0$ such that $b(t) - a(t) \geq w$ for $t \in \mathbb{R}_+$, then from (2.131) we get

$$^cD_+^\alpha V(t, \varphi_1(0), \varphi_2(0), \varphi_3(0)) \leq -wV(t, \varphi_1(0), \varphi_2(0), \varphi_3(0)), \ t \neq t_k, \ t_k > 0,$$

whenever $V(t + \theta, \varphi_1(\theta), \varphi_2(\theta), \varphi_3(\theta)) \leq V(t, \varphi_1(0), \varphi_2(0), \varphi_3(0))$ for $-r \leq \theta \leq 0$.

Therefore, from Corollary 2.4 it follows that the zero solution of the chaotic system (2.130) is globally Mittag–Leffler stable with respect to the function $h(t, x_1, x_2, x_3) = x_1 + x_2 + x_3$.

In Example 2.16, we have used a single Lyapunov function $V \in V_0$. To demonstrate the advantage of employing several Lyapunov functions, let us consider the following example.

Example 2.17 Consider the system

$$
\begin{cases}
{}_0^c D_t^\alpha x(t) = e^{-t} x(t - r(t)) + \sin(t) y(t), \ t \neq t_k, \\
{}_0^c D_t^\alpha y(t) = \sin(t) x(t) - e^{-t} y(t - r(t)), \ t \neq t_k, \\
\Delta x(t) = a_k x(t) + b_k y(t), \ t = t_k, \ k = 1, 2, ..., \\
\Delta y(t) = b_k x(t) + a_k y(t), \ t = t_k, \ k = 1, 2, ...,
\end{cases} \tag{2.133}
$$

where $x, y \in \mathbb{R}$, $t \geq 0$, $0 < \alpha < 1$, $0 \leq r(t) \leq r$, $r(t)$ is continuous on $\mathbb{R}$, $a_k = \frac{1}{2}(c_k + d_k)$, $b_k = \frac{1}{2}(c_k - d_k)$, $-1 < c_k \leq 0$, $-1 < d_k \leq 0$, $k = 1, 2, ..., 0 < t_1 < t_2 < ...$, and $\lim\limits_{k \to \infty} t_k = \infty$.

Suppose that we choose a scalar Lyapunov function $V(t, x, y) = |x + y| + |x - y|$. Then for $t \geq 0$, $\varphi, \phi \in PC[[-r, 0], \mathbb{R}]$ and $t \neq t_k$, for the upper right-hand derivative of V of order α with respect to (2.113), we have

$$
{}^c D_+^\alpha V(t, \varphi(0), \phi(0)) = {}^c D_+^\alpha |\varphi(0) + \phi(0)| + {}^c D_+^\alpha |\varphi(0) - \phi(0)|
$$

$$
\leq (e^{-t} + |\sin(t)|) V(t, \varphi(0), \phi(0)),
$$

whenever $V(t + \theta, \varphi(\theta), \phi(\theta)) \leq V(t, \varphi(0), \phi(0))$ for $-r \leq \theta \leq 0$.

Also,

$$
V(t_k^+, \varphi(0) + a_k \varphi(0) + b_k \phi(0), \phi(0) + b_k \varphi(0) + a_k \phi(0))
$$

$$
= |(1 + a_k)\varphi(0) + b_k \phi(0) + (1 + a_k)\phi(0) + b_k \varphi(0)|
$$

$$
+ |(1 + a_k)\varphi(0) + b_k \phi(0) - (1 + a_k)\phi(0) - b_k \varphi(0)|
$$

$$
\leq \gamma_k V(t_k, \varphi(0), \phi(0)), \ k = 1, 2, ...,
$$

where $\gamma_k = 1 + |a_k| + |b_k| > 1$, $k = 1, 2, ...$.

Consider the following comparison fractional scalar equation

$$
\begin{cases}
{}_0^c D_t^\alpha u(t) = \left[e^{-t} + |\sin(t)| \right] u(t), \ t \neq t_k, \ t > 0, \\
\Delta u(t_k) = \gamma_k u(t_k), \ k = 1, 2, ...,
\end{cases}
$$

where $u \in \mathbb{R}_+$. The zero $u = 0$ is an equilibrium of the above equation, but it is not stable (see [Stamova 2014a]) and, consequently we cannot deduce any information about the stability of the system (2.133) from Theorem 2.38, even in the cases when the system (2.133) is stable.

Now, let us take the function $V = (V_1, V_2)$, where the functions V_1 and V_2 are defined by $V_1(t, x, y) = |x + y|$, $V_2(t, x, y) = |x - y|$. Then for $t \geq 0$ the vector inequality

$$
{}^c D_+^\alpha V(t, \varphi(0), \phi(0)) \leq F(t, V(t, \varphi(0), \phi(0))), \ t \neq t_k, \ k = 1, 2, ...
$$

is satisfied whenever $V(t+\theta,\varphi(\theta),\phi(\theta)) \le V(t,\varphi(0),\phi(0))$ for $-r \le \theta \le 0$ with $F = (F_1, F_2)$, where

$$F_1(t,u_1,u_2) = [e^{-t} + \sin(t)]u_1,$$

$$F_2(t,u_1,u_2) = [e^{-t} - \sin(t)]u_2,$$

and

$$V_1(t_k^+, \varphi(0) + \Delta\varphi(0), \phi(0) + \Delta\phi(0))$$

$$\le |1 + c_k|V_1(t_k, \varphi(0), \phi(0)) \le V_1(t_k, \varphi(0), \phi(0)), \; k = 1,2,...,$$

$$V_2(t_k^+, \varphi(0) + \Delta\varphi(0), \phi(0) + \Delta\phi(0))$$

$$\le |1 + d_k|V_2(t_k, \varphi(0), \phi(0)) \le V_2(t_k, \varphi(0), \phi(0)), \; k = 1,2,....$$

Hence, the conditions of Theorem 2.39 are satisfied, and according to Theorem 2.13, the zero solution of the comparison system

$$\begin{cases} {}_0^cD_t^\alpha u_1(t) = [e^{-t} + \sin(t)]u_1(t), \; t \ne t_k, \\ {}_0^cD_t^\alpha u_2(t) = [e^{-t} - \sin(t)]u_2(t), \; t \ne t_k, \\ \Delta u_1(t_k) = c_k u_1(t_k), \; \Delta u_2(t_k) = d_k u_2(t_k), \; k = 1,2,... \end{cases}$$

is uniformly stable (see [Stamova 2014a]), then the zero solution of system (2.133) is uniformly stable with respect to the function $h(t,x,y) = \sqrt{x^2 + y^2}$.

Example 2.18 Let $0 < r(t) \le r$ and $r(t)$ is continuous on $\mathbb{R}$. Consider the system

$$\begin{cases} {}_0^cD_t^\alpha x(t) = [20ax(t) - 10y(t) - 1]x(t) \\ \qquad + 2x(t-r(t))[y(t-r(t)) - 2ax(t-r(t))], \; t \ne t_k, \\ {}_0^cD_t^\alpha y(t) = y^2(t-r(t)) - 4ax(t) - 5y^2(t), \; t \ne t_k, \qquad (2.134) \\ \Delta x(t_k) = \left(\frac{1}{2^k} - 1\right)x(t_k), \; \Delta y(t_k) = \left(\frac{1}{4^k} - 1\right)y(t_k), \; k = 1,2,..., \end{cases}$$

where $x,y \in \mathbb{R}, t \ge 0, 0 < \alpha < 1, a \in \mathbb{R}, 0 < t_1 < t_2 < ...$ and $\lim_{k\to\infty} t_k = \infty$.

Let $h(t,x,y) = y - 4ax$ and $V(t,x,y) = |h|$. Then for $t \ge 0$, $\varphi, \phi \in PC[[-r,0],\mathbb{R}]$ and $t \ne t_k$, for the upper right-hand derivative of V of order α with respect to (2.134), we have

$$^cD_+^\alpha V(t,\varphi(0),\phi(0)) = \text{sgn}(\phi(0) - 4a\varphi(0)){}_0^cD_t^\alpha\Big(\phi(0) - 4a\varphi(0)\Big)$$

$$\le -5V^2(t,\varphi(0),\phi(0)) + \sup_{-r\le s\le 0} V^2(t+\theta,\varphi(\theta),\phi(\theta)) \le -4V^2(t,\varphi(0),\phi(0)),$$

whenever $V(t+\theta,\varphi(\theta),\phi(\theta)) \le V(t,\varphi(0),\phi(0))$ for $-r \le \theta \le 0$.

Also,

$$V\left(t_k^+, \varphi(0) + \left(\frac{1}{2^k} - 1\right)\varphi(0), \phi(0) + \left(\frac{1}{4^k} - 1\right)\phi(0)\right)$$

$$= \left|\frac{1}{4^k}\phi(0) - 4a\frac{1}{2^k}\varphi(0)\right| \le \frac{1}{2^k}V(t_k, \varphi(0), \phi(0)).$$

Consider the fractional-order impulsive system

$$\begin{cases} {}_0^c D_t^\alpha u(t) = -4u^2, t \neq t_k, t \ge 0, \\ \Delta u(t_k) = \left(\frac{1}{2^k} - 1\right)u(t_k), k = 1, 2, \dots. \end{cases} \tag{2.135}$$

Since by Theorem 2.20, the zero solution of system (2.135) is uniformly globally asymptotically stable (see [Stamova 2014a]), then by Theorem 2.39 (b) the zero solution of system (2.134) is uniformly globally asymptotically stable with respect to the function h.

Moreover, using the same ideas as in theorems 2.16 and 2.17, we can proof that the solutions of system (2.135) are uniformly bounded and uniformly ultimately bounded. Then from Theorem 2.42 it follows that the solutions of system (2.134) are uniformly bounded and uniformly ultimately bounded with respect to the function h.

2.8   Stability of Integral Manifolds

The method of integral manifolds is a very powerful instrument for investigating of various qualitative properties of differential equations. For example, the method is effectively applied in studying of singularly perturbed ordinary differential equations (see [Sakamoto 1990], [Stamov 1996], [Stamov 1997]), the systems with time lag (see [Halanay 1966], [Strygin and Fridman 1984]), linearization problems (see [Palmer 1975], [Papaschinopoulos 1997]), etc. Also, the existence of integral manifolds is one of the most interesting problems for different types of functional differential equations and impulsive functional differential equations. See, for example, [Corduneanu and Ignatyev 2005], [Mitropol'skii, Fodchuk and Klevchuk 1986], [Naito 1970], [Perestyuk and Cherevko 2002], [Stamov 2009b], [Stamov and Stamova 2001] and the references therein.

One of the most efficient tools for the study of the existence and stability of integral manifolds is provided by the Lyapunov second method. For integer-order differential equations this technique has been applied by several authors (see [Bernfeld, Corduneanu and Ignatyev 2003], [Ghorbel and Spong 2000], [Stamov 2009b], [Stamov and Stamova 2001]). The goal of this section is to extend such results for impulsive functional differential fractional-order equations of the type (1.6).

Definition 2.13 An arbitrary manifold M in the extended phase space $[t_0 - r, \infty) \times \Omega$ of (1.6) is said to be an integral manifold of (1.6), if for any solution $x(t) = x(t; t_0, \varphi_0)$, $(t, \varphi_0(t - t_0)) \in M$, $t \in [t_0 - r, t_0]$ implies $(t, x(t)) \in M$ for $t \geq t_0$.

Let M be an integral manifold of (1.6). In the next, we shall use the following notations:

$$M(t) = \{x \in \Omega : (t, x) \in M, \, t \in [t_0, \infty)\};$$

$$M_0(t) = \{x \in \Omega : (t, x) \in M, t \in [t_0 - r, t_0]\};$$

$$d(x, M(t)) = \inf_{y \in M(t)} ||x - y|| \text{ is the distance between } x \in \Omega \text{ and } M(t);$$

$$M(t, \varepsilon) = \{x \in \Omega : d(x, M(t)) < \varepsilon\} \, (\varepsilon > 0) \text{ is an } \varepsilon\text{-neighborhood of } M(t);$$

$$d_0(\phi, M_0(t)) = \sup_{t \in [t_0 - r, t_0]} d(\phi(t - t_0), M_0(t)), \, \phi \in PC[[-r, 0], \Omega];$$

$M_0(t, \varepsilon) = \{\phi \in PC[[-r, 0], \Omega] : d_0(\phi, M_0(t)) < \varepsilon\}$ is an ε-neighborhood of $M_0(t)$;

$$\overline{B}_\rho = \{x \in \mathbb{R}^n : ||x|| \leq \rho\}; \overline{B}_\rho(\mathscr{PC}) = \{\phi \in \mathscr{PC} : ||\phi||_r \leq \rho\}, \rho > 0.$$

We shall use the following definition:

Definition 2.14 The integral manifold M is said to be:
  (a) *stable* with respect to system (1.6) if

$$(\forall t_0 \in \mathbb{R})(\forall \rho > 0)(\forall \varepsilon > 0)(\exists \delta = \delta(t_0, \rho, \varepsilon) > 0)$$

$$(\forall \varphi_0 \in \overline{B}_\rho(\mathscr{PC}) \cap M_0(t, \delta))(\forall t \geq t_0) : x(t; t_0, \varphi_0) \in M(t, \varepsilon);$$

  (b) *uniformly stable* with respect to system (1.6) if the number δ in (a) depends only on ε;
  (c) *attractive* with respect to system (1.6), if

$$(\forall t_0 \in \mathbb{R})(\forall \rho > 0)(\exists \eta > 0)(\forall \varepsilon > 0)(\forall \varphi_0 \in \overline{B}_\rho(\mathscr{PC}) \cap M_0(t, \eta))$$

$$(\exists T > 0)(\forall t \geq t_0 + T) : x(t; t_0, \varphi_0) \in M(t, \varepsilon);$$

  (d) *asymptotically stable* with respect to system (1.6) if M is stable and attractive.

Note that Definition 2.14 can be reduced to the following particular cases:
  1. Lyapunov stability of the zero solution of (1.6), if

$$M = [t_0 - r, \infty) \times \{x \in \Omega : x_i \equiv 0, \, i = 1, ..., n\}.$$

2. Lyapunov stability of a non-null solution $x^*(t)$, $x^* = (x_1^*, ..., x_n^*)$ of (1.6), if

$$M = [t_0 - r, \infty) \times \{x \in \Omega : x_i \equiv x_i^*, \ i = 1, ..., n\}.$$

We shall introduce the Mittag–Leffler stability definition for the integral manifold M with respect to system (1.6) which generalizes definitions given in [Li, Chen and Podlubny 2010].

Definition 2.15 The integral manifold M is said to be Mittag–Leffler stable, if there exist constants $\mu > 0$ and $d > 0$ such that

$$x(t; t_0, \varphi_0) \in M(t, \{m(\varphi_0) E_\beta(-\mu(t - t_0)^\beta)\}^d), \ t \geq t_0,$$

where $0 < \beta < 1$, $m(0) = 0$, $m(\varphi) \geq 0$, and $m(\varphi)$ is Lipschitz with respect to $\varphi \in \overline{B_\rho}(\mathscr{PC}) \cap M_0(t, \delta)$.

It is our aim in this section to contribute to the development of the qualitative theory of fractional-order impulsive functional differential equations, presenting Lyapunov-type existence and stability theorems for integral manifolds. We shall use piecewise continuous Lyapunov functions $V : [t_0, \infty) \times \Omega \to \mathbb{R}_+$, $V \in V_0$ for which the following condition is true:

H2.7 $V(t, x) = 0$ for $(t, x) \in M$, $t \geq t_0$ and $V(t, x) > 0$ for $(t, x) \in \{[t_0, \infty) \times \Omega\} \setminus M$.

For such functions, we shall say that they have the manifold M as a kernel.

Firstly, the main existence results will be given.

Theorem 2.44 Let M be a manifold in the extended phase space for the system (1.6), there exists a function $V \in V_0$ with a kernel the manifold M and a constant $\beta, 0 < \beta < 1$, such that

$$V(t^+, \varphi(0) + I_k(\varphi)) \leq V(t, \varphi(0)), \ t = t_k, \ t_k > t_0, \tag{2.136}$$

and the inequality

$$^cD_+^\beta V(t, \varphi(0)) \leq 0 \tag{2.137}$$

is valid whenever $V(t + \theta, \varphi(\theta)) \leq V(t, \varphi(0))$ for $-r \leq \theta \leq 0$, $t \in [t_0, \infty)$, $\varphi \in PC[[-r, 0], \Omega]$.

Then M is an integral manifold of (1.6).

Proof. Suppose that the set M is not an integral manifold of the system (1.6). It follows that there exists a moment t', $t' > t_0$ so that if $(t, \varphi_0(t - t_0)) \in M$ for $t \in [t_0 - r, t_0]$, $(t, x(t; t_0, \varphi_0)) \in M$ when $t_0 < t \leq t'$ and $(t, x(t; t_0, \varphi_0)) \notin M$ when $t > t'$.

Obviously we have two possible cases for the position of the moment t':

1. If $t' \neq t_k$, $k = j, j+1, ..., j \geq 1$, then there exists $t'' \in (t_k, t_{k+1})$, $t'' > t'$ such that $(t'', x(t''; t_0, \varphi_0)) \notin M$ and $V(t'', x(t''; t_0, \varphi_0)) > 0$. On the other hand

by Corollary 1.6 we have $V(t'', x(t''; t_0, \varphi_0)) \leq \sup\limits_{-r \leq \theta \leq 0} V(t_0^+, \varphi_0(\theta)) = 0$; a
contradiction, proving our claim.

2. If $t' = t_k$, $k = j, j+1, ..., j \geq 1$, again by (2.136), Corollary 1.6 and the
definition of V we get $V(t'^+, x(t'^+; t_0, \varphi_0)) \leq \sup_{-r \leq \theta \leq 0} V(t_0^+, \varphi_0(\theta)) = 0$.
The received contradiction proves our claim.

The proof of the next theorem is analogous.

Theorem 2.45 Let M be a manifold in the extended phase space for the system
(1.6), there exist a function $V \in V_0$ with a kernel the manifold M, a constant β,
$0 < \beta < 1$, such that

$$V(t^+, \varphi(0) + I_k(\varphi)) \leq V(t, \varphi(0)), \ t = t_k, \ t_k > t_0,$$

and the inequality

$$^cD_+^\beta V(t, \varphi(0)) \leq -w(||\varphi(0)||), \ w \in K$$

is valid whenever $V(t + \theta, \varphi(\theta)) \leq V(t, \varphi(0))$ for $-r \leq \theta \leq 0$, $t \in [t_0, \infty)$, $\varphi \in PC[[-r, 0], \Omega]$.
Then M is an integral manifold of (1.6).

Let M be an integral manifold for system (1.6). In this section applying
Lyapunov functions and Razumikhin technique, we shall establish some
stability, asymptotic stability and Mittag–Leffler stability criteria for the integral
manifold M with respect to (1.6).

Theorem 2.46 Let there exist a function $V \in V_0$ with a kernel the manifold M
for which (2.136) holds, $\beta, 0 < \beta < 1$, such that

$$w_1(d(x, M(t))) \leq V(t, x), \ w_1 \in K, \ (t, x) \in [t_0, \infty) \times \Omega, \tag{2.138}$$

and the inequality (2.137) is valid whenever $V(t + \theta, \varphi(\theta)) \leq V(t, \varphi(0))$ for
$-r \leq \theta \leq 0, t \in [t_0, \infty), \varphi \in PC[[-r, 0], \Omega]$.
Then the integral manifold M is stable with respect to system (1.6).

Proof. Let $t_0 \in \mathbb{R}, \varepsilon > 0, \rho > 0$. From the properties of the function V, it follows
that there exists a constant $\delta = \delta(t_0, \rho, \varepsilon) > 0$ such that if $x \in \overline{B}_\rho \cap M(t_0, \delta)$, then
$\sup\limits_{|x| < \delta} V(t_0, x) < w_1(\varepsilon)$.

Let $\varphi_0 \in \overline{B}_\rho(\mathscr{PC}) \cap M_0(t, \delta)$. Then

$$d_0(\varphi_0, M_0(t)) = \sup\limits_{t \in [t_0-r, t_0]} d(\varphi_0(t-t_0), M_0(t)) < \delta,$$

hence

$$\sup\limits_{-r \leq \theta \leq 0} V(t_0^+, \varphi_0(\theta)) < w_1(\varepsilon). \tag{2.139}$$

Let $x(t) = x(t; t_0, \varphi_0)$ be the solution of the IVP (1.6), (1.7). Since the conditions of Corollary 1.6 are met, then

$$V(t, x(t; t_0, \varphi_0)) \leq \sup_{-r \leq \theta \leq 0} V(t_0^+, \varphi_0(\theta)), \ t \in [t_0, \infty). \tag{2.140}$$

From (2.138), (2.140) and (2.139), there follow the inequalities

$$w_1(d(x(t; t_0, \varphi_0), M(t))) \leq V(t, x(t; t_0, \varphi_0))$$

$$\leq \sup_{-r \leq \theta \leq 0} V(t_0^+, \varphi_0(\theta)) < w_1(\varepsilon), \ t \in [t_0, \infty).$$

Hence, $x(t; t_0, \varphi_0) \in M(t, \varepsilon)$ for $t \geq t_0$, i.e. the integral manifold M is stable with respect to system (1.6).

Theorem 2.47 Let the conditions of Theorem 2.46 hold, and a function $w_2 \in K$ exists such that

$$V(t, x) \leq w_2(d(x, M(t))), \ (t, x) \in [t_0, \infty) \times \Omega. \tag{2.141}$$

Then the integral manifold M is uniformly stable with respect to system (1.6).

Proof. Let $\varepsilon > 0$. Choose $\delta = \delta(\varepsilon) > 0$ so that $w_2(\delta) < w_1(\varepsilon)$. Let $\varphi_0 \in \overline{B_\rho}(\mathscr{PC}) \bigcap M_0(t, \delta)$. Using successively (2.138), (2.140) and (2.141), we obtain

$$w_1(d(x(t; t_0, \varphi_0), M(t))) \leq V(t, x(t; t_0, \varphi_0)) \leq \sup_{-r \leq \theta \leq 0} V(t_0^+, \varphi_0(\theta))$$

$$\leq w_2(d(\varphi_0, M_0(t))) < w_2(\delta) < w_1(\varepsilon),$$

for $t \in [t_0, \infty)$. Hence, $x(t; t_0, \varphi_0) \in M(t, \varepsilon)$ for $t \geq t_0$.

This proves that the integral manifold M is uniformly stable with respect to system (1.6).

Theorem 2.48 Let there exist a function $V \in V_0$ with a kernel the manifold M for which (2.136) and (2.138) holds, $\beta, 0 < \beta < 1$, such that the inequality

$$^cD_+^\beta V(t, \varphi(0)) \leq -wV(t, \varphi(0)), \ w = const > 0 \tag{2.142}$$

is valid whenever $V(t + \theta, \varphi(\theta)) \leq V(t, \varphi(0))$ for $-r \leq \theta \leq 0, \ t \in [t_0, \infty), \ \varphi \in PC[[-r, 0], \Omega]$.

Then the integral manifold M is asymptotically stable with respect to system (1.6).

Proof. Since the conditions of Theorem 2.46 are met, the integral manifold M is stable with respect to system (1.6). We shall show that it is an attractive manifold with respect to system (1.6).

Let $t_0 \in \mathbb{R}$, $\varepsilon > 0$, $\rho > 0$. From (2.136). (2.142) and Corollary 1.5, we have

$$V(t, x(t; t_0, \varphi_0)) \leq \sup_{-r \leq \theta \leq 0} V(t_0^+, \varphi_0(\theta)) E_\beta(-w(t - t_0)^\beta), \ t \in [t_0, \infty). \quad (2.143)$$

Let $\eta = const > 0$ be such that $\varphi_0 \in \overline{B}_\rho(\mathscr{PC}) \bigcap M_0(t, \eta)$ and $x(t) = x(t; t_0, \varphi_0)$ be the solution of the IVP (1.6), (1.7). We set $N = N(t_0, \eta, \rho) = \sup_{-r \leq \theta \leq 0} \{V(t_0^+, \phi(\theta)) : \phi \in \overline{B}_\rho(\mathscr{PC}) \bigcap M_0(t, \eta)\}$. Choose $T > 0$ so that

$$T > \left[\frac{-L_\beta\left(\frac{w_1(\varepsilon)}{N(t_0, \eta, \rho)}\right)}{w} \right]^{1/\beta},$$

where $L_\beta(z)$ is the inverse function of the Mittag-Leffler function $E_\beta(z)$, defined as the solution of the equation $E_\beta(L_\beta(z)) = z$.

Then, from (2.138) and (2.143) for $t \geq t_0 + T$, the following inequalities hold

$$w_1(d(x(t; t_0, \varphi_0), M(t))) \leq V(t, x(t; t_0, \varphi_0))$$

$$\leq \sup_{-r \leq \theta \leq 0} V(t_0^+, \varphi_0(\theta)) E_\beta(-w(t - t_0)^\beta) < w_1(\varepsilon).$$

Hence, $x(t; t_0, \varphi_0) \in M(t, \varepsilon)$ for $t \geq t_0 + T$ and the integral manifold M is an attractive set with respect to system (1.6).

We shall next consider a result which gives asymptotic stability of the integral manifold M with respect to system (1.6). We shall use two Lyapunov like functions.

Theorem 2.49 Assume that there exist functions V, $W \in V_0$ with a kernel the manifold M such that (2.138) holds,

$$w_2(d(x, M(t))) \leq W(t, x), \ w_2 \in K, \ (t, x) \in [t_0, \infty) \times \Omega, \quad (2.144)$$

$$\sup\{^cD_+^\beta W(t, \varphi(0)) : \ (t, \varphi) \in [t_0, \infty) \times PC[[-r, 0], \Omega]\} = N_1 < \infty, \quad (2.145)$$

$$V(t^+, \varphi(0) + I_k(\varphi)) \leq V(t, \varphi(0)), \ t = t_k, \ t_k > t_0, \quad (2.146)$$

$$W(t^+, \varphi(0) + I_k(\varphi)) \leq W(t, \varphi(0)), \ t = t_k, \ t_k > t_0, \quad (2.147)$$

and the inequality

$$^cD_+^\beta V(t, \varphi(0)) \leq -w_3(W(t, \varphi(0))), \quad (2.148)$$

is valid whenever $V(t + \theta, \varphi(\theta)) \leq V(t, \varphi(0))$ for $-r \leq \theta \leq 0$, where $t \in [t_0, \infty)$, $\varphi \in PC[[-r, 0], \Omega]$, $0 < \beta < 1$, $w_3 \in K$.

Then the integral manifold M is asymptotically stable with respect to system (1.6).

Proof. From Theorem 2.46, it follows that the integral manifold M is stable with respect to system (1.6).

Let $t_0 \in \mathbb{R}$. Let $\rho = const > 0$ be such that $M_0(t,\rho) \subset \Omega$ for $t \in [t_0 - r, t_0]$ and $M(t,\rho) \subset \Omega$ for $t \geq t_0$. For an arbitrary $t \geq t_0$, we put

$$V_{t,\rho}^{-1} = \{x \in \Omega : V(t^+, x) \leq w_1(\rho)\}.$$

From (2.138), we have that for each $t \geq t_0$ the following inclusions are valid

$$V_{t,\rho}^{-1} \subset M(t,\rho) \subset \Omega.$$

From (2.146), (2.147) and (2.148), we obtain that if $\varphi_0 \in V_{t_0,\rho}^{-1}$, where

$$V_{t_0,\rho}^{-1} = \{\phi \in PC[[-r,0],\Omega] : \sup_{-r \leq \theta \leq 0} V(t_0^+, \phi(\theta)) \leq w_1(\rho)\},$$

then $x(t;t_0,\varphi_0) \in V_{t,\rho}^{-1}$ for $t \in [t_0,\infty)$.

Let $\varphi_0 \in V_{t_0,\rho}^{-1}$. We shall prove that

$$\lim_{t \to \infty} d(x(t;t_0,\varphi_0), M(t)) = 0.$$

Suppose that this is not true. Then there exist $\varphi_0 \in V_{t_0,\rho}^{-1}$, $\kappa > 0$, $l > 0$, and a sequence $\{\xi_k\} \subset (t_0,\infty)$ such that for $k = 1,2,\ldots$ the following inequalities are valid

$$\xi_{k+1} - \xi_k \geq \kappa$$

and

$$d(x(\xi_k;t_0,\varphi_0), M(\xi_k)) \geq l.$$

From the last inequality and (2.144), we have

$$W(\xi_k, x(\xi_k;t_0,\varphi_0)) \geq w_2(l), \quad k = 1,2,\ldots. \tag{2.149}$$

Choose the constant $\gamma : 0 < \gamma < \min\{\kappa, \left[\Gamma(1+\beta)\frac{w_2(l)}{2N_1}\right]^{1/\beta}\}$ and from (2.145) and (2.149), we obtain

$$W(t,x(t)) = W(\xi_k, x(\xi_k)) + \frac{1}{\Gamma(\beta)} \int_{\xi_k}^{t} {}^cD_+^\beta W(s,x(s))(t-s)^{\beta-1} ds$$

$$\geq w_2(l) - \frac{N_1}{\Gamma(1+\beta)}(\xi_k - t)^\beta \geq w_2(l) - \frac{N_1\gamma^\beta}{\Gamma(1+\beta)} > \frac{w_2(l)}{2}$$

for $t \in [\xi_k - \gamma, \xi_k]$. From the above estimate and from (2.148), we conclude that

$$V(\xi_n, x(\xi_n; t_0, \varphi_0)) \leq$$

$$\leq V(t_0, \varphi_0(0)) - \frac{1}{\Gamma(\beta)} \sum_{k=1}^{n} \int_{\xi_k - \gamma}^{\xi_k} w_3(W(s, x(s; t_0, \varphi_0)))(\xi_k - s)^{\beta - 1} ds$$

$$\leq V(t_0, \varphi_0(0)) - \frac{1}{\Gamma(1+\beta)} w_3 \left(\frac{w_2(l)}{2} \right) \gamma^\beta n \to -\infty \text{ as } n \to \infty,$$

which contradicts (2.138).

Consequently, $\lim_{t \to \infty} d(x(t; t_0, \varphi_0), M(t)) = 0$ and since $V_{t_0, \rho}^{-1}$ is contained in $M_0(t, \rho)$, then the integral manifold M is an attractive set with respect to system (1.6).

In Theorem 2.49 two auxiliary functions that have the manifold M as a kernel were used. The function $W(t, x)$ may have a special form. In the case when $W(t, x) = d(x, M(t))$, we deduce the following corollary of Theorem 2.49.

Corollary 2.5 Assume that there exists a function $V \in V_0$ with a kernel the manifold M such that (2.138), (2.146) hold,

$$d(\varphi(0) + I_k(\varphi), M(t)) \leq d(\varphi(0), M(t)), \quad \varphi \in PC[[-r, 0], \Omega], \ t = t_k, \ t_k > t_0,$$

and the inequality

$$^c D_+^\beta V(t, \varphi(0)) \leq -w_3(d(\varphi(0), M(t)))$$

is valid whenever $V(t + \theta, \varphi(\theta)) \leq V(t, \varphi(0))$ for $-r \leq \theta \leq 0$, where $t \in [t_0, \infty)$, $\varphi \in PC[[-r, 0], \Omega], 0 < \beta < 1, w_3 \in K$.

Then the integral manifold M is asymptotically stable with respect to system (1.6).

In the case when $W(t, x) = V(t, x)$, we deduce the following corollary of Theorem 2.49.

Corollary 2.6 Assume that there exists a function $V \in V_0$ with a kernel the manifold M such that (2.138), (2.146) hold, and the inequality

$$^c D_+^\beta V(t, \varphi(0)) \leq -w_3(V(t, \varphi(0)))$$

is valid whenever $V(t + \theta, \varphi(\theta)) \leq V(t, \varphi(0))$ for $-r \leq \theta \leq 0$, where $t \in [t_0, \infty)$, $\varphi \in PC[[-r, 0], \Omega], 0 < \beta < 1, w_3 \in K$.

Then the integral manifold M is asymptotically stable with respect to system (1.6).

Theorem 2.50 Assume that there exists a function $V \in V_0$ with a kernel the manifold M, for which (2.146) hold, for any $\rho > 0$ there exists $\gamma(\rho) > 0$ such that

$$d(x,M(t)) \le V(t,x) \le \gamma(\rho)d(x,M(t)), \ (t,x) \in [t_0,\infty) \times \Omega, \qquad (2.150)$$

and the inequality (2.142) is valid whenever $V(t+\theta, \varphi(\theta)) \le V(t,\varphi(0))$ for $-r \le \theta \le 0$, $\varphi \in PC[[-r,0],\Omega]$, then the integral manifold M is Mittag–Leffler stable with respect to system (1.6).

Proof. Let $t_0 \in \mathbb{R}$. Let $\rho = const > 0$ be such that $M_0(t,\rho) \subset \Omega$ for $t \in [t_0-r,t_0]$ and $M(t,\rho) \subset \Omega$ for $t \ge t_0$. Let $\varphi_0 \in \overline{B}_\rho(\mathscr{PC}) \cap M_0(t,\delta)$ and $x(t) = x(t;t_0,\varphi_0)$ be the solution of IVP (1.6), (1.7). From (2.142) and (2.146), it follows by Corollary 1.5, that for $t \ge t_0$ the following inequality is valid

$$V(t,x(t;t_0,\varphi_0)) \le \sup_{-r \le \theta \le 0} V(t_0^+, \varphi_0(\theta))E_\beta(-w(t-t_0)^\beta).$$

From the above inequality and (2.150), we obtain

$$d(x(t;t_0,\varphi_0),M(t)) \le \sup_{-r \le \theta \le 0} V(t_0^+, \varphi_0(\theta))E_\beta(-w(t-t_0)^\beta)$$

$$\le \gamma(\rho)d_0(\varphi_0,M_0(t))E_\beta(-w(t-t_0)^\beta), \ t \ge t_0.$$

Let $m = \gamma(\rho)d_0(\varphi_0,M_0(t))$. Then

$$d(x(t;t_0,\varphi_0),M(t)) \le mE_\beta(-w(t-t_0)^\beta), \ t \ge t_0,$$

where $m \ge 0$ and $m = 0$ holds only if $d_0(\varphi_0,M_0(t)) = 0$, which implies that the integral manifold M is Mittag–Leffler stable with respect to system (1.6).

Since the fractional functional differential system

$$_{t_0}^c D_t^\alpha x(t) = f(t,x_t), \qquad (2.151)$$

where $f : [t_0,\infty) \times C[[-r,0],\Omega] \to \mathbb{R}^n$, $0 < \alpha < 1$ is a special case of (1.6), the following theorems follow directly from theorems 2.44 –2.48 and 2.50.

Theorem 2.51 Let M be a manifold in the extended phase space for the system (1.151), there exists a function $V \in C_0$ with a kernel the manifold M and β, $0 < \beta < 1$, such that the inequality (2.137) is valid whenever $V(t+\theta, \varphi(\theta)) \le V(t,\varphi(0))$ for $-r \le \theta \le 0, t \in [t_0,\infty)$, $\varphi \in C[[-r,0],\Omega]$.
Then M is an integral manifold of (1.151).

Theorem 2.52 Let M be a manifold in the extended phase space for the system (1.151), there exists a function $V \in C_0$ with a kernel the manifold M and β, $0 < \beta < 1$, such that the inequality

$$^c D_+^\beta V(t,\varphi(0)) \le -w(||\varphi(0)||), \ w \in K$$

is valid whenever $V(t + \theta, \varphi(\theta)) \leq V(t, \varphi(0))$ for $-r \leq \theta \leq 0$, $t \in [t_0, \infty)$, $\varphi \in C[[-r, 0], \Omega]$.

Then M is an integral manifold of (1.151).

Theorem 2.53 Let there exists a function $V \in C_0$ with a kernel M and β, $0 < \beta < 1$, such that (2.138) holds, and the inequality (2.137) is valid whenever $V(t + \theta, \varphi(\theta)) \leq V(t, \varphi(0))$ for $-r \leq \theta \leq 0$, $t \in [t_0, \infty)$, $\varphi \in C[[-r, 0], \Omega]$.

Then the integral manifold M is stable with respect to system (1.151).

Theorem 2.54 Let the conditions of Theorem 2.53 are met, and a function $w_2 \in K$ exists such that (2.141) hold.

Then the integral manifold M is uniformly stable with respect to system (1.151).

Theorem 2.55 If in Theorem 2.53 condition (2.137) is replaced by the condition (2.142) whenever $V(t + \theta, \varphi(\theta)) \leq V(t, \varphi(0))$ for $-r \leq \theta \leq 0$, where $t \in [t_0, \infty)$, $\varphi \in C[[-r, 0], \Omega]$, $0 < \beta < 1$, $w = const > 0$, then the integral manifold M is asymptotically stable with respect to system (1.151).

Theorem 2.56 Assume that there exists a function $V \in C_0$ with a kernel the manifold M, for any $\rho > 0$ there exists $\gamma(\rho) > 0$ such that (2.150) holds, and the inequality (2.142) is valid whenever $V(t + \theta, \varphi(\theta)) \leq V(t, \varphi(0))$ for $-r \leq \theta \leq 0$, then the integral manifold M is Mittag-Leffler stable with respect to system (1.151).

The proofs of the next theorems is similar to the proofs of Corollary 2.5 and Corollary 2.6, respectively.

Theorem 2.57 Assume that there exists a function $V \in C_0$ with a kernel the manifold M such that (2.138) holds, and the inequality

$$^{c}D_{+}^{\beta}V(t, \varphi(0)) \leq -w_3(d(\varphi(0), M(t)))$$

is valid whenever $V(t + \theta, \varphi(\theta)) \leq V(t, \varphi(0))$ for $-r \leq \theta \leq 0$, where $t \in [t_0, \infty)$, $\varphi \in C[[-r, 0], \Omega]$, $0 < \beta < 1$, $w_3 \in K$.

Then the integral manifold M is asymptotically stable with respect to system (1.151).

Theorem 2.58 Assume that there exists a function $V \in C_0$ with a kernel the manifold M such that (2.138) holds, and the inequality

$$^{c}D_{+}^{\beta}V(t, \varphi(0)) \leq -w_3(V(t, \varphi(0)))$$

is valid whenever $V(t + \theta, \varphi(\theta)) \leq V(t, \varphi(0))$ for $-r \leq \theta \leq 0$, where $t \in [t_0, \infty)$, $\varphi \in C[[-r, 0], \Omega]$, $0 < \beta < 1$, $w_3 \in K$.

Then the integral manifold M is asymptotically stable with respect to system (1.151).

Now, we shall consider several examples to illustrate our theoretical findings.

Example 2.19 Let $x \in \mathbb{R}_+$, $r > 0$, $\sigma \in C[\mathbb{R}_+, \mathbb{R}_+]$, $t - \sigma(t) \to \infty$ as $t \to \infty$.
Consider the following impulsive scalar quasilinear model

$$
\begin{cases}
{}_0^c D_t^\alpha x(t) = -ax(t) + bx(t-r) + e(t)x(t-\sigma(t)), \, t \neq t_k, t \geq 0, \\
\Delta x(t_k) = c_k x(t_k), \quad t_k > 0, \, k = 1, 2, ...,
\end{cases}
\tag{2.152}
$$

where $0 < \alpha < 1$, a, $b > 0$, $e(t) \geq 0$ is a continuous function, $-1 < c_k \leq 0$, $k = 1, 2, ...$, $0 < t_1 < t_2 < ... < t_k < t_{k+1} < ...$ and $\lim_{k \to \infty} t_k = \infty$.

Let $t_{-1} = \min\{-r, \inf_{t \geq 0}\{t - \sigma(t)\}\} < 0$ and $\varphi_0 \in C[[t_{-1}, 0], \mathbb{R}_+]$.

Consider the set $M = \{(t, 0) : t \in [t_{-1}, \infty)\}$, $d(x, M(t)) = |x|$, and $d_0(\phi, M_0(t)) = |\phi|_{t_{-1}} = \sup_{\theta \in [t_{-1}, 0]} |\phi(\theta)|$, $\phi \in C[[t_{-1}, 0], \mathbb{R}_+]$.

We define a Lyapunov function

$$
V(t, x) = \frac{1}{2} x^2.
$$

Suppose that:

(i) there exists a constant $\eta > 0$ such that

$$
e(t) < \eta, \, t \in \mathbb{R}_+;
$$

(ii) $b + \eta < a$, $t \in \mathbb{R}_+$.

Then for $t \neq t_k$, $k = 1, 2, ...$, $\varphi \in PC[[t_{-1}, 0]$ for the upper right-hand derivative ${}^c D_+^\alpha V(t, \varphi(0))$ along the solutions of (2.151) using Lemma 1.8, we have

$$
{}^c D_+^\alpha V(t, \varphi(0)) \leq \varphi(0)\,{}^c D_+^\alpha \varphi(0)
$$

$$
\leq \varphi(0)[-a\varphi(0) + b\varphi(-r) + \eta\varphi(-\sigma(0))] \leq 0,
\tag{2.153}
$$

whenever $V(t + \theta, \varphi(\theta)) \leq V(t, \varphi(0))$ for $t_{-1} \leq \theta \leq 0$, $t \in [0, \infty)$.

Also,

$$
V(t_k^+, x(t_k) + c_k x(t_k))
$$

$$
= \frac{1}{2}(1 + c_k)^2 x^2(t_k) \leq V(t_k, x(t_k)), \, k = 1, 2, ...
\tag{2.154}
$$

Thus:

(i) from Theorem 2.44 it follows that the set $M = \{(t, 0) : t \in [t_{-1}, \infty)\}$ is an integral manifold for (2.152);

(ii) from Theorem 2.46 it follows that the integral manifold M is stable with respect to (2.152).

Moreover, if there exists a constant $w > 0$ such that $b + \eta \leq a - w$ for $t \geq 0$, then from (2.153) we have

$$^cD_+^{\alpha}V(t, \varphi(0)) \leq -wV(t, \varphi(0)), \quad t \neq t_k \tag{2.155}$$

for $t \geq 0$ and for all $\varphi \in PC[[t_{-1}, 0]$ such that $V(t + \theta, \varphi(\theta)) \leq V(t, \varphi(0)), t_{-1} \leq \theta \leq 0$.

From (2.154), (2.155) and Theorem 2.50 it follows that the integral manifold M is Mittag–Leffler stable with respect to (2.152). Therefore, the integral manifold M is asymptotically stable with respect to (2.152).

Example 2.20　Consider the following delayed cellular neural network with dynamical thresholds of $\alpha-$ fractional order

$$^c_0D_t^{\alpha}x(t) = -x(t) + Af(x(t) - Bx(t - \tau(t)) - C), \quad t \geq 0, \tag{2.156}$$

where $0 < \alpha < 1$, $A > 0$, B and C are nonnegative constants, $f : \mathbb{R} \to \mathbb{R}$, $\tau(t)$ corresponds to the transmission delay and satisfies $0 \leq \tau(t) \leq \tau$ ($\tau = const$), $t - \tau(t) \to \infty$ as $t \to \infty$.

For $\alpha = 1$ some stability criteria for the equilibrium are investigated in [Zhong, Lin and Jiong 2001]. Gopalsamy and Leung (see [Gopalsamy and Leung 1997]) established a sufficient condition for global asymptotic stability of the equilibrium $x^* = 0$ for the case $\alpha = 1$, $f = \tanh$ and $C = 0$.

Let there exists an equilibrium $x^* \in \mathbb{R}$ of the model (2.156) such that

$$x^* = Af(x^* - Bx^* - C).$$

Consider the set $M = \{(t, x^*) : t \in [-\tau, \infty)\}$. Denote $d(x, M(t)) = |x - x^*|$ and

$$d_0(\phi, M_0(t)) = ||\phi - x^*||,$$

for a function $\phi \in C[[-\tau, 0], \mathbb{R}]$.

We define a Lyapunov function

$$V(t, x) = |x - x^*|.$$

Let $t \geq 0$. If there exists a constant $L > 0$ such that

$$|f(u) - f(v)| \leq L|u - v|$$

for all $u, v \in \mathbb{R}$, $u \neq v$, and $f(0) = 0$, then for the upper right-hand derivative $^cD_+^{\alpha}V(t, \varphi(0))$ along the solution of (2.156), we get

$$^cD_+^{\alpha}V(t, \varphi(0)) \leq -|\varphi(0) - x^*| + AL|\varphi(0) - x^*| + ABL|\varphi(-\tau(t)) - x^*|.$$

From the above estimate, we have

$$^cD_+^{\alpha}V(t, \varphi(0)) \leq -(1 - AL(1 + B))V(t, \varphi(0))$$

for any solution $\varphi(t)$ of (2.156) which satisfies the Razumikhin condition

$$V(t+\theta,\varphi(\theta)) \leq V(t,\varphi(0)), \quad -\tau \leq \theta \leq 0.$$

If the system parameters A and B satisfy

$$0 < AL(1+B) < 1,$$

then there exits a real number $\lambda > 0$ such that

$$1 - AL(1+B) \geq \lambda,$$

and it follows that

$$^cD_+^\alpha V(t,\varphi(0)) \leq -\lambda V(t,\varphi(0))$$
$$= -\lambda|\varphi(0) - x^*| = -\lambda d(\varphi(0), M(t)), \, t \geq 0.$$

Thus:

(i) from Theorem 2.51 it follows that the set $M = \{(t,x^*) : t \in [-\tau,\infty)\}$ is an integral manifold for (2.156);

(ii) from Theorem 2.56 it follows that the integral manifold M is Mittag–Leffler stable with respect to (2.156).

Since the Mittag–Leffler stability implies asymptotic stability of the integral manifold M, which contains the model's equilibrium, this means that the domain of attraction of the equilibrium point is the whole space and the convergence is in real time.

Example 2.21 Let $t_0 \in \mathbb{R}_+$. We consider the following fractional quasilinear functional differential model

$$^c_0D_t^\alpha x(t) = Ax(t) + Bx(t - h(t)) + f_1(x(t), x(t - h(t))), \tag{2.157}$$

where $0 < \alpha < 1$, $t \geq t_0$, A and B are constant matrices of type $(n \times n)$, $f_1 : \mathbb{R}^n \times \mathbb{R}^n \to \mathbb{R}^n$, $h \in C[[t_0,\infty), \mathbb{R}_+]$, $r = \sup_{t \geq t_0} h(t)$.

Let $\varphi_0 \in PC[[-r,0], \mathbb{R}_+^n]$. Denote by $x(t) = x(t; t_0, \varphi_0)$ the solution of system (2.157) satisfying the initial conditions

$$x(t+\theta) = \varphi_0(\theta) \geq 0, \, \theta \in [-r,0); \, x(t_0^+) = \varphi_0(0) > 0. \tag{2.158}$$

We introduce the following conditions:

H2.8 The matrix A is a real symmetric matrix with (distinct) negative eigenvalues.

H2.9 $f_1 \in C[\mathbb{R}^n \times \mathbb{R}^n, \mathbb{R}^n]$ and there exists a constant $\eta > 0$ such that

$$\|f_1(x,\tilde{x})\| \leq \eta\|\tilde{x}\|, \, x,\tilde{x} \in \mathbb{R}^n.$$

Remark 2.12 Condition H2.8 guarantees the asymptotic stability of the zero solution of the integer-order linear system

$$\dot{x}(t) = Ax(t).$$

Lemma 2.3 Let the condition H2.9 holds.

Then for $t_0 < t \leq t_0 + r$, for the solution $x(t) = x(t; t_0, \varphi_0)$ of problem (2.157), (2.158) the following inequality holds

$$||x(t)|| < \left[1 + \frac{1}{\Gamma(\alpha+1)}(||B|| + \eta)r^\alpha\right]||\varphi_0||_r E_\alpha(||A||r^\alpha), \qquad (2.159)$$

where E_α is the corresponding Mittag–Leffler function, and $||A||$ is the matix norm induced by the norm in $\mathbb{R}^n$.

Proof. Since the solution $x(t) = x(t; t_0, \varphi_0)$ of problem (2.157), (2.158) satisfies

$$x_{t_0} = \varphi_0,$$

$$x(t) = \varphi_0(0)$$

$$+ \frac{1}{\Gamma(\alpha)} \int_{t_0}^t (t-s)^{\alpha-1}[Ax(s) + Bx(s - h(s)) + f_1(x(s), x(s - h(s)))]ds,$$

by H2.9 it follows that for $t_0 < t \leq t_0 + r$ we have

$$||x(t)|| \leq ||\varphi_0(0)||$$

$$+ \frac{1}{\Gamma(\alpha)} \int_{t_0}^t (t-s)^{\alpha-1}||A|| \, ||x(s)||ds + \frac{1}{\Gamma(\alpha+1)}||\varphi_0||_r(||B|| + \eta)r^\alpha$$

$$\leq ||\varphi_0||_r + \frac{1}{\Gamma(\alpha)} \int_{t_0}^t (t-s)^{\alpha-1}||A|| \, ||x(s)||ds + \frac{1}{\Gamma(\alpha+1)}||\varphi_0||_r(||B|| + \eta)r^\alpha.$$

Hence, Corollary 1.7 yields the estimate

$$||x(t)|| \leq \left[1 + \frac{1}{\Gamma(\alpha+1)}(||B|| + \eta)r^\alpha\right]||\varphi_0||_r E_\alpha(||A||(t-t_0)^\alpha)$$

$$\leq \left[1 + \frac{1}{\Gamma(\alpha+1)}(||B|| + \eta)r^\alpha\right]||\varphi_0||_r E_\alpha(||A||r^\alpha) \qquad (2.160)$$

for $t_0 < t \leq t_0 + r$.

We introduce the following notation:

$$\bar{a} = \left\{\left[1 + \frac{1}{\Gamma(\alpha+1)}(||B|| + \eta)r^\alpha\right]||\varphi_0||_r E_\alpha(||A||r^\alpha)\right\}^2. \qquad (2.161)$$

Consider the manifold

$$M = [t_0 - r, \infty) \times \{x \in \mathbb{R}^n : x \leq 0\}.$$

Theorem 2.59 Assume that:

1. Conditions H2.8 and H2.9 hold.
2. The function f_1 and the matrix B are such that

$$\lambda_{\max}(A) + ||B|| + \eta \leq 0,$$

where $\lambda_{\max}(A)$ is the greatest eigenvalue of the matrix A.

Then:

1. M is an integral manifold for the system (2.157).
2. If

$$\lambda_{\max}(A) + ||B|| + \eta \leq -w, \tag{2.162}$$

where w is a positive constant, the integral manifold M is globally asymptotically stable with respect to system (2.157).

Proof of Assertion 1 Consider the function

$$V(t,x) = \begin{cases} \frac{1}{2}x^T x, & x > 0, \\ 0, & x \leq 0. \end{cases}$$

From Lemma 1.8 and H2.9, for the upper right-hand derivative $^cD_+^\alpha V(t,\varphi(0))$ along the solutions of system (2.157), we have

$$^cD_+^\alpha V(t,\varphi(0)) = \begin{cases} \frac{1}{2}\varphi^T(0)\varphi(0), & \varphi(0) > 0, \\ 0, & \varphi(0) \leq 0 \end{cases}$$

$$\leq \begin{cases} (\varphi^T(0))^c D_+^\alpha \varphi(0), & \varphi(0) > 0, \\ 0, & \varphi(0) \leq 0 \end{cases}$$

$$\leq \begin{cases} \varphi^T(0)A\varphi(0) + (||B|| + \eta)||\varphi(-r)||||\varphi(0)||, & \varphi(0) > 0, \\ 0, & \varphi(0) \leq 0, \end{cases}$$

where $\varphi \in C[[-r,0],\mathbb{R}^n]$.

Since A is a real symmetric matrix, it follows that for $x \in \mathbb{R}^n$, $x \neq 0$, the following inequalities hold

$$\lambda_{\min}(A)||x||^2 \leq x^T A x \leq \lambda_{\max}(A)||x||^2. \tag{2.163}$$

From (2.163), we obtain

$$^cD_+^\alpha V(t,\varphi(0)) \leq \begin{cases} \lambda_{\max}(A)||\varphi(0)||^2 + (||B|| + \eta)||\varphi(-r)||||\varphi(0)||, & \varphi(0) > 0, \\ 0, & \varphi(0) \leq 0. \end{cases}$$

Using the inequality $2|a||b| \leq a^2 + b^2$, we get

$$^cD_+^\alpha V(t,\varphi(0))$$

$$\leq \begin{cases} \lambda_{\max}(A)||\varphi(0)||^2 + (||B|| + \eta)\left(\frac{||\varphi(-r)||^2}{2} + \frac{||\varphi(0)||^2}{2}\right), & \varphi(0) > 0, \\ 0, & \varphi(0) \leq 0. \end{cases}$$

$$(2.164)$$

If $V(t + \theta, \varphi(\theta)) = \varphi^T(\theta)\varphi(\theta) \leq \varphi^T(0)\varphi(0) = V(t, \varphi(0))$, $-r \leq \theta \leq 0$, $t \geq t_0$, then the following inequalities hold

$$||\varphi(\theta)||^2 \leq ||\varphi(0)||^2.$$

$$(2.165)$$

From (2.164), (2.165) and condition 2 of Theorem 2.59, we have

$$^cD_+^\alpha V(t, \varphi(0)) \leq \begin{cases} (\lambda_{\max}(A) + ||B|| + \eta)||\varphi(0)||^2, & \varphi(0) > 0, \\ \\ 0, & \varphi(0) \leq 0 \end{cases} \leq 0,$$

whenever $V(t + \theta, \varphi(\theta)) \leq V(t, \varphi(0))$, $-r \leq \theta \leq 0$, $t \geq t_0$.

Since all the conditions of Theorem 2.51 are met, then the set $M = [t_0 - r, \infty) \times \{x \in \mathbb{R}^n : x \leq 0\}$ is an integral manifold of (2.157).

Proof of Assertion 2 By Lemma 2.3 for $t_0 < t \leq t_0 + r$ the solution $x(t)$ of (2.157), (2.158) satisfies the inequality (2.159).

Let $\bar{a} > 0$ be the constant defined by (2.161). For $t \in [t_0 - r, \infty)$ introduce the notations:

$v^{\bar{a}} = \{x \in \mathbb{R}^n : V(t, x) < \bar{a}\}$ and $\partial v^{\bar{a}} = \{x \in \mathbb{R}^n : V(t, x) = \bar{a}\}$.

Then, from (2.159) it follows that $x(t) = x(t; t_0, \varphi_0) \in v^{\bar{a}}$ for $t_0 - r \leq t \leq t_0 + r$.

We shall prove that $x(t; t_0, \varphi_0) \in v^{\bar{a}}$ for $t > t_0 + r$, too. Suppose that this is not true. This implies the existence of $T > t_0 + r$, such that $x(t) \in v^{\bar{a}}$ for $t_0 - r \leq t < T$ and $x(T) \in \partial v^{\bar{a}}$. Note that in view of the choice of the Lyapunov function, $x(T) > 0$.

Consider again the upper right-hand Caputo fractional derivative of the Lyapunov function $V(t, x) = \frac{1}{2}x^T x$ with respect to system (2.157). For $t = T$ and $x > 0$, we derive the estimate

$$^cD_+^\alpha V(T, x(T)) \leq \left[\lambda_{\max}(A)||x(T)|| + (||B|| + \eta)||x(T - h(T))||\right]||x(T)||.$$

From the above estimate, (2.163) and the condition 2 of Theorem 2.59, we obtain $^cD_+^\alpha V(T, x(T)) < 0$.

Hence by Corollary 1.2, we have $x(t) \in v^{\bar{a}}$ for all $t \geq t_0 - r$.

Thus, we have $||x(t)|| < \sqrt{\bar{a}}$ for all $t \geq t_0$. If $\bar{a} > 0$ is chosen so small that $\bar{a} \leq \varepsilon^2$, then $||x(t)|| < \varepsilon$ for all $t \geq t_0$. Therefore, the integral manifold M is stable with respect to system (2.157).

Also, from the condition (2.162) we have

$$^cD_+^\alpha V(t, \varphi(0)) \leq \begin{cases} (\lambda_{\max}(A) + ||B|| + \eta)||\varphi(0)||^2, & \varphi(0) > 0, \\ \\ 0, & \varphi(0) \leq 0 \end{cases}$$

$$\leq \begin{cases} -wV(t, \varphi(0)), \ \varphi(0) > 0, \\ \\ 0, \ \varphi(0) \leq 0, \end{cases}$$

whenever $V(t + \theta, \varphi(\theta)) \leq V(t, \varphi(0))$, $-r \leq \theta \leq 0$, $t \geq t_0$.

Thus, all conditions of Theorem 2.58 are met for $\Omega \equiv \mathbb{R}^n$, and the integral manifold M is globally asymptotically stable with respect to system (2.157).

2.9 Notes and Comments

The Lyapunov stability theorems 2.1–2.5 for fractional order functional differential systems in Section 2.1 are adopted from [Stamova 2016]. Similar technique is applied also in [Baleanu, Sadati, Ghaderi, Ranjbar, Abdeljawad and Jarad 2010] and [Liu and Jiang 2014]. Theorems 2.6 and 2.7 are new. Theorems 2.8–2.11 for impulsive fractional systems are due to Stamova (see [Stamova 2015]). Theorems 2.12–2.14 are taken from Stamova and Stamov (see [Stamova and Stamov G 2014b]).

The investigation of the boundedness properties of the solutions of different classes of differential equations has played a significant role in the existence of periodic and almost periodic solutions and it has many applications in biological population management and control. The boundedness Theorems 2.15–2.17 for fractional systems of the type (1.1) are adopted from [Stamova 2016]. Interesting boundedness results for fractional-order systems without delays by similar technique are given in [Yakar, Çiçek and Gücen 2011].

The global stability results in Section 2.3 for impulsive differential equations of Caputo fractional order with impulse effect at fixed moments of time are adapted from [Stamova 2014a]. These results extend the corresponding theory of Caputo fractional order systems to the impulsive case.

The Mittag–Leffler stability results (Theorem 2.21 and Theorem 2.22) for fractional functional differential systems (1.1) listed in Section 2.4 were adopted from [Stamova 2016]. Theorems 2.23–2.25 are taken from [Stamova 2015]. Close to them are the results in [Sadati, Baleanu, Ranjbar, Ghaderi and Abdeljawad 2010].

The content in Section 2.5 is adapted from [Stamova and Henderson 2016]. Similar results for non-impulsive systems can be found in [Çiçek, Yaker and Gücen 2014].

The results in Section 2.6 are given by Stamova and Stamov in [Stamova and Stamov G 2013].

The idea of stability of sets was initiated by Yoshizawa in [Yoshizawa 1966]. The results on stability and boundedness of solutions of the impulsive delay fractional differential equations with respect to a set discussed in Section 2.7 are new. Our results are quite general in their applicability to different impulsive delayed systems of fractional order. The advantages of using vector Lyapunov function method are demonstrated.

Theorems 2.44 and 2.45 for the existence of integral manifolds of the impulsive functional fractional differential equations are adapted from [Stamov and Stamova 2014b]. The existence results for integral manifolds for fractional functional differential equations, and the stability results for integral manifolds of impulsive fractional functional differential equations in Section 2.8 are new. Similar results for zero solutions of impulsive functional fractional differential equations are given in [Stamova 2014a].

Chapter 3

Almost Periodicity

Periodicity is one of the important fundamental concepts in various mathematical models governed by differential equations in a periodic environment in as diverse fields as economics, model ecosystems, physics and engineering. However, the problem of exact periodicity of the solutions is usually too strong and has limited applicability. Upon considering long-term dynamical behavior, it has been discovered that the periodic parameters often turn out to experience certain perturbations that may lead to changing character. Thus, the investigation of almost periodic behavior is considered to be more accordant with reality.

The concept of almost periodicity is with deep historical roots. The idea that the motion of a planet can be described by almost periodic functions is developed by the Greek astronomers, as well as by Ptolemy and Copernicus. The main aspects of the historical development of this problem can be found in [Neugebauer 1957] and [Sternberg 1969]. The formal notion of an almost periodic function was introduced in the 1920s by Bohr and Bochner. The fundamentals of the theory of almost periodic functions can be found in their papers [Bochner 1927], [Bochner 1933], [Bochner and Neumann 1935], [Bohr 1925], [Bohr 1926]. Since then, the almost periodicity has been extensively studied for differential equations and dynamical systems and a number of impressive results are reached. Part of them are in [Fink 1974], [Fink and Seifert 1969], [Levitan and Zhikov 1983], [Seifert 1966], [Yoshizawa 1966].

Later, Markoff recognized that almost periodicity and stability qualitative properties for a differential equation are closely related. In his paper [Markoff 1933] for the fist time it was considered that strong stable bounded solutions are almost periodic.

117

The beginning of the study of almost periodic piecewise continuous functions is in the sixteenth years of XX century. Then, parallel with the investigation of the impulsive differential equations these new notions have been introduced (see [Halanay and Wexler 1971], [Samoilenko and Perestyuk 1995]). From 1980s, the theory of almost periodic solutions to impulsive differential equations has been well developed, see [Stamov 2012], and the references therein.

A variety of direct and indirect methods have been applied in the investigation of the existence, uniqueness and stability of almost periodic solutions for numerous classes of differential equations. Among them, the Lyapunov direct method is one of the most efficient techniques (see [Yoshizawa 1966]).

For fractional-order systems the non-existence of exact periodic solutions has been investigated by several authors (see [Kaslik and Sivasundaram 2012b] and the references therein). It has been shown that the fractional-order derivative of a periodic function can not be a periodic function of the same period, underlining a very remarkable difference between integer and fractional-order derivatives and explaining the absence of periodic solutions in a class of fractional-order dynamical systems. Since no periodic solutions are possible for a wider class of such equations, the existence of almost periodic solutions is a possibility.

In the present chapter, using different techniques, we shall state existence, uniqueness and stability results for almost periodic solutions of impulsive and functional differential systems of fractional order.

In *Section* 3.1 an impulsive fractional differential equation in a Banach space with an infinitesimal generator of an analytic semigroup will be considered. Applying the fractional powers of this generator sufficient conditions for the existence and exponential stability of an almost periodic solution will be proved.

Section 3.2 is dedicated to the application of the Lyapunov method for the existence and uniqueness of almost periodic solutions of fractional-order equations. Results on almost periodic solutions of impulsive fractional differential equations, impulsive functional fractional differential equations, impulsive integro-differential equations of fractional order are given.

In *Section* 3.3 we shall study uncertain differential systems of fractional order. The existence theorems for almost periodic solutions of such systems will be obtained by means of the fractional Lyapunov direct method.

3.1 Almost Periodic Solutions

Let $(X, ||.||_X)$ be an abstract Banach space and $t_0 \in \mathbb{R}$. In this section we shall investigate the existence of almost periodic solutions for an impulsive fractional differential equation of the form

$$\,^{c}_{t_0}D^{\alpha}_t x(t) + Ax(t) = F(t,x(t)) + \sum_{k=\pm1,\pm2,...} H_k(x(t))\delta(t - t_k), \qquad (3.1)$$

where:

(i) $0 < \alpha < 1$, $t \in \mathbb{R}$, $-A: \mathbb{D}(A) \subset X \to X$ is a linear infinitesimal generator of an analytic C_0-semigroup $S(t)$ satisfying the exponential stability;

(ii) $F(t,x): \mathbb{D}(\mathbb{R} \times X) \to X$ is a continuous function with respect to $t \in \mathbb{R}$ and with respect to $x \in X$;

(iii) $H_k: \mathbb{D}(H_k) \subset X \to X$ are continuous impulsive operators;

(iv) $\delta(.)$ is the Dirac's delta-function, $\{t_k\} \in \mathscr{B}$.

Let $x_0 \in X$. Denote by $x(t) = x(t;t_0,x_0)$ the solution of (3.1) with the initial condition

$$x(t_0^+) = x_0. \qquad (3.2)$$

The solution $x(t) = x(t;t_0,x_0)$ of problem (3.1), (3.2) is a piecewise continuous function (see [Bainov, Kostadinov and Myshkis 1988], [Stamov and Stamova 2014a], [Wang, Fečkan and Zhou 2011]) with points of discontinuity of the first kind at the moments t_k, $k = \pm1,\pm2,...$ at which it is continuous from the left, i.e. the following relations are valid:

$$x(t_k^-) = x(t_k), \; x(t_k^+) = x(t_k) + H_k(x(t_k)), \; k = \pm1,\pm2,....$$

With respect to the norm

$$||\varphi||_{PC} = \max\left\{ \sup_{t\in\mathbb{R}}||\varphi(t^+)||_X, \sup_{t\in\mathbb{R}}||\varphi(t^-)||_X \right\},$$

it is easy to see that $PC[\mathbb{R},X]$ is a Banach space (see [Bainov, Kostadinov and Myshkis 1988]).

Since the solutions of (3.1) belong to the space $PC[\mathbb{R},X]$, in this section, we shall use Definition 1.9 for almost periodicity of the set of sequences $\{t_k^l\}$, $t_k^l = t_{k+l} - t_k$, $k,l = \pm1,\pm2,...$, $\{t_k\} \in \mathscr{B}$.

The next definition is for almost periodic functions in a Banach space of the form $PC[\mathbb{R},X]$.

Definition 3.1 The function $\phi \in PC[\mathbb{R},X]$ is said to be *almost periodic*, if:

(a) the set of sequences $\{t_k^l\}$, $t_k^l = t_{k+l} - t_k$, $k,l = \pm1,\pm2,...$, $\{t_k\} \in \mathscr{B}$ is uniformly almost periodic;

(b) for any $\varepsilon > 0$ there exists a real number $\delta > 0$ such that, if the points t' and t'' belong to one and the same interval of continuity of $\phi(t)$ and satisfy the inequality $|t' - t''| < \delta$, then $||\phi(t') - \phi(t'')||_X < \varepsilon$;

(c) for any $\varepsilon > 0$ there exists a relatively dense set T such that, if $\tau \in T$, then $\|\phi(t+\tau) - \phi(t)\|_X < \varepsilon$ for all $t \in \mathbb{R}$ satisfying the condition $|t - t_k| > \varepsilon$, $k = \pm 1, \pm 2, \ldots$.

The elements of T are called $\varepsilon - almost\ periods$.

Denote by $PCB[\mathbb{R}, X]$ the subspace of $PC[\mathbb{R}, X]$ formed of all bounded functions in $PC[\mathbb{R}, X]$ and together with (3.1) we consider the corresponding linear non-homogeneous impulsive differential equation

$$^c_{t_0}D^\alpha_t x(t) + Ax(t) = f(t) + \sum_{k=\pm 1, \pm 2, \ldots} h_k \delta(t - t_k), \qquad (3.3)$$

where $f \in PCB[\mathbb{R}, X]$ and $h_k : \mathbb{D}(h_k) \subset X \to X$.

Introduce the following conditions:

H3.1 The set of sequences $\{t^l_k\}$, $t^l_k = t_{k+l} - t_k$, $k, l = \pm 1, \pm 2, \ldots$, $\{t_k\} \in \mathscr{B}$ is uniformly almost periodic and there exists $\bar{\theta} > 0$ such that $\inf_k t^1_k = \bar{\theta} > 0$.

H3.2 The function $f(t)$ is almost periodic and locally Hölder continuous with points of discontinuity at the moments t_k, $k = \pm 1, \pm 2, \ldots$ at which it is continuous from the left.

H3.3 The sequence $\{h_k\}$, $k = \pm 1, \pm 2, \ldots$ of impulsive operators is almost periodic.

Lemma 3.1 ([Samoilenko and Perestyuk 1995]) Let the condition H3.1 holds. Then:
  (i) Uniformly with respect to $t \in \mathbb{R}$ there exists a constant $p > 0$ such that

$$\lim_{T \to \infty} \frac{i(t, t+T)}{T} = p;$$

  (ii) For each $p > 0$ there exists a positive integer N such that each interval of a length p has no more than N elements of the sequence $\{t_k\}$, i.e.,

$$i(s, t) \leq N(t - s) + N,$$

where $i(s, t)$ is the number of points t_k in the interval (s, t).

Lemma 3.2 ([Samoilenko and Perestyuk 1995]) Let the conditions H3.1–H3.3 hold.
  Then for each $\varepsilon > 0$ there exist ε_1, $0 < \varepsilon_1 < \varepsilon$ and relatively dense sets T^* of real numbers and Q^* of integer numbers, such that the following relations are fulfilled:
  (i) $\|f(t+\tau) - f(t)\|_X < \varepsilon$, $t \in \mathbb{R}$, $\tau \in T^*$, $|t - t_k| > \varepsilon$, $k = \pm 1, \pm 2, \ldots$;
  (ii) $\|h_{k+q} - h_k\|_X < \varepsilon$, $q \in Q^*$, $k = \pm 1, \pm 2, \ldots$;
  (iii) $|t^q_k - \tau| < \varepsilon_1$, $q \in Q^*$, $\tau \in T^*$, $k = \pm 1, \pm 2, \ldots$.

Definition 3.2 ([Wang, Fečkan and Zhou 2011]) By a *mild solution* of the system (3.2), (3.3) we mean a function $x \in PC[J,X]$, $J \subseteq \mathbb{R}$ which satisfies the following integral equation

$$x(t) = \begin{cases} \mathscr{T}(t-t_0)x_0 + \int_{t_0}^{t}(t-s)^{\alpha-1}\mathscr{S}(t-s)f(s)ds, \ t \in [t_0,t_1], \\ \\ \mathscr{T}(t-t_0)x_0 + \int_{t_0}^{t}(t-s)^{\alpha-1}\mathscr{S}(t-s)f(s)ds \\ \qquad\qquad + \mathscr{T}(t-t_1)h_1, \ t \in (t_1,t_2], \\ \dots\dots\dots\dots\dots\dots\dots\dots\dots\dots\dots\dots, \\ \\ \mathscr{T}(t-t_0)x_0 + \int_{t_0}^{t}(t-s)^{\alpha-1}\mathscr{S}(t-s)f(s)ds \\ \qquad\qquad + \sum_{t_0<t_k<t} \mathscr{T}(t-t_k)h_k, \ t \in (t_k,t_{k+1}], \\ \dots\dots\dots\dots\dots\dots\dots\dots\dots\dots\dots\dots, \end{cases} \tag{3.4}$$

where

$$\mathscr{T}(t) = \int_0^\infty \xi_\alpha(\sigma)S(t^\alpha\sigma)d\sigma, \quad \mathscr{S}(t) = \alpha\int_0^\infty \sigma\xi_\alpha(\sigma)S(t^\alpha\sigma)d\sigma,$$

$$\xi_\alpha(\sigma) = \frac{1}{\alpha}\sigma^{-1-\frac{1}{\alpha}}\varpi_\alpha(\sigma^{-\frac{1}{\alpha}}) \geq 0,$$

$$\varpi_\alpha(\sigma) = \frac{1}{\pi}\sum_{n=1}^\infty (-1)^{n-1}\sigma^{-n\alpha-1}\frac{\Gamma(n\alpha+1)}{n!}\sin(n\pi\alpha), \ \sigma \in (0,\infty),$$

ξ_α is a probability density function defined on $(0,\infty)$, that is

$$\xi_\alpha \geq 0, \ \theta \in (0,\infty) \ \text{and} \ \int_0^\infty \xi_\alpha(\sigma)d\sigma = 1.$$

Let the operator $-A$ in (3.1) and (3.3) be an infinitesimal operator of an analytic semigroup $S(t)$ in the Banach space X and $0 \in \rho(A)$, $\rho(A)$ be the resolvent set of A. For any $\beta > 0$, we define the fractional power $A^{-\beta}$ of the operator A by

$$A^{-\beta} = \frac{1}{\Gamma(\beta)}\int_0^\infty t^{\beta-1}S(t)dt,$$

where $A^{-\beta}$ is bounded, bijective and $A^\beta = (A^{-\beta})^{-1}$, $\beta > 0$ is a closed linear operator such that $\mathbb{D}(A^\beta) = \mathscr{R}(A^{-\beta})$ where $\mathscr{R}(A^{-\beta})$ is the range of $A^{-\beta}$. The operator A^0 is the identity operator in X and for $0 \leq \beta \leq 1$, the space $X_\beta = \mathbb{D}(A^\beta)$ with norm $||x||_\beta = ||A^\beta x||_X$ is a Banach space (see [Dalec'kii and Krein 1974], [Diethelm 2010], [El-Borai 2004], [El-Borai and Debbouche 2009]).

Lemma 3.3 ([Pazy 1983]) Let—A be an infinitesimal operator of an analytic semigroup $S(t)$.

Then:

(i) $S(t) : X \to \mathbb{D}(A^\beta)$ for every $t > 0$, and $\beta \geq 0$;

(ii) For every $x \in \mathbb{D}(A^\beta)$, the following equality $S(t)A^\beta x = A^\beta S(t)x$ holds;

(iii) For every $t > 0$, the operator $A^\beta S(t)$ is bounded and

$$||A^\beta S(t)||_X \leq K_\beta t^{-\beta} e^{-\lambda t}, \ K_\beta > 0, \ \lambda > 0;$$

(iv) For $0 < \beta \leq 1$ and $x \in \mathbb{D}(A^\beta)$ we have

$$||S(t)x - x||_X \leq C_\beta t^\beta ||A^\beta x||_X, \ C_\beta > 0.$$

When $-A$ generates a semigroup with a negative exponent, we deduce that if $x(t)$ is a bounded solution of (3.3) on $\mathbb{R}$, then we take the limit as $t_0 \to -\infty$ and using (3.4), we obtain

$$x(t) = \int_{-\infty}^{t} (t-s)^{\alpha-1} \mathscr{S}(t-s)f(s)ds + \sum_{t_k < t} \mathscr{T}(t-t_k)h_k. \tag{3.5}$$

Now, we shall prove the main results on almost periodic solutions of (3.1).

Lemma 3.4 Assume that:

1. Conditions H3.1–H3.3 are fulfilled;
2. $-A$ is the infinitesimal generator of an analytic semigroup $S(t)$.

Then:

1. There exists a unique mild almost periodic solution $x(t) \in PCB[\mathbb{R}, X]$ of (3.3).
2. The mild almost periodic solution $x(t)$ is exponentially stable.

Proof of Assertion 1 First we shall show that the right-hand side of (3.5) is well defined.

From conditions H3.2, H3.3 it follows that $f(t)$ and $\{h_k\}$ are bounded, and let

$$\max\{||f(t)||_{PC}, ||h_k||_X\} \leq M_0, \ M_0 > 0.$$

In view of Lemma 3.3 and the definition of the norm in X_β, we obtain

$$||x(t)||_\beta = \int_{-\infty}^{t} (t-s)^{\alpha-1} ||A^\beta \mathscr{S}(t-s)||_X ||f(s)||_{PC} ds$$

$$+ \sum_{t_k < t} ||A^\beta \mathscr{T}(t-t_k)||_X ||h_k||_X$$

$$\leq K_\beta M_0 \Big[\alpha \int_{-\infty}^t \int_0^\infty \sigma^{1-\beta} \xi_\alpha(\sigma)(t-s)^{-\alpha\beta+\alpha-1} e^{-\lambda\sigma(t-s)^\alpha} d\sigma ds$$

$$+ \sum_{t_k < t} \int_0^\infty \sigma^{-\beta} \xi_\alpha(\sigma)(t-t_k)^{-\alpha\beta} e^{-\lambda\sigma(t-t_k)^\alpha} d\sigma \Big]. \tag{3.6}$$

Then

$$\|x(t)\|_\alpha \leq K_\beta M_0 \Big[\alpha \int_0^\infty \int_0^\infty \sigma^{1-\beta} \xi_\alpha(\sigma) \eta^{-\alpha\beta+\alpha-1} e^{-\lambda\sigma\eta^\alpha} d\sigma d\eta$$

$$+ \sum_{t_k < t} \int_0^\infty \sigma^{-\beta} \xi_\alpha(\sigma)(t-t_k)^{-\alpha\beta} e^{-\lambda\sigma(t-t_k)^\alpha} d\sigma \Big], \tag{3.7}$$

where $\eta = t - s$.

However, for (3.7) we observe that

$$\alpha \int_0^\infty \int_0^\infty \sigma^{1-\beta} \xi_\alpha(\sigma) \eta^{-\alpha\beta+\alpha-1} e^{-\lambda\sigma\eta^\alpha} d\sigma d\eta$$

$$= \alpha \int_0^\infty \xi_\alpha(\sigma) \int_0^\infty \sigma^{1-\beta} \eta^{-\alpha\beta+\alpha-1} e^{-\lambda\sigma\eta^\alpha} d\eta d\sigma$$

$$= \frac{1}{\lambda^{1-\beta}} \int_0^\infty \xi_\alpha(\sigma) \int_0^\infty (\lambda\sigma\eta^\alpha)^{-\beta} e^{-\lambda\sigma\eta^\alpha} d\lambda\sigma\eta^\alpha d\sigma$$

$$= \frac{\Gamma(1-\beta)}{\lambda^{1-\beta}}. \tag{3.8}$$

On the other hand, by the help of H3.1 and Lemma 3.1, the sum in (3.7) can be estimated as follows

$$\sum_{t_k < t} \int_0^\infty \sigma^{-\beta} \xi_\alpha(\sigma)(t-t_k)^{-\alpha\beta} e^{-\lambda\sigma(t-t_k)^\alpha} d\sigma$$

$$= \int_0^\infty \xi_\alpha(\sigma) \Big[\sum_{0 < t-t_k \leq 1} (\sigma(t-t_k)^\alpha)^{-\beta} e^{-\lambda\sigma(t-t_k)^\alpha}$$

$$+ \sum_{j=1}^\infty \sum_{j < t-t_k \leq j+1} (\sigma(t-t_k)^\alpha)^{-\beta} e^{-\lambda\sigma(t-t_k)^\alpha} \Big] d\sigma$$

$$\leq \int_0^\infty \xi_\alpha(\sigma) \Big[\frac{2N}{m^\beta} + \frac{2N}{e^\lambda - 1} \Big] d\sigma = 2N \Big[\frac{1}{m^\beta} + \frac{1}{e^\lambda - 1} \Big], \tag{3.9}$$

where $m = \min\{ \sigma(t-t_k)^\alpha, \ 0 < t - t_k \leq 1 \}$.

In virtue of (3.8), (3.9) and the identity

$$\Gamma(\alpha)\Gamma(1-\alpha)\sin\pi\alpha = \pi, \ 0 < \alpha < 1,$$

we have

$$\|x(t)\|_\alpha \leq K_\beta M_0 \Big[\frac{\pi}{\Gamma(\beta)\sin\pi\beta\lambda^{1-\beta}} + 2N\Big(\frac{1}{m^\beta} + \frac{1}{e^\beta - 1} \Big) \Big],$$

and $x(t) \in PCB[\mathbb{R}, X]$.

Let $\varepsilon > 0$, $\tau \in T^*$, $q \in Q^*$, where the sets T^* and Q^* are defined as in Lemma 3.2.

Then

$$\|x(t+\tau) - x(t)\|_\beta = \|A^\beta (x(t+\tau) - x(t))\|_{PC}$$

$$\leq \int_{-\infty}^t \|A^\beta \mathscr{S}(t-s)\|_X \|f(s+\tau) - f(s)\|_{PC} ds$$

$$+ \sum_{t_k < t} \|A^\beta \mathscr{T}(t-\tau_k)\|_X \|h_{k+q} - h_k\|_X \leq M_\beta \varepsilon,$$

where $|t - t_k| > \varepsilon$ and

$$M_\beta = M_0 \left[\frac{\pi}{\Gamma(\beta) \sin \pi \beta \lambda^{1-\beta}} + 2N \left(\frac{1}{m^\beta} + \frac{1}{e^\beta - 1} \right) \right].$$

The last inequality implies that the function $x(t)$ is almost periodic. The uniqueness of this solution follows from conditions H3.1–H3.3 (see [Wang, Fečkan and Zhou 2011]).

Proof of Assertion 2 Let $\tilde{x} \in PCB[\mathbb{R}, X]$ be an arbitrary mild solution of (3.3) and $y = \tilde{x} - x$. Then $y \in PCB[\mathbb{R}, X]$ and

$$y = \mathscr{T}(t - t_0) y(t_0). \tag{3.10}$$

The proof follows from (3.10), the estimates from Lemma 3.3 and the fact that $i(t_0, t) - p(t - t_0) = o(t)$ for $t \to \infty$.

Introduce the following conditions:

H3.4 The function $F(t,x)$ is almost periodic with respect to $t \in \mathbb{R}$ uniformly on $x \in \Omega$, (Ω is a compact set from X), and there exist constants $L_1 > 0$, $1 > \kappa > 0, 1 > \beta > 0$, such that

$$\|F(t_1, x_1) - F(t_2, x_2)\|_X \leq L_1 (|t_1 - t_2|^\kappa + \|x_1 - x_2\|_\beta),$$

where $(t_i, x_i) \in \mathbb{R} \times X_\beta$, $i = 1, 2$.

H3.5 The sequence of functions $\{H_k(x)\}$ is almost periodic with respect to $k = \pm 1, \pm 2, \ldots$ uniformly at $x \in \Omega$, and there exist constants $L_2 > 0$, $0 < \beta < 1$, such that

$$\|H_k(x_1) - H_k(x_2)\|_X \leq L_2 \|x_1 - x_2\|_\beta,$$

where x_1, $x_2 \in X_\beta$.

Theorem 3.1 Assume that:

1. Conditions H3.1, H3.4, H3.5 hold.
2. — A is the infinitesimal generator of an analytic semigroup $S(t)$.

3. The functions $F(t,x)$ and $H_k(x)$ are bounded and

$$\max\left\{||F(t,x)||_X, \ ||H_k(x)||_X\right\} \le M,$$

$t \in \mathbb{R}$, $k = \pm 1, \pm 2, ...$, $x \in \Omega$, $M > 0$.

Then if $L > 0$, $L = \max\{L_1, L_2\}$, is sufficiently small it follows that:

(i) There exists a unique mild almost periodic solution $x(t) \in PCB[\mathbb{R}, X]$ of (3.1).

(ii) The mild almost periodic solution $x(t)$ is exponentially stable.

Proof. We denote by $\mathscr{D} \subset PCB[\mathbb{R}, X]$ the set of all almost periodic functions $\varphi \in PCB[\mathbb{R}, X]$ with discontinuities of the first type at the points t_k, $k = \pm 1, \pm 2, ...$, $\{t_k\} \in \mathscr{B}$ satisfying the inequality $||\varphi||_{PC} < h$, $h > 0$. In $\mathscr{D}$, we define the operator Θ in the following way

$$\Theta\varphi(t) = \int_{-\infty}^{t} (t-s)^{\alpha-1} A^\beta \mathscr{S}(t-s) F(s, A^{-\beta}\varphi(s)) ds$$

$$+ \sum_{t_k < t} A^\beta \mathscr{T}(t-t_k) H_k(A^{-\beta}\varphi(t_k)). \tag{3.11}$$

Proceeding in the same way as in the proof of Assertion 1 of Lemma 3.4, we can show that Θ is well defined and $\Theta\varphi(t)$ is an almost periodic function.

First, we shall show that Θ is a contracting operator in $\mathscr{D}$. Let $\varphi_1(t)$, $\varphi_2(t) \in \mathscr{D}$. Then, we obtain

$$||\Theta\varphi_1(t) - \Theta\varphi_2(t)||_X$$

$$\le \int_{-\infty}^{t} (t-s)^{\alpha-1} ||A^\beta \mathscr{S}(t-s)||_X ||F(t, A^{-\beta}\varphi_1(t)) - F(s, A^{-\beta}\varphi_2(t))||_X ds$$

$$+ \sum_{t_k < t} ||A^\beta \mathscr{T}(t-t_k)||_X ||H_k(A^{-\beta}\varphi_1(t_k)) - H_k(A^{-\beta}\varphi_2(t_k))||_X$$

$$\le L K_\beta ||\varphi_1(t) - \varphi_2(t)||_X \left[\alpha \int_{-\infty}^{t} \int_0^\infty \sigma^{1-\beta} \xi_\alpha(\sigma)(t-s)^{-\alpha\beta+\alpha-1} e^{-\lambda\sigma(t-s)^\alpha} d\sigma ds \right.$$

$$\left. + \sum_{t_k < t} \int_0^\infty \sigma^{-\beta} \xi_\alpha(\sigma)(t-t_k)^{-\alpha\beta} e^{-\lambda\sigma(t-t_k)^\alpha} d\sigma \right].$$

By following similar arguments as those used in (3.7), we have

$$||\Theta\varphi_1(t) - \Theta\varphi_2(t)||_X$$

$$\le L K_\beta \left[\frac{\pi}{\Gamma(\beta)\sin(\pi\beta)\lambda^{1-\beta}} + 2N\left(\frac{1}{m^\beta} + \frac{1}{e^\beta - 1}\right) \right] ||\varphi_1(t) - \varphi_2(t)||_X.$$

Therefore, if L is chosen in the form

$$L \le \left[K_\beta \left[\frac{\pi}{\Gamma(\beta)\sin(\pi\beta)\lambda^{1-\beta}} + 2N\left(\frac{1}{m^\beta} + \frac{1}{e^\beta - 1}\right) \right] \right]^{-1},$$

then the operator Θ is a contracting operator in $\mathscr{D}$.

Consequently, there exists $\varphi \in \mathscr{D}$ such that

$$\varphi(t) = \int_{-\infty}^{t} (t-s)^{\alpha-1} A^{\beta} \mathscr{S}(t-s) F(t, A^{-\beta}\varphi(s)) ds$$

$$+ \sum_{\tau_k < t} A^{\beta} \mathscr{T}(t-t_k) H_k(A^{-\beta}\varphi(t_k)). \tag{3.12}$$

On the other hand, since A^{β} is closed and $\varphi \in \mathscr{D}$, then $P = \sup_{t \in \mathbb{R}} ||F(t, A^{-\beta}\varphi(t))|| < \infty$, and we get

$$A^{-\beta}\varphi(t) = \int_{-\infty}^{t} (t-s)^{\alpha-1} \mathscr{S}(t-s) F(s, A^{-\beta}\varphi(s)) ds$$

$$+ \sum_{t_k < t} \mathscr{T}(t-t_k) H_k(A^{-\beta}\varphi(t_k)). \tag{3.13}$$

Now we shall show that $F(t, A^{-\beta}\varphi(s))$ is locally Hölder continuous. Let $h \in (0, \theta)$, where θ is the constant from H3.1, $t \in (t_k, t_{k+1} - h]$. By Lemma 3.3 we deduce that $0 < q < 1 - \beta$ implies

$$||S((h) - I) A^{\beta} S(t-s)|| \leq C_q h^q ||A^{\beta+q} S(t-s)||. \tag{3.14}$$

Again for $h \in (0, \theta)$, we have

$$||S((t+h-s)^{\alpha}\theta)|| = ||S((t+h-s)^{\alpha}\theta - (t-s)^{\alpha}\theta^* - h^{\alpha}\theta^*) S(h^{\alpha}\theta^*) Q((t-s)^{\alpha}\theta^*)||$$

$$\leq M^* ||S(h^{\alpha}\theta^*) S((t-s)^{\alpha}\theta^*)||, \tag{3.15}$$

where $M^* > 0$, $\theta^* = \dfrac{\theta}{2}$.

In virtue of Lemma 3.3, (3.14), (3.15), (see [El-Borai 2004], [El-Borai and Debbouche 2009]), and standard properties of the almost periodicity, we have that there exists $M_1 > 0$, such that

$$M_1 = \sup_{t \in \mathbb{R}} ||F(t, A^{-\beta}\varphi(t))||_{PC}.$$

Then

$$||\varphi(t+h) - \varphi(t)||_{PC}$$

$$\leq \left\| \alpha M^* \int_{-\infty}^{t} \int_{0}^{\infty} \theta \left[(t+h-s)^{\alpha-1} - (t-s)^{\alpha-1} \right] \xi(\theta) \right.$$

$$\left. \left(S(h^{\alpha}\theta^*) - I \right) A^{\beta} S((t-s)^{\alpha}\theta^*) F(s, A^{-\beta}\varphi(s)) d\theta ds \right\|_{PC}$$

$$+ \left\| \alpha \int_t^{t+h} \int_0^\infty \theta(t+h-s)^{\alpha-1}\xi(\theta)A^\beta S((t+h-s)^\alpha\theta)F(s,A^{-\beta}\varphi(s))d\theta ds \right\|_{PC}$$

$$\leq \alpha M^* C^q h^{\alpha q} M_1 M_{\beta+q} \int_{-\infty}^t \int_0^\infty \left| \theta \left[(t+h-s)^{\alpha-1} - (t-s)^{\alpha-1} \right] \xi(\theta) \right.$$

$$\left. \theta^{*\beta}(t-s)^{-\alpha(\beta+q)}\theta^* - (\beta+q)e^{-\lambda(t-s)^\alpha\theta^*} \right] |d\theta ds$$

$$+\alpha M_1 M_\beta \int_t^{t+h} \int_0^\infty \left| \theta(t+h-s)^{\alpha-1}\xi_\alpha(\theta)(t+h-s)^{-\alpha\beta}\theta^{-\beta}e^{-\lambda(t+h-s)^\alpha}\theta \right| d\theta ds.$$

$$(3.16)$$

From Lemma 3.3 it follows that there exists a constant $C > 0$ such that

$$\|\varphi(t+h) - \varphi(t)\|_\alpha \leq Ch^q.$$

On the other hand from H3.4 it follows that $F(t,A^{-\beta}\varphi(t))$ is locally Hölder continuous and from H3.5 and conditions of Theorem 3.1, we deduce that $H_k(A^{-\beta}\varphi(t_k))$ is a bounded almost periodic sequence.

Let $\varphi(t)$ be the solution of (3.12). Consider the equation

$$^c_{t_0}D_t^\alpha x(t) + Ax(t) = F(t,A^{-\beta}\varphi(t)) + \sum_{t_k > t_0} H_k(A^{-\beta}\varphi(t_k))\delta(t-t_k). \quad (3.17)$$

In view of Lemma 3.4, it follows that there exists a unique mild exponentially stable solution in the form

$$\psi(t) = \int_{-\infty}^t (t-s)^{\alpha-1}\mathscr{S}(t-s)F(s,A^{-\beta}\varphi(s))ds$$

$$+ \sum_{t_k < t} \mathscr{T}(t-t_k)H_k(A^{-\beta}\varphi(t_k)). \quad (3.18)$$

Then

$$A^\beta\psi(t) = \int_{-\infty}^t A^\beta(t-s)^{\alpha-1}\mathscr{S}(t-s)F(s,A^{-\beta}\varphi(s))ds$$

$$+ \sum_{t_k < t} A^\beta\mathscr{T}(t-t_k)H_k(A^{-\beta}\varphi(t_k)) = \varphi(t). \quad (3.19)$$

The last equality shows that $\psi(t) = A^{-\beta}\varphi(t)$ is a mild solution of (3.1), and the uniqueness follows from the uniqueness of the solutions of (3.12) and (3.17), and Lemma 3.4.

Example 3.1 As an example we shall consider a two-dimensional fractional impulsive predator-prey system with diffusion, when biological parameters assumed to be changed in an almost periodical manner. The system is affected by impulses, which can be considered as a control.

Assuming that the system is confined to a fixed bounded space domain $\Omega \subset \mathbb{R}^n$ with smooth boundary $\partial\Omega$, non-uniformly distributed in the domain

$\overline{\Omega} = \Omega \times \partial\Omega$ and subjected to short-term external influence at fixed moment of time t_k, $k = \pm 1, \pm 2, \ldots$. The functions $u(t,x)$ and $v(t,x)$ determine the densities of predator and pray, respectively, $\Delta = \frac{\partial^2}{\partial x_1^2} + \frac{\partial^2}{\partial x_2^2} + \ldots + \frac{\partial^2}{\partial x_n^2}$ is the Laplace operator and $\frac{\partial}{\partial n}$ is the outward normal derivative.

The system is written in the form

$$
\begin{cases}
{}_{t_0}^{c}D_t^{\alpha}u(t,x(t)) = \mu_1\Delta u \\
\qquad\qquad + u\left[a_1(t,x) - b(t,x)u - \frac{c_1(t,x)v}{r(t,x)v+u}\right], \; t \neq t_k, \\
{}_{t_0}^{c}D_t^{\alpha}v(t,x(t)) = \mu_2\Delta v \\
\qquad\qquad + v\left[-a_2(t,x) + \frac{c_2(t,x)u}{r(t,x)u+v}\right], \; t \neq t_k, \\
u(t_k^+,x) = u(t_k,x)I_k(x,u(t_k,x),v(t_k,x)), \; k = \pm 1, \pm 2, \ldots, \\
v(t_k^+,x) = v(t_k,x)J_k(x,u(t_k,x),v(t_k,x)), \; k = \pm 1, \pm 2, \ldots, \\
\left.\dfrac{\partial u}{\partial n}\right|_{\partial\Omega} = 0, \; \left.\dfrac{\partial v}{\partial n}\right|_{\partial\Omega} = 0.
\end{cases} \qquad (3.20)
$$

The boundary condition characterizes the absence of migration, $\mu_1 > 0$, $\mu_2 > 0$ are diffusion coefficients. We assume that, the predator functional response has the form of the ratio function $\dfrac{c_1 v}{rv + u}$. The ratio function $\dfrac{c_2 u}{rv + u}$ represents the conversion of prey to predator, a_1, a_2, c_1 and c_2 are positive functions that stand for prey intrinsic growth rate, capturing rate of the predator, death rate of the predator and conversion rate, respectively, $\dfrac{a_1(t,x)}{b(t,x)}$ gives the carrying capacity of the prey, and $r(t,x)$ is the half saturation function.

We note that for $\alpha = 1$ the existence of almost periodic solutions of (3.20) has been investigated in [Stamov 2012].

Let the moments of impulsive perturbations t_k be such that

$$t_k = k + \alpha_k, \; k = \pm 1, \pm 2, \ldots,$$

where $\{\alpha_k\}$, $\alpha_k \in \mathbb{R}$, $k = \pm 1, \pm 2, \ldots$ is an almost periodic sequence such that

$$\sup_{k=\pm 1, \pm 2, \ldots} |\alpha_k| = \alpha < \frac{1}{2}.$$

Then, we have

$$t_{k+1} - t_k \geq 1 - 2\alpha > 0,$$

and $\lim\limits_{k \to \pm\infty} t_k = \pm\infty$.

Let $\varepsilon > 0$ and p be an $\dfrac{\varepsilon}{2}$-almost period of the sequence $\{\alpha_k\}$. Then, for all integers k and l it follows

$$\left| t^l_{k+p} - t^l_k \right| = \left| \alpha_{k+l+p} - \alpha_{k+l} \right| + \left| \alpha_{k+p} - \alpha_k \right| < \varepsilon.$$

The last inequality shows that the set of sequences $\{t^l_k\}$ is uniformly almost periodic, i.e. condition H3.1 holds for the system (3.20).

As for the equation (3.1), we assume that:

H3.6 The functions $a_i(t,x)$, $c_i(t,x)$, $i = 1,2$, $b(t,x)$ and $r(t,x)$ are almost periodic with respect to t, uniformly at $x \in \overline{\Omega}$, positive-valued on $\mathbb{R} \times \overline{\Omega}$ and locally Hölder continuous with points of discontinuity at the moments t_k, $k = \pm 1, \pm 2, \ldots$, at which they are continuous from the left.

H3.7 The sequences of functions $\{I_k(x,u,v)\}$, $\{J_k(x,u,v)\}$, $k = \pm 1, \pm 2, \ldots$ are almost periodic with respect to k, uniformly at $x, u, v \in \overline{\Omega}$.

Set $w = (u,v)$, and

$$-A = \begin{bmatrix} \lambda - \mu_1 \Delta & 0 \\ 0 & \lambda - \mu_2 \Delta \end{bmatrix},$$

$$F(t,w) = \begin{bmatrix} u\left[a_1(t,x) - b(t,x)u - \dfrac{c_1(t,x)v}{r(t,x)v + u} \right] + \lambda u \\ v\left[-a_2(t,x) + \dfrac{c_2(t,x)u}{r(t,x)u + v} \right] + \lambda v \end{bmatrix},$$

$$H_k(w(t_k)) = \begin{bmatrix} u(t_k,x)I_k(x,u(t_k,x),v(t_k,x)) - u(t_k,x) \\ v(t_k,x)J_k(x,u(t_k,x),v(t_k,x)) - v(t_k,x) \end{bmatrix},$$

where $\lambda > 0$.

Then, the system (3.20) moves to the equation

$$^c_{t_0}D^\alpha_t w(t) + Aw(t) = F(t,w(t)) + \sum_{k=\pm 1, \pm 2, \ldots} H_k(w(t))\delta(t - t_k). \qquad (3.21)$$

It is well-known (see [Stamov 2012]), that the operator A is sectorial, and $Re\sigma_1(A) \leq -\lambda$, where $\sigma_1(A)$ is the spectrum of A. Now, the analytic semigroup of the operator A is e^{-At}, and

$$A^{-\beta} = \dfrac{1}{\Gamma(\beta)} \int_0^\infty t^{\beta-1} e^{-At} dt.$$

Theorem 3.2 Let for the equation (3.21) the following conditions hold.

1. Conditions H3.6 and H3.7 are met.
2. For the functions $F(t,w)$ there exist constants $L_1 > 0$, $1 > \kappa > 0$, $1 > \beta > 0$ such that

$$||F(t_1,w_1) - F(t_2,w_2)||_X \le L_1 \left(|t_1 - t_2|^\kappa + ||w_1 - w_2||_\beta\right),$$

where $(t_i, w_i) \in \mathbb{R} \times X_\beta$, $i = 1, 2$.

3. For the set of functions $\{H_k(w)\}$, $k = \pm 1, \pm 2, \dots$ there exist constants $L_2 > 0$, $1 > \beta > 0$ such that

$$||H_k(w_1) - H_k(w_2)||_X \le L_2 ||w_1 - w_2||_\beta.$$

where w_1, $w_2 \in X_\beta$.

4. The functions $F(t,w)$ and $H_k(w)$ are bounded for $t \in \mathbb{R}, w \in X_\beta$ and $k = \pm 1, \pm 2, \dots$.

Then, if $L = max\{L_1, L_2\}$ is sufficiently small, it follows:

1. There exists a unique mild almost periodic solution $x \in PCB[\mathbb{R}, X]$ *of* (3.20).
2. The mild almost periodic solution $x(t)$ is asymptotically stable.

Proof. From conditions H3.6, and H3.7 and conditions of the theorem, it follows that all conditions of Theorem 3.1 hold. Then, for (3.21), and consequently for (3.20), there exists a unique mild almost periodic solution, which is asymptotically stable.

3.2 Lyapunov Method for Almost Periodic Solutions

In the present section, by means of the fractional comparison lemmas and the Lyapunov method, the existence and uniqueness of almost periodic solutions of impulsive fractional differential equations and impulsive fractional functional differential equations will be investigated.

3.2.1 Impulsive Fractional Differential Systems

First, we shall consider the class of impulsive differential equations involving Caputo fractional derivatives of the type

$$\begin{cases} {}^c_{t_0}D^\alpha_t x(t) = f(t, x(t)), \ t \ne t_k, \\ \Delta x(t_k) = I_k(x(t_k)), \ k = \pm 1, \pm 2, \dots, \end{cases} \quad (3.22)$$

where $f : \mathbb{R} \times \mathbb{R}^n \to \mathbb{R}^n$, $0 < \alpha < 1$, $\{t_k\} \in \mathscr{B}$, $\Delta x(t_k) = x(t_k^+) - x(t_k)$, $I_k : \mathbb{R}^n \to \mathbb{R}^n$, $k = \pm 1, \pm 2, \dots$.

Let $t_0 \in \mathbb{R}$ and $x_0 \in \mathbb{R}^n$. Denote by $x(t) = x(t;t_0,x_0)$ the solution of system (3.22), satisfying the initial condition

$$x(t_0^+) = x_0, \qquad (3.23)$$

and by $J^+(t_0,x_0)$ - the maximal interval of type $[t_0, \gamma)$ in which the solution $x(t;t_0,x_0)$ is defined. We shall also consider such solutions of system (3.22) for which the continuability to the left of t_0 is guaranteed.

The solution $x(t) = x(t;t_0,x_0)$ of the IVP (3.22), (3.23) is a piecewise continuous function with points of discontinuity of the first kind at the moments t_k, $k = \pm 1, \pm 2, \ldots$ at which it is continuous from the left, i.e. the following relations are valid:

$$x(t_k^-) = x(t_k), \ x(t_k^+) = x(t_k) + I_k(x(t_k)), \ k = \pm 1, \pm 2, \ldots.$$

We introduce the following conditions:

H3.8 The function $f(t,x)$ is almost periodic in $t \in \mathbb{R}$ uniformly with respect to $x \in \mathbb{R}^n$.

H3.9 The sequence $\{I_k(x)\}$, $k = \pm 1, \pm 2, \ldots$ is almost periodic uniformly with respect to $x \in \mathbb{R}^n$.

Let the conditions H3.1, H3.8 and H3.9 hold, and let $\{s_m'\}$ be an arbitrary sequence of real numbers. Then, there exists a subsequence $\{s_n\}$, $s_n = s_{m_n}'$ such that the system (3.22) moves to the system

$$\begin{cases} {}_{t_0}^c D_t^\alpha x(t) = f^s(t,x(t)), \ t \neq t_k^s, \\ \Delta x(t_k^s) = I_k^s(x(t_k^s)), \ k = \pm 1, \pm 2, \ldots \end{cases} \qquad (3.24)$$

and the set of systems in the form (3.24) we shall denote by $H(f, I_k, t_k)$.

In the proof of the main results here we shall use the following lemma.

Lemma 3.5 ([Samoilenko and Perestyuk 1995]) The set of sequences $\{t_k^l\}$ is uniformly almost periodic, if and only if from each infinite sequence of shifts $\{t_k - s_n\}$, $k, l = \pm 1, \pm 2, \ldots$, $n = 1, 2, \ldots$, $s_n \in \mathbb{R}$ we can choose a subsequence convergent in $\mathscr{B}$.

Let $\bar{x}_1 \in \mathbb{R}^n$. Denote by $x_1(t) = x_1(t;t_0,\bar{x}_1)$ the solution of system (3.22), satisfying the initial condition

$$x_1(t_0^+) = \bar{x}_1.$$

Definition 3.3 The solution $x_1(t)$ of system (3.22) is said to be *globally quasi-equi-asymptotically stable*, if

$$(\forall \lambda > 0)(\forall \varepsilon > 0)(\forall t_0 \in \mathbb{R})(\exists T > 0)(\forall x_0 \in \mathbb{R}^n : ||x_0 - \bar{x}_1|| < \lambda)$$

$$(\forall t \geq t_0 + T) : ||x(t;t_0,x_0) - x_1(t;t_0,\bar{x}_1)|| < \varepsilon.$$

The next definition is a generalization of the corresponding definitions in [Lakshmikantham and Leela 1969] and [Yoshizawa 1966].

Definition 3.4 The solution $x_1(t)$ of (3.22) is said to be *globally perfectly uniform-asymptotically stable*, if it is uniformly stable, the number T in Definition 3.3 is independent of $t_0 \in \mathbb{R}$ and the solutions of (3.22) are uniformly bounded.

In the proof of the main results we shall use the following comparison lemma similar to Lemma 1.5.

Lemma 3.6 Assume that:

1. The function $g : \mathbb{R} \times \mathbb{R}_+ \to \mathbb{R}$ is continuous in each of the sets $(t_{k-1}, t_k] \times \mathbb{R}_+$, $k = \pm 1, \pm 2, \dots$.
2. $B_k \in C[\mathbb{R}_+, \mathbb{R}_+]$ and $\tilde{\psi}_k(u) = u + B_k(u) \geq 0$, $k = \pm 1, \pm 2, \dots$ are non-decreasing with respect to u.
3. The maximal solution $u^+(t; t_0, u_0)$ of the scalar problem

$$\begin{cases} {}^c_{t_0} D^\alpha_t u(t) = g(t, u(t)), \ t \neq t_k, \\ u(t_0^+) = u_0 \geq 0, \\ \Delta u(t_k) = B_k(u(t_k)), \ k = \pm 1, \pm 2, \dots \end{cases}$$

   is defined in the interval $[t_0, \infty)$.
4. The function $V \in V_2$ is such that for $t \in [t_0, \infty)$,

$$V(t^+, x + I_k(x), y + I_k(y)) \leq \tilde{\psi}_k(V(t, x, y)), \ t = t_k, \ k = \pm 1, \pm 2, \dots,$$

$${}^c D^\alpha_+ V(t, x, y) \leq g(t, V(t, x, y)), \ t \neq t_k, \ k = \pm 1, \pm 2, \dots.$$

   Then $V(t_0^+, x_0, y_0) \leq u_0$ implies

$$V(t, x(t; t_0, x_0), y(t; t_0, y_0)) \leq u^+(t; t_0, u_0), \ t \in [t_0, \infty),$$

where $x = x(t; t_0, x_0), y = y(t; t_0, y_0)$ are two solutions of (3.22), $x_0, y_0 \in \mathbb{R}^n$, existing on $[t_0, \infty)$.

In the case when $g(t, u) = qu$ for $(t, u) \in \mathbb{R} \times \mathbb{R}_+$, where $q \in \mathbb{R}$ is a constant, and $\tilde{\psi}_k(u) = u$ for $u \in \mathbb{R}_+$, $k = \pm 1, \pm 2, \dots$, we deduce the following corollary from Lemma 3.6.

Corollary 3.1 Assume that the function $V \in V_2$ is such that:

$$V(t^+, x + I_k(x), y + I_k(y)) \leq V(t, x, y), \ x, y \in \mathbb{R}^n, t = t_k, \ t_k > t_0,$$

$${}^c D^\alpha_+ V(t, x, y) \leq qV(t, x, y), \ x, y \in \mathbb{R}^n, t \neq t_k, \ t_k > t_0,$$

Then

$$V(t,x(t;t_0,x_0),y(t;t_0,y_0)) \leq V(t_0^+,x_0,y_0)E_\alpha(q(t-t_0)^\alpha), \, t \in [t_0,\infty).$$

In the proofs of our main theorems in this section we shall use piecewise continuous Lyapunov functions $V : \mathbb{R} \times \mathbb{R}^n \times \mathbb{R}^n \to \mathbb{R}_+$, $V \in V_2$ for which the following condition is true:

H3.10 $V(t,0,0) = 0, \, t \in \mathbb{R}$.

Theorem 3.3 Assume that:

1. Conditions H3.1, H3.8 and H3.9 are met.
2. There exist functions $V \in V_2$ and $w_1, w_2 \in K$ such that H3.10 is true,

$$w_1(||x-y||) \leq V(t,x,y) \leq w_2(||x-y||), \, (t,x,y) \in \mathbb{R} \times \mathbb{R}^n \times \mathbb{R}^n, \quad (3.25)$$

and $w_1(u) \to \infty$ as $u \to \infty$,

$$V(t_k^+, x+I_k(x), y+I_k(y)) \leq V(t_k,x,y), \, (x,y) \in \mathbb{R}^n \times \mathbb{R}^n, \, k = \pm 1, \pm 2, ...,$$
$$(3.26)$$
$$^cD_+^\alpha V(t,x,y) \leq -wV(t,x,y), \, t = t_k, \, (x,y) \in \mathbb{R}^n \times \mathbb{R}^n, \, k = \pm 1, \pm 2, ...,$$
$$(3.27)$$

$w > 0$.

3. There exists a solution $x(t) = x(t;t_0,x_0)$ of (3.22) such that $x(t) \in B_q$, where $t \geq t_0$, $q > 0$.

Then for the system (3.22) there exists a unique almost periodic solution $\tilde{x}(t)$ such that:

1. $\tilde{x}(t) \in \overline{B}_{q_1} < q$;
2. $\tilde{x}(t)$ is globally perfectly uniform-asymptotically stable;
3. $H(\tilde{x},t_k) \subset H(f,I_k,t_k)$.

Proof. Let $t_0 \in \mathbb{R}$ and let $\{s_j\}$ be any sequence of real numbers such that $s_j \to \infty$ as $j \to \infty$ and $\{s_j\}$ moves the system (3.22), (3.23) to a system at $H(f,I_k,t_k)$.

For any real number γ, let $j_0 = j_0(\gamma)$ be the smallest value of j, such that $s_{j_0} + \gamma \geq t_0$. Since $||x(t;t_0,x_0)|| \leq q_1$, $q_1 < q$ for all $t \geq t_0$, then $||x(t+s_j;t_0,x_0)|| \leq q_1$ for $t \geq \gamma$, $j \geq j_0$.

Let U, $U \subset (\gamma,\infty)$ be compact. Since for the function $w_1 \in K$ we have $w_1(u) \to \infty$ as $u \to \infty$, then for any $\varepsilon > 0$, we can choose an integer $n_0(\varepsilon,\gamma) \geq j_0(\gamma)$, so large that for $l \geq j \geq n_0(\varepsilon,\gamma)$ and $t \in \mathbb{R}$, $t \neq t_k$, $k = \pm 1, \pm 2, ...$, it follows

$$w_2(2q_1)E_\alpha\left(-w(\gamma+s_j-t_0)^\alpha\right) \leq \frac{w_1(\varepsilon)}{2}, \quad (3.28)$$

$$\|f(t+s_l,x) - f(t+s_j,x)\| \le \frac{w_1(\varepsilon)w}{2L_V}. \tag{3.29}$$

where $E_\alpha(.)$ is the corresponding Mittag-Leffler function.

Now, following [Lakshmikantham, Leela and Sambandham 2008] and [Lakshmikantham, Leela and Vasundhara Devi 2009], we shall set

$$S(x,\chi,h,\alpha) = \sum_{h=1}^{n} (-1)^{h+1} \binom{\alpha}{h} x(t-h\chi).$$

Then for small $\chi > 0$, we have

$$V(\sigma,x(\sigma),x(\sigma+s_j-s_l)) - V(\sigma-\chi,S_1,S_2) = V(\sigma,x(\sigma),x(\sigma+s_j-s_l))$$

$$-V(\sigma-\chi,x(\sigma)-\chi^\alpha f(\sigma,x(\sigma)),x(\sigma+s_j-s_l)-\chi^\alpha f(\sigma+s_j-s_l,x(\sigma+s_j-s_l))$$

$$+V(\sigma-\chi,x(\sigma)-\chi^\alpha f(\sigma,x(\sigma)),x(\sigma+s_j-s_l)-\chi^\alpha f(\sigma+s_j-s_l,x(\sigma+s_j-s_l))$$

$$-V(\sigma-\chi,S_1,S_2),$$

where $S_1 = S(x(\sigma),\chi,h,\alpha)$ and $S_2 = S(x(\sigma+s_j-s_l),\chi,h,\alpha)$.

Then for the fractional derivative in the Caputo sense of the order α, using (3.29), we obtain

$$^cD_+^\alpha V(\sigma,x(\sigma),x(\sigma+s_j-s_l)) = \lim_{\substack{\chi\to 0 \\ n\chi=t-t_0}} \sup \frac{1}{\chi^\alpha} \Big[V(\sigma,x(\sigma),x(\sigma+s_j-s_l))$$

$$-V(\sigma-\chi,x(\sigma)-\chi^\alpha f(\sigma,x(\sigma)),x(\sigma+s_j-s_l)-\chi^\alpha f(\sigma+s_j-s_l,x(\sigma+s_j-s_l))$$

$$+V(\sigma-\chi,x(\sigma)-\chi^\alpha f(\sigma,x(\sigma)),x(\sigma+s_j-s_l)-\chi^\alpha f(\sigma+s_p-s_l,x(\sigma+s_j-s_l))$$

$$-V(\sigma-\chi,S_1,S_2)\Big]$$

$$\le -wV(\sigma,x(\sigma),x(\sigma+s_j-s_l)) + L_V\|f(\sigma+s_j-s_l,x(\sigma+s_j-s_l)) - f(\sigma,x(\sigma))\|.$$

Set $\zeta = \sigma - s_j$. Then,

$$f(\sigma+s_l-s_j,x(\sigma+s_l-s_j)) - f(\sigma,x(\sigma))$$

$$= f(\zeta+s_l,x(\zeta+s_l)) - f(\zeta+s_j,x(\zeta+s_j))$$

and by (3.28)

$$^cD_+^\alpha V(\sigma,x(\sigma),x(\sigma+s_j-s_l)) \le -wV(\sigma,x(\sigma),x(\sigma+s_j-s_l)) + \frac{w_1(\varepsilon)w}{2}. \tag{3.30}$$

On the other hand, from $\sigma = t_k - (s_l - s_j)$ it follows

$$V(\sigma,x(\sigma)+I_k(x(\sigma)),x(\sigma+s_l-s_j)+I_k(x(\sigma+s_l-s_j)))$$

$$\le V(\sigma,x(\sigma),x(\sigma+s_l-s_j)). \tag{3.31}$$

Then from (3.30) and (3.31) it follows that the conditions of Corollary 3.1 are fulfilled and consequently, for $j, l \geq n_0(\varepsilon, \gamma)$ and any $t \in U$,

$$V(t + s_j, x(t + s_j), x(t + s_l))$$

$$\leq E_\alpha\left(-w(t + s_j - s_l)^\alpha\right)V(t_0^+, x(t_0 + s_j), x(t_0 + s_l)) + \frac{w_1(\varepsilon)}{2} \leq w_1(\varepsilon).$$

Finally from (3.25) and (3.26) it follows that for any $t \in U$, we get

$$\|x(t + s_j) - x(t + s_l)\| < \varepsilon.$$

Consequently, there exists a function $\tilde{x}(t)$, such that $x(t + s_j) - \tilde{x}(t) \to \infty$ for $j \to \infty$. Since γ is arbitrary, it follows that $\tilde{x}(t)$ is defined uniformly on $t \in \mathbb{R}$.
Next, we shall show that $\tilde{x}(t)$ is a solution of (3.24).
Since $x(t; t_0, x_0)$ is a solution of (3.22), (3.23), we have

$$\|{}^c_{t_0}D_t^\alpha x(t + s_j) - {}^c_{t_0}D_t^\alpha x(t + s_l)\| \leq \|f(t + s_j, x(t + s_j)) - f(t + s_j, x(t + s_j))\|$$

$$+ \|f(t + s_l, x(t + s_j)) - f(t + s_l, x(t + s_l))\|,$$

for $t + s_\kappa \neq t_k$, $\kappa = j, l$; $k = \pm 1, \pm 2, \ldots$.
As $x(t + s_j) \in \overline{B}_{q_1}$ for large s_j for each compact subset of $\mathbb{R}$ there exists an $n_1(\varepsilon) > 0$ such that if $l \geq j \geq n_1(\varepsilon)$, then

$$\|f(t + s_j, x(t + s_j)) - f(t + s_l, x(t + s_j))\| < \frac{\varepsilon}{2}.$$

Since $x(t + s_\kappa) \in B_{\gamma(q)}$, $\kappa = j, l$ and from Lemma 3.5 it follows that there exists $n_2(\varepsilon) > 0$ such that if $l \geq j \geq n_2(\varepsilon)$, then

$$\|f(t + s_l, x(t + s_j)) - f(t + s_l, x(t + s_l))\| < \frac{\varepsilon}{2}.$$

For $l \geq j \geq n(\varepsilon)$, $n(\varepsilon) = \max\{n_1(\varepsilon), n_2(\varepsilon)\}$, we obtain

$$\|{}^c_{t_0}D_t^\alpha x(t + s_j) - {}^c_{t_0}D_t^\alpha x(t + s_l)\| \leq \varepsilon,$$

where $t + s_\kappa \neq t_k^s$ which shows that $\lim\limits_{j \to \infty} {}^c_{t_0}D_t^\alpha x(t + s_j)$ exists uniformly on all compact subsets of $\mathbb{R}$.
Let now $\lim\limits_{j \to \infty} {}^c_{t_0}D_t^\alpha x(t + s_j) = {}^c_{t_0}D_t^\alpha \tilde{x}(t)$, and

$${}^c_{t_0}D_t^\alpha \tilde{x}(t) = \lim\limits_{j \to \infty}\left[f(t + s_j, x_{t + s_j}) - f(t + s_j, \tilde{x}(t)) + f(t + s_j, \tilde{x}(t))\right]$$

$$= f^s(t, \tilde{x}(t)), \tag{3.32}$$

where $t \neq t_k^s$, $t_k^s = \lim\limits_{j \to \infty} t_{k + j(s)}$.

On the other hand, for $t + s_j = t_k^s$ it follows

$$\tilde{x}(t_k^{s+}) - \tilde{x}(t_k^{s-}) = \lim_{j \to \infty} (x(t_k^s + s_j + 0) - x(t_k^s + s_j - 0))$$

$$= \lim_{j \to \infty} I_k^s(x(t_k^s + s_j)) = I_k^s(\tilde{x}(t_k^s)). \tag{3.33}$$

From (3.32) and (3.33), we get that $\tilde{x}(t)$ is a solution of (3.24).

We shall show that $\tilde{x}(t)$ is an almost periodic function.

Let the sequence $\{s_j\}$ moves the system (3.22) to $H(f, I_k, t_k)$. For any $\varepsilon > 0$ there exists $m_0(\varepsilon) > 0$ such that if $l \geq j \geq m_0(\varepsilon)$, then

$$E_\alpha(-w(s_p)^\alpha) w_2(2q_1) < \frac{w_1(\varepsilon)}{4},$$

and

$$\|f(\sigma + s_j, x(\sigma + s_j)) - f(\sigma + s_l, x(\sigma + s_l))\| < \frac{w_1(\varepsilon)}{4L_V}.$$

For each fixed $t \in \mathbb{R}$ let $\tau_\varepsilon = \frac{w_1(\varepsilon)}{4L_V}$ be a translation number of f such that $t + \tau_\varepsilon \geq 0$.

Consider the function

$$V(\tau_\varepsilon + \sigma, \tilde{x}(\sigma), \tilde{x}(\sigma + s_l - s_j)),$$

where $t \leq \sigma \leq t + s_j$.

Then, the same way as above, we get

$$^cD_+^\alpha V(\tau_\varepsilon + \sigma, \tilde{x}(\sigma), \tilde{x}(\sigma + s_l - s_j))$$

$$\leq -wV(\tau_\varepsilon + \sigma, \tilde{x}(\sigma), \tilde{x}(\sigma + s_l - s_j)) + L_V \|f^s(\sigma, \tilde{x}(\sigma)) - f^s(\tau_\varepsilon + \sigma, \tilde{x}(\sigma))\|$$

$$+ L_V \|f^s(\sigma + s_l - s_j, \tilde{x}(\sigma + s_l - s_j)) - f^s(\tau_\varepsilon + \sigma, \tilde{x}(\sigma + s_l - s_j))\|$$

$$\leq -wV(\tau_\varepsilon + \sigma, \tilde{x}(\sigma), \tilde{x}(\sigma + s_l - s_j)) + \frac{3w_1(\varepsilon)}{4}. \tag{3.34}$$

On the other hand,

$$V(\tau_\varepsilon + t_k^s, \tilde{x}(t_k^s) + I_k^s(\tilde{x}(t_k^s)), \tilde{x}(t_k^s + s_l - s_j) + I_k^s(\tilde{x}(t_k^s + s_l - s_p)))$$

$$\leq V(\tau_\varepsilon + t_k^s, \tilde{x}(t_k^s), \tilde{x}(t_k^s + s_l - s_j)). \tag{3.35}$$

From (3.34), (3.36) and Corollary 3.1 it follows

$$V(\tau_\varepsilon + t + s_j, \tilde{x}(t + s_j), \tilde{x}(t + s_l))$$

$$\leq E_\alpha(-w(s_j)^\alpha) V(\tau_\varepsilon + t, \tilde{x}(t), \tilde{x}(t + s_j - s_l)) + \frac{3w_1(\varepsilon)}{4} < w_1(\varepsilon). \tag{3.36}$$

Then from (3.36) for $l \geq j \geq m_0(\varepsilon)$, we have

$$||\tilde{x}(t+s_j) - \tilde{x}(t+s_l)|| < \varepsilon. \tag{3.37}$$

Now, from definitions of the sequence $\{s_j\}$ and for $l \geq j \geq m_0(\varepsilon)$ it follows that $\rho(t_k + s_j, t_k + s_l) < \varepsilon$.

Then from (3.37) and the last inequality we obtain that the sequence $\tilde{x}(t+s_j)$ is convergent uniformly to the function $\tilde{x}(t)$.

The assertions 1 and 3 of the theorem follow immediately. We shall prove the assertion 2.

Let $\overline{\omega}(t)$ be an arbitrary solution of (3.24). Set

$$u(t) = \overline{\omega}(t) - \tilde{x}(t),$$

$$g^s(t, u(t)) = f^s(t, u(t) + \tilde{x}(t)) - f^s(t, \tilde{x}(t)),$$

$$B_k^s(u) = I_k^s(u + \tilde{x}) - I_k^s(\tilde{x}).$$

Now we consider the system

$$\begin{cases} {}^c_{t_0}D_t^\alpha u = g^s(t, u(t)), \ t \neq t_k^s, \\ \Delta u(t_k^s) = B_k^s(u(t_k^s)), \ k = \pm 1, \pm 2, ..., \end{cases} \tag{3.38}$$

and let $W(t, u(t)) = V(t, \tilde{x}(t), \tilde{x}(t) + u(t))$. Then, from Lemma 3.6 it follows that the zero solution $u(t) = 0$ of (3.38) is globally perfectly uniform-asymptotically stable, and consequently $\tilde{x}(t)$ is globally perfectly uniform-asymptotically stable.

3.2.2 Impulsive Fractional Integro-Differential Systems

In the part of Section 3.2 we shall consider the following impulsive integro-differential system with a Caputo fractional operator

$$\begin{cases} {}^c_{t_0}D_t^\alpha x(t) = f(t, x(t), \int_{t_0}^t \mathcal{K}(t, \tau, x(\tau))d\tau), \ t \neq t_k, \\ \Delta x(t_k) = I_k(x(t_k)), \ k = \pm 1, \pm 2, ..., \end{cases} \tag{3.39}$$

where $0 < \alpha < 1$, $f \in C[\mathbb{R} \times \mathbb{R}^n \times \mathbb{R}^n, \mathbb{R}^n]$, $\mathcal{K} \in [\mathbb{R} \times \mathbb{R} \times \mathbb{R}^n, \mathbb{R}^n]$, $\{t_k\} \in \mathcal{B}$, $\Delta x(t_k) = x(t_k^+) - x(t_k)$, $I_k : \mathbb{R}^n \to \mathbb{R}^n$, $k = \pm 1, \pm 2, ...$

Let $x_0 \in \mathbb{R}^n$. Denote by $x(t) = x(t; t_0, x_0)$ the solution of system (3.39), satisfying the initial condition

$$x(t_0^+) = x_0.$$

Note that, the existence and uniqueness of the solutions of impulsive integro-differential systems of fractional order have been studied by several authors. See, for example [Anguraj and Maheswari 2012], [Gao, Yang and

Liu 2013], [Suganya, Mallika Arjunan and Trujillo 2015], [Xie 2014] and the references therein.

By the specific character of system (3.39) we need the next definition.

Definition 3.5   The function $\mathscr{K} \in C[\mathbb{R} \times \mathbb{R} \times \mathbb{R}^n, \mathbb{R}^n]$ is said to be *integro-almost periodic* in $t \in \mathbb{R}$ uniformly for $x \in PC[\mathbb{R}, \mathbb{R}^n]$, if for every sequence of real numbers $\{s'_m\}$ there exists a subsequence $\{s_n\}$, $s_n = s'_{m_n}$, such that the sequence

$$\left\{ \int_{t_0}^{t+s_n} \mathscr{K}(t+s_n, \tau, x(\tau)) d\tau \right\}$$

converges uniformly with respect to $n \to \infty$.

Introduce the following conditions:

H3.11   The function $\mathscr{K}(t, \tau, x)$ is integro-almost periodic in $t \in \mathbb{R}$ uniformly on $\tau \in \mathbb{R}$ and $x \in PC[\mathbb{R}, \mathbb{R}^n]$.

H3.12   The function $f(t, x, \int_{t_0}^{t} \mathscr{K}(t, \tau, x) d\tau)$ is almost periodic in $t \in \mathbb{R}$ uniformly with respect to $x \in \mathbb{R}^n$.

Let conditions H3.1, H3.9, H3.11 and H3.12 hold, and let $\{s'_m\}$ be an arbitrary sequence of real numbers. Then, there exists a subsequence $\{s_n\}$, $s_n = s'_{m_n}$ such that the system (3.39) moves to the system

$$\begin{cases} {}^{c}_{t_0}D^{\alpha}_t x(t) = f^s(t, x(t), \int_{t_0}^{t} \mathscr{K}^s(t, \tau, x(\tau)) d\tau), \ t \neq t^s_k, \\ \Delta x(t^s_k) = I^s_k(x(t^s_k)), \ k = \pm 1, \pm 2, ... \end{cases} \tag{3.40}$$

We shall denote the set of systems in the form (3.40) by $H(f, \mathscr{K}, I_k, t_k)$.

Theorem 3.4   Assume that:
1. Conditions H3.1, H3.9. H3.11 and H3.12 are met.
2. There exist functions $V \in V_2$ and $w_1, w_2 \in K$ such that H3.10, (3.25)–(3.27) are true.
3. There exists a solution $x(t) = x(t; t_0, x_0)$ of (3.39) such that $x(t) \in B_q$, where $t \geq t_0$, $q > 0$.

Then for the system (3.39) there exists a unique almost periodic solution $\tilde{x}(t)$ such that:
1. $\tilde{x}(t) \in \overline{B}_{q_1}$, $q_1 < q$;
2. $\tilde{x}(t)$ is globally perfectly uniform-asymptotically stable;
3. $H(\tilde{x}, t_k) \subset H(f, \mathscr{K}, I_k, t_k)$.

Proof. Let $t_0 \in \mathbb{R}$ and let $\{s_i\}$ be any sequence of real numbers such that $s_i \to \infty$ as $i \to \infty$ and $\{s_i\}$ moves the system (3.39) to a system at $H(f, \mathscr{K}, I_k, t_k)$.

For any real number γ, let $i_0 = i_0(\gamma)$ be the smallest value of i, such that $s_{i_0} + \gamma \geq t_0$. Since $x(t) \in \overline{B}_{q_1}$, $q_1 < q$ for all $t \geq t_0$, then $x(t + s_i) \in \overline{B}_{q_1}$ for $t \geq \gamma$, $i \geq i_0$.

Let U, $U \subset (\gamma, \infty)$ be compact. Then, for any $\varepsilon > 0$, choose an integer $n_0(\varepsilon, \gamma) \geq i_0(\gamma)$, so large that for $l \geq i \geq n_0(\varepsilon, \gamma)$ and $t > t_0$, $t \neq t_k$, it follows

$$w_2(2q_1)E_\alpha\left(-w(\gamma + s_i - t_0)^\alpha\right) \leq \frac{w_1(\varepsilon)}{2}, \tag{3.41}$$

$$\left\|f(t + s_l, x, \int_{t_0}^{t+s_l} \mathscr{K}(t + s_l, \tau, x)d\tau)\right.$$

$$\left. -f(t + s_i, x, \int_{t_0}^{t+s_i} \mathscr{K}(t + s_i, \tau, x)d\tau)\right\| \leq \frac{w_1(\varepsilon)w}{2L_V}, \tag{3.42}$$

where $E_\alpha(.)$ is the corresponding Mittag-Leffler function.

Set $m(\sigma) = V(\sigma, x(\sigma), x(\sigma + s_i - s_l))$ and let again

$$S(x, \chi, h, \alpha) = \sum_{h=1}^{n} (-1)^{h+1}\binom{\alpha}{h}x(t - h\chi).$$

Then for small $\chi > 0$, we have

$$V(\sigma, x(\sigma), x(\sigma + s_i - s_l)) - V(\sigma - \chi, S_1, S_2) = V(\sigma, x(\sigma), x(\sigma + s_i - s_l))$$

$$-V\left(\sigma - \chi, x(\sigma) - \chi^\alpha f(\sigma, x(\sigma), \int_{t_0}^\sigma \mathscr{K}(\sigma, \tau, x(\tau))d\tau),\right.$$

$$\left. x(\sigma + s_i - s_l) - \chi^\alpha f(\sigma + s_i - s_l, x(\sigma + s_i - s_l), \int_{t_0}^{\sigma + s_i - s_l} \mathscr{K}(\sigma + s_i - s_l, \tau, x(\tau))d\tau)\right)$$

$$+V\left(\sigma - \chi, x(\sigma) - \chi^\alpha f(\sigma, x(\sigma), \int_{t_0}^\sigma \mathscr{K}(\sigma, \tau, x(\tau))d\tau),\right.$$

$$\left. x(\sigma + s_i - s_l) - \chi^\alpha f(\sigma + s_i - s_l, x(\tau + s_i - s_l), \int_{t_0}^{\sigma + s_i - s_l} \mathscr{K}(\sigma + s_i - s_l, \tau, x(\tau))d\tau)\right)$$

$$-V(\sigma - \chi, S_1, S_2),$$

where $S_1 = S(x(\sigma), \chi, h, \alpha)$ and $S_2 = S(x(\sigma + s_i - s_l), \chi, h, \alpha)$.

Then by using the definition of the fractional order derivative of V in the Caputo's sense and (3.42) we obtain

$${}^cD_+^\alpha m(\sigma) = \lim_{\substack{\chi \to 0 \\ n\chi = t - t_0}} \sup \frac{1}{\chi^\alpha}\left[V(\sigma, x(\sigma), x(\sigma + s_i - s_l))\right.$$

$$-V\left(\sigma - \chi, x(\sigma) - \chi^\alpha f(\sigma, x(\sigma), \int_{t_0}^\sigma \mathscr{K}(\sigma, \tau, x(\tau))d\tau),\right.$$

$$\left. x(\sigma + s_i - s_l) - \chi^\alpha f(\sigma + s_i - s_l, x(\sigma + s_i - s_l), \int_{t_0}^{\sigma + s_i - s_l} \mathscr{K}(\sigma + s_i - s_l, \tau, x(\tau))d\tau)\right)$$

$$+V\left(\sigma-\chi,x(\sigma)-\chi^{\alpha}f(\sigma,x(\sigma),\int_{t_0}^{\sigma}\mathscr{K}(\sigma,\tau,x(\tau))d\tau),\right.$$

$$x(\sigma+s_i-s_l)-\chi^{\alpha}f(\sigma+s_i-s_l,x(\sigma+s_i-s_l),\int_{t_0}^{\sigma+s_i-s_l}\mathscr{K}(\sigma+s_i-s_l,\tau,x(\tau))d\tau\right)$$

$$\left.-V(\sigma-\chi,S_1,S_2)\right]$$

$$\leq -wV(\sigma,x(\sigma),x(\sigma+s_i-s_l))$$

$$+L_V||f(\sigma+s_i-s_l,x(\sigma+s_i-s_l),\int_{t_0}^{\sigma+s_i-s_l}\mathscr{K}(\sigma+s_i-s_l,\tau,x(\tau)d\tau))$$

$$-f(\sigma,x(\sigma)),\int_{t_0}^{\sigma}\mathscr{K}(\sigma,\tau,x(\tau)d\tau))||.$$

Set $\zeta=\sigma-s_i$. Then,

$$f(\sigma+s_i-s_l,x(\sigma+s_i-s_l),\int_{t_0}^{\sigma+s_i-s_l}\mathscr{K}(t+s_i-s_l,\tau,x(\tau)d\tau))$$

$$-f(\sigma,x(\sigma)),\int_{t_0}^{\sigma}\mathscr{K}(t,\tau,x(\tau)d\tau))$$

$$=f(\zeta+s_l,x(\zeta+s_l),\int_{t_0}^{\zeta+s_l}\mathscr{K}(\zeta+s_l,\tau,x(\tau)d\tau))$$

$$-f(\zeta+s_i,x(\zeta+s_i),\int_{t_0}^{\zeta+s_i}\mathscr{K}(\zeta+s_i,\tau,x(\tau)d\tau))$$

and by (3.42)

$$^cD_+^{\alpha}V(\sigma,x(\sigma),x(\sigma+s_i-s_l))$$

$$\leq -wV(\sigma,x(\sigma),x(\sigma+s_i-s_l))+\frac{w_1(\varepsilon)w}{2}. \tag{3.43}$$

On the other hand, from $\sigma=t_k-(s_l-s_i)$ it follows

$$V(\sigma,x(\sigma)+I_k(x(\sigma)),x(\sigma+s_l-s_i)+I_k(x(\sigma+s_l-s_i)))$$

$$\leq V(\sigma,x(\sigma),x(\sigma+s_l-s_i)). \tag{3.44}$$

Then from (3.43) and (3.44) and Corollary 3.1 for $i,l\geq n_0(\varepsilon,\gamma)$ and any $t\in U$, we have that

$$V(t+s_i,x(t+s_i),x(t+s_l))$$

$$\leq E_{\alpha}\left(-w(t+s_i-s_l)^{\alpha}\right)V(t_0,x(t_0+s_i),x(t_0+s_l))+\frac{w_1(\varepsilon)}{2}\leq w_1(\varepsilon).$$

Finally, from (3.25) and (3.26) for any $t\in U$, we get

$$||x(t+s_i)-x(t+s_l)||<\varepsilon.$$

Consequently, there exists a function $\tilde{x}(t)$, such that $x(t+s_i) - \tilde{x}(t) \to \infty$ for $i \to \infty$. Since γ is arbitrary, it follows that $\tilde{x}(t)$ is defined uniformly on $t \in \mathbb{R}$.

Next, we shall show that $\tilde{x}(t)$ is a solution of (3.40).

Since $x(t;t_0,x_0)$ is a solution of (3.39) with an initial condition $x(t_0^+) = x_0$, we have

$$||{}_{t_0}^{c}D_t^{\alpha}x(t+s_i) - {}_{t_0}^{c}D_t^{\alpha}x(t+s_l)||$$

$$\leq ||f(t+s_i,x(t+s_i), \int_{t_0}^{t+s_i} \mathscr{K}(t+s_i,\tau,x)d\tau$$

$$-f(t+s_l,x(t+s_i), \int_{t_0}^{t+s_l} \mathscr{K}(t+s_l,\tau,x(\tau))d\tau)||$$

$$+||f(t+s_l,x(t+s_i), \int_{t_0}^{t+s_l} \mathscr{K}(t+s_l,\tau,x(\tau))d\tau$$

$$-f(t+s_l,x(t+s_l), \int_{t_0}^{t+s_l} \mathscr{K}(t+s_l,\tau,x(\tau))d\tau)||,$$

for $t+s_\kappa \neq t_k$, $\kappa = i,l$; $k = \pm 1, \pm 2, \ldots$.

As $x(t+s_i) \in \overline{B}_{q_1}$ for large s_i and for each compact subset of $\mathbb{R}$ there exists $n_1(\varepsilon) > 0$ such that if $l \geq i \geq n_1(\varepsilon)$, then

$$||f(t+s_i,x(t+s_i), \int_{t_0}^{t+s_i} \mathscr{K}(t+s_i,\tau,x(\tau))d\tau$$

$$-f(t+s_l,x(t+s_i), \int_{t_0}^{t+s_l} \mathscr{K}(t+s_l,\tau,x(\tau))d\tau)|| < \frac{\varepsilon}{2}.$$

Since $x(t+s_\kappa) \in B_{\gamma(q)}$, $\kappa = i,l$, then it follows that there exists $n_2(\varepsilon) > 0$ such that if $l \geq i \geq n_2(\varepsilon)$, then

$$||f(t+s_l,x(t+s_i), \int_{t_0}^{t+s_l} \mathscr{K}(t+s_l,\tau,x(\tau))d\tau$$

$$-f(t+s_l,x(t+s_l), \int_{t_0}^{t+s_l} \mathscr{K}(t+s_l,\tau,x(\tau))d\tau)|| < \frac{\varepsilon}{2}.$$

For $l \geq i \geq n(\varepsilon)$, $n(\varepsilon) = \max\{n_1(\varepsilon), n_2(\varepsilon)\}$, we obtain

$$||{}_{t_0}^{c}D_t^{\alpha}x(t+s_i) - {}_{t_0}^{c}D_t^{\alpha}x(t+s_l)|| \leq \varepsilon,$$

where $t+s_\kappa \neq t_k^s$, which shows that $\lim_{i \to \infty} {}_{t_0}^{c}D_t^{\alpha}x(t+s_i)$ exists uniformly on all compact subsets of $\mathbb{R}$.

Let now $\lim_{i \to \infty} {}_{t_0}^{c}D_t^{\alpha}x(t+s_i) = {}_{t_0}^{c}D_t^{\alpha}\tilde{x}(t)$, and

$$_{t_0}^{c}D_t^{\alpha}\tilde{x}(t) = \lim_{i \to \infty} \left[f(t+s_i,x(t+s_i), \int_{t_0}^{t+s_i} \mathscr{K}(t+s_i,\tau,x(\tau))d\tau\right]$$

$$-f(t+s_i, \tilde{x}(t), \int_{t_0}^{t+s_l} \mathscr{K}(t+s_i, \tau, \tilde{x}(\tau))d\tau)$$

$$+f(t+s_i, \tilde{x}(t), \int_{t_0}^{t+s_p} \mathscr{K}(t+s_i, \tau, \tilde{x}(\tau))d\tau)]$$

$$= f^s(t, \tilde{x}(t), \int_{t_0}^{t} K^s(t, \tau, \tilde{x}(\tau))d\tau), \tag{3.45}$$

where $t \neq t_k^s$, $t_k^s = \lim_{i \to \infty} t_{k+i(s)}$.

On the other hand, for $t + s_i = t_k^s$ it follows

$$\tilde{x}(t_k^{s+}) - \tilde{x}(t_k^{s-}) = \lim_{i \to \infty}(x(t_k^s + s_i + 0) - x(t_k^s + s_i - 0))$$

$$= \lim_{i \to \infty} I_k^s(x(t_k^s + s_i)) = I_k^s(\tilde{x}(t_k^s)). \tag{3.46}$$

From (3.45) and (3.46), we get that $\tilde{x}(t)$ is a solution of (3.40).

We shall show that $\tilde{x}(t)$ is an almost periodic function.

Let the sequence $\{s_i\}$ moves the system (3.39) to $H(f, \mathscr{K}, I_k, t_k)$. For any $\varepsilon > 0$ there exists $m_0(\varepsilon) > 0$ such that if $l \geq i \geq m_0(\varepsilon)$, then

$$E_\alpha(-w(s_i)^\alpha)b(2q_1) < \frac{w_1(\varepsilon)}{4},$$

and

$$\|f(\sigma+s_i, x(\sigma+s_i), \int_{t_0}^{\sigma+s_i} \mathscr{K}(\sigma+s_i, \tau, x(\tau))d\tau)$$

$$-f(\sigma+s_l, x(\sigma+s_l), \int_{t_0}^{\sigma+s_l} \mathscr{K}(\sigma+s_l, \tau, x(\tau))d\tau)\| < \frac{w_1(\varepsilon)}{4L_V}.$$

For each fixed $t \in \mathbb{R}$ let $\tau_\varepsilon = \frac{w_1(\varepsilon)}{4L_V}$ be a translation number of f such that $t + \tau_\varepsilon \geq 0$.

Consider the function

$$V(\tau_\varepsilon + \sigma, \tilde{x}(\sigma), \tilde{x}(\sigma+s_l - s_i)),$$

where $t \leq \sigma \leq t + s_i$.

Then the same way like above, we have

$${}^cD_+^\alpha V(\tau_\varepsilon + \sigma, \tilde{x}(\sigma), \tilde{x}(\sigma + s_l - s_i))$$

$$\leq -wV(\tau_\varepsilon + \sigma, \tilde{x}(\sigma), \tilde{x}(\sigma + s_l - s_i))$$

$$+L_V\|f^s(\sigma, \tilde{x}(\sigma), \int_{t_0}^{\sigma} \mathscr{K}^s(\sigma, \tau, \tilde{x}(\tau))d\tau)$$

$$-f^s(\tau_\varepsilon + \sigma, \tilde{x}(\sigma), \int_{t_0}^{\tau_\varepsilon + \sigma} \mathscr{K}^s(\tau_\varepsilon + \sigma, \tau, \tilde{x}(\tau))d\tau)\|$$

$$+L_V||f^s(\sigma+s_l-s_i,\tilde{x}(\sigma+s_l-s_i),\int_{t_0}^{\sigma+s_l-s_i}\mathscr{K}^s(\sigma+s_l-s_i,\tau,\tilde{x}(\tau))d\tau)$$

$$-f^s(\tau_\varepsilon+\sigma,\tilde{x}(\sigma+s_l-s_i),\int_{t_0}^{\tau_\varepsilon+\sigma}\mathscr{K}^s(\tau_\varepsilon+\sigma,\tau,\tilde{x}(\tau))d\tau)||$$

$$\leq -wV(\tau_\varepsilon+\sigma,\tilde{x}(\sigma),\tilde{x}(\sigma+s_l-s_i))+\frac{3w_1(\varepsilon)}{4}. \tag{3.47}$$

On the other hand,

$$V(\tau_\varepsilon+t_k^s,\tilde{x}(t_k^s)+I_k^s(\tilde{x}(t_k^s)),\tilde{x}(t_k^s+s_l-s_i)+I_k^s(\tilde{x}(t_k^s+s_l-s_i)))$$

$$\leq V(\tau_\varepsilon+t_k^s,\tilde{x}(t_k^s),\tilde{x}(t_k^s+s_l-s_i)). \tag{3.48}$$

From (3.47), (3.48) and Corollary 3.1 it follows

$$V(\tau_\varepsilon+t+s_i,\tilde{x}(t+s_i),\tilde{x}(t+s_l))$$

$$\leq E_\alpha(-w(s_i)^\alpha)V(\tau_\varepsilon+t,\tilde{x}(t),\tilde{x}(t+s_i-s_l))+\frac{3w_1(\varepsilon)}{4}<w_1(\varepsilon). \tag{3.49}$$

Then from (3.49) for $l\geq i\geq m_0(\varepsilon)$, we have

$$||\tilde{x}(t+s_i)-\tilde{x}(t+s_l)||<\varepsilon. \tag{3.50}$$

Now, from definitions of the sequence $\{s_i\}$ and for $l\geq i\geq m_0(\varepsilon)$ it follows that $\rho(t_k+s_i,t_k+s_l)<\varepsilon$.

Then from (3.50) and the last inequality we obtain that the sequence $\tilde{x}(t+s_i)$ is convergent uniformly to the function $\tilde{x}(t)$.

The assertions 1 and 3 of the theorem follow immediately. We shall again prove the assertion 2.

Let $\overline{\omega}(t)$ be an arbitrary solution of (3.40).

Set

$$u(t)=\overline{\omega}(t)-\tilde{x}(t),$$

$$g_{\mathscr{K}}^s(t,u(t),\int_{t_0}^t\mathscr{K}^s(t,\tau,u(\tau))d\tau)$$

$$=f^s(t,u(t)+\tilde{x}(t),\int_{t_0}^t\mathscr{K}^s(t,\tau,u(\tau)+\tilde{x}(\tau))d\tau)$$

$$-f^s(t,\tilde{x}(t),\int_{t_0}^t\mathscr{K}^s(t,\tau,\tilde{x}(\tau))d\tau),$$

$$B_k^s(u)=I_k^s(u+\tilde{x})-I_k^s(\tilde{x}).$$

Now, we consider the system

$$\begin{cases} {}_{t_0}^cD_t^\alpha u(t)=g_{\mathscr{K}}^s(t,u(t)), \ t\neq t_k^s, \\ \Delta u(t_k^s)=B_k^s(u(t_k^s)), \ k=\pm1,\pm2,..., \end{cases} \tag{3.51}$$

and let $W(t,u(t)) = V(t,\tilde{x}(t),\tilde{x}(t)+u(t))$. Then, from Lemma 3.6 it follows that the zero solution $u(t) = 0$ of (3.51) is globally perfectly uniform-asymptotically stable, and consequently $\omega(t)$ is globally perfectly uniform-asymptotically stable.

3.2.3 Impulsive Fractional Functional Differential Systems

Finally, we shall consider the class of impulsive functional differential equations involving Caputo fractional derivatives of the type

$$\begin{cases} {}^c_{t_0}D^\alpha_t x(t) = f(t,x_t),\, t \neq t_k, \\ \Delta x(t) = I_k(x(t)),\, t = t_k,\, k = \pm 1, \pm 2, ..., \end{cases} \tag{3.52}$$

where $f : \mathbb{R} \times \mathscr{PC} \to \mathbb{R}^n$, $I_k : \mathbb{R}^n \to \mathbb{R}^n$, $k = \pm 1, \pm 2, ...$, $\{t_k\} \in \mathscr{B}$.

Let the function $\varphi_0 \in \mathscr{PC}$. Denote by $x(t) = x(t;t_0,\varphi_0)$ the solution of system (3.52), satisfying the initial conditions

$$\begin{cases} x(t;t_0,\varphi_0) = \varphi_0(t-t_0),\, t_0 - r \leq t \leq t_0, \\ x(t_0^+;t_0,\varphi_0) = \varphi_0(0). \end{cases}$$

Introduce the following conditions:

H3.13 The function $f(t,\varphi)$ is almost periodic in $t \in \mathbb{R}$ uniformly with respect to $\varphi \in \mathscr{PC}$.

H3.14 The function $\varphi_0 \in \mathscr{PC}$ is almost periodic.

Let the conditions H3.1, H3.9, H3.13 and H3.14 hold, and let $\{s'_m\}$ be an arbitrary sequence of real numbers. Then, there exists a subsequence $\{s_n\}$, $s_n = s'_{m_n}$ such that the system (3.52) moves to the system

$$\begin{cases} {}^c_{t_0}D^\alpha_t x(t) = f^s(t,x_t),\, t \neq t^s_k, \\ \Delta x(t^s_k) = I^s_k(x(t^s_k)),\, k = \pm 1, \pm 2, ... \end{cases} \tag{3.53}$$

We shall denote the systems in the form (3.53) by $H(f,I_k,t_k)$.

Let $\varphi_1 \in \mathscr{PC}$. Denote by $x_1(t) = x_1(t;t_0,\varphi_1)$ the solution of system (3.52), satisfying the initial conditions

$$\begin{cases} x_1(t;t_0,\varphi_1) = \varphi_1(t-t_0),\, t_0 - r \leq t \leq t_0, \\ x_1(t_0^+;t_0,\varphi_1) = \varphi_1(0). \end{cases}$$

The next two definitions are generalizations of Definition 3.3 and Definition 3.4, respectively. The global quasi-equi-asymptotic stability and global perfect uniform-asymptotic stability for system (3.52) are defined as follows.

Definition 3.6 The solution $x_1(t)$ of system (3.52) is said to be *globally quasi-equi-asymptotically stable*, if

$$(\forall \lambda > 0)(\forall \varepsilon > 0)(\forall t_0 \in \mathbb{R})(\exists T > 0)(\forall \varphi_0 \in \mathscr{PC} : ||\varphi_0 - \varphi_1||_r < \lambda)$$

$$(\forall t \geq t_0 + T) : ||x(t;t_0,\varphi_0) - x_1(t;t_0,\varphi_1)|| < \varepsilon.$$

Definition 3.7 The solution $x_1(t)$ of (3.52) is said to be *globally perfectly uniform-asymptotically stable*, if it is uniformly stable, the number T in Definition 3.6 is independent of $t_0 \in \mathbb{R}$, and the solutions of (3.52) are uniformly bounded.

For the system (3.52) we shall use the following comparison results for a function from the class V_2 which is similar to Lemma 1.6.

Lemma 3.7 Assume that:
1. Conditions 1–3 of Lemma 3.6 are met.
2. The function $V \in V_2$ is such that for $t \in [t_0,\infty)$, $\varphi,\phi \in \mathscr{PC}$,

$$V(t^+,\varphi(0)+I_k(\varphi),\phi(0)+I_k(\phi)) \leq \tilde{\psi}_k(V(t,\varphi(0),\phi(0))), \ t=t_k, \ t_k > t_0$$

and the inequality

$$^cD_+^\alpha V(t,\varphi(0),\phi(0)) \leq g(t,V(t,\varphi(0),\phi(0))), t \neq t_k, t \in [t_0,\infty)$$

is valid whenever $V(t+\theta,\varphi(\theta),\phi(\theta)) \leq V(t,\varphi(0),\phi(0))$ for $-r \leq \theta \leq 0$. Then $\sup_{-r\leq\theta\leq0} V(t_0+\theta,\varphi_0(\theta),\phi_0(\theta)) \leq u_0$ implies

$$V(t,x(t;t_0,\varphi_0),y(t;t_0,\varphi_0)) \leq u^+(t;t_0,u_0), t \in [t_0,\infty),$$

where $x(t;t_0,\varphi_0),y(t;t_0,\varphi_0)$ are two solutions of (3.52), $\varphi_0,\phi_0 \in \mathscr{PC}$, existing on $[t_0,\infty)$.

In the case when $g(t,u) = qu$ for $(t,u) \in \mathbb{R} \times \mathbb{R}_+$, where $q \in \mathbb{R}$ is a constant, and $\tilde{\psi}_k(u) = u$ for $u \in \mathbb{R}_+$, $k = \pm1,\pm2,...$, the following corollary holds.

Corollary 3.2 Assume that the function $V \in V_2$ is such that for $t \in [t_0,\infty)$, $\varphi,\phi \in \mathscr{PC}$,

$$V(t^+,\varphi(0)+I_k(\varphi),\phi(0)+I_k(\phi)) \leq V(t,\varphi(0),\phi(0)), \ t=t_k, \ t_k > t_0$$

and the inequality

$$^cD_+^\alpha V(t,\varphi(0),\phi(0)) \leq qV(t,\varphi(0),\phi(0)), t \neq t_k, t \in [t_0,\infty)$$

is valid whenever $V(t+\theta,\varphi(\theta),\phi(\theta)) \leq V(t,\varphi(0),\phi(0))$ for $-r \leq \theta \leq 0$. Then $\sup_{-r\leq\theta\leq0} V(t_0+\theta,\varphi_0(\theta),\phi_0(\theta)) \leq u_0$ implies

$$V(t,x(t;t_0,\varphi_0),y(t;t_0,\varphi_0)) \leq \sup_{-r\leq\theta\leq0} V(t_0^+,\varphi_0(\theta),\phi_0(\theta))E_\alpha(q(t-t_0)^\alpha), t \in [t_0,\infty).$$

Theorem 3.5 Let the following conditions hold.

1. Conditions H3.1, H3.9, H3.13 and H3.14 are met.
2. There exist functions $V \in V_2$ and $w_1, w_2 \in K$ such that H3.10 is true,

$$w_1(||x-y||) \leq V(t,x,y) \leq w_2(||x-y||), \quad (t,x,y) \in \mathbb{R} \times \mathbb{R}^n \times \mathbb{R}^n, \quad (3.54)$$

and $w_1(u) \to \infty$ as $u \to \infty$,

$$V(t^+, \varphi(0) + I_k(\varphi), \phi(0) + I_k(\phi)) \leq V(t, \varphi(0), \phi(0)), \quad (3.55)$$

$t = t_k$, $k = \pm 1, \pm 2, \ldots$, and the inequality

$$^cD_+^\alpha V(t, \varphi(0), \phi(0)) \leq -wV(t, \varphi(0), \phi(0)), \quad (3.56)$$

$w > 0$, $t \neq t_k$, $k = \pm 1, \pm 2, \ldots$, is valid whenever $V(t + \theta, \varphi(\theta), \phi(\theta)) \leq V(t, \varphi(0), \phi(0))$ for $-r \leq \theta \leq 0$, $t \in \mathbb{R}$, $\varphi, \phi \in \mathscr{PC}$.

3. There exists a solution $x(t; t_0, \varphi_0)$ of (3.52) such that

$$||x(t; t_0, \varphi_0)|| < q,$$

where $t \geq t_0$, $q > 0$.

Then for the system (3.52) there exists a unique almost periodic solution $\tilde{x}(t)$ such that:

1. $||\tilde{x}(t)|| \leq q_1$, $q_1 < q$;
2. $\tilde{x}(t)$ is globally perfectly uniform-asymptotically stable;
3. $H(\tilde{x}, t_k) \subset H(f, I_k, t_k)$.

Proof. Let $t_0 \in \mathbb{R}$ and let $\{s_i\}$ be any sequence of real numbers such that $s_i \to \infty$ as $i \to \infty$ and $\{s_i\}$ moves the system (3.52) to a system at $H(f, I_k, t_k)$.

For any real number γ, let $i_0 = i_0(\gamma)$ be the smallest value of i, such that $s_{i_0} + \gamma \geq t_0$. Since $||x(t; t_0, \varphi_0)|| \leq q_1$, $q_1 < q$ for all $t \geq t_0$, then $||x(t + s_i; t_0, \varphi_0)|| \leq q_1$ for $t \geq \gamma$, $i \geq i_0$.

Let U, $U \subset (\gamma, \infty)$ be compact. Since for the function $w_1 \in K$ we have $w_1(u) \to \infty$ as $u \to \infty$, then for any $\varepsilon > 0$, we can choose an integer $n_0(\varepsilon, \gamma) \geq i_0(\gamma)$, so large that for $l \geq i \geq n_0(\varepsilon, \gamma)$ and $t \in \mathbb{R}$, $t \neq t_k$, $k = \pm 1, \pm 2, \ldots$, it follows $V(t + \theta, \varphi(\theta + s_i), \varphi(\theta + s_l)) \leq V(t, \varphi(0), \varphi(0))$ for $-r \leq \theta \leq 0$, $\varphi \in \mathscr{PC}$, and

$$w_2(2q_1)E_\alpha\left(-w(\gamma + s_i - t_0)^\alpha\right) \leq \frac{w_1(\varepsilon)}{2}, \quad (3.57)$$

$$||f(t + s_l, x_t) - f(t + s_i, x_t)|| \leq \frac{w_1(\varepsilon)w}{2L_V}, \quad (3.58)$$

where E_α is the corresponding Mittag-Leffler function.

Now, we shall again set

$$m(\sigma) = V(\sigma, \varphi(0), \varphi(s_i - s_l))$$

and similar to the proof of Theorem 3.3, we have

$$V(\sigma, \varphi(0), \varphi(s_i - s_l)) - V(\sigma - \chi, S_1(\varphi), S_2(\varphi))$$
$$= V(\sigma, \varphi(0), \varphi(s_i - s_l))$$
$$-V(\sigma - \chi, \varphi(0) - \chi^\alpha f(\sigma, \varphi(\sigma)), \varphi(s_i - s_l) - \chi^\alpha f(\sigma + s_i - s_l, \varphi(\sigma + s_i - s_l)))$$
$$+V(\sigma - \chi, \varphi(0) - \chi^\alpha f(\sigma, \varphi(\sigma)), \varphi(s_i - s_l) - \chi^\alpha f(\sigma + s_i - s_l, \varphi(\sigma + s_i - s_l)))$$
$$-V(\sigma - \chi, S_1(\varphi), S_2(\varphi)),$$

where $S_1(\varphi) = \sum_{h=1}^n (-1)^{h+1} \binom{\alpha}{h} \varphi(-h\chi)$ and $S_2(\varphi) = \sum_{h=1}^n (-1)^{h+1} \binom{\alpha}{h} \varphi(s_i - s_l - h\chi)$.

Then, by using the definition of the fractional order derivative of V in the Caputo sense, for small $\chi > 0$, we have

$$^cD_+^\alpha m(\sigma) = \lim_{\substack{\chi \to 0 \\ n\chi = t - t_0}} \sup \frac{1}{\chi^\alpha} \Big[V(\sigma, \varphi(0), \varphi(s_i - s_l))$$

$$-V(\sigma - \chi, \varphi(0) - \chi^\alpha f(\sigma, \varphi(\sigma)), \varphi(s_i - s_l) - \chi^\alpha f(\sigma + s_i - s_l, \varphi(\sigma + s_i - s_l)))$$
$$+V(\sigma - \chi, \varphi(0) - \chi^\alpha f(\sigma, \varphi(\sigma)), \varphi(s_i - s_l) - \chi^\alpha f(\sigma + s_i - s_l, \varphi(\sigma + s_i - s_l)))$$
$$-V(\sigma - \chi, S_1(\varphi), S_2(\varphi)) \Big]$$

$$\leq -wV(\sigma, \varphi(0), \varphi(s_i - s_l)) + L_V \| f(\sigma + s_i - s_l, x_{\sigma + s_i - s_l}) - f(\sigma, x_\sigma) \|.$$

Set $\zeta = \sigma - s_i$. Then,

$$f(\sigma + s_l - s_i, x_{\sigma + s_l - s_i}) - f(\sigma, x_\sigma)$$
$$= f(\zeta + s_l, x_{\zeta + s_l}) - f(\zeta + s_i, x_{\zeta + s_i})$$

and by (3.57) we get

$$^cD_+^\alpha V(\sigma, \varphi(0), \varphi(s_i - s_l)) \leq -wV(\sigma, \varphi(0), \varphi(s_i - s_l)) + \frac{w_1(\varepsilon)w}{2}. \quad (3.59)$$

Now for $\sigma = t_k - (s_l - s_i)$ and (3.55), we have

$$V(\sigma, \varphi(0) + I_k(\varphi(\sigma)), \varphi(\sigma + s_l - s_i) + I_k(\varphi(\sigma + s_l - s_i)))$$
$$\leq V(\sigma, \varphi(0), \varphi(\sigma + s_l - s_i)). \quad (3.60)$$

Then, from (3.57) and (3.60), by Corollary 3.2, it follows

$$V(t + s_i, x(t + s_i), x(t + s_l))$$

$$\leq E_\alpha(-w(t + s_i - s_l)^\alpha) \sup_{-r \leq \theta \leq 0} V(t_0^+, x(\theta + s_i), x(\theta + s_l)) + \frac{w_1(\varepsilon)}{2} \leq w_1(\varepsilon)$$

for $i, l \geq n_0(\varepsilon, \gamma)$ and any $t \in U$.

Finally, from (3.54) and (3.55) it follows that for any $t \in U$, we get

$$||x(t+s_i) - x(t+s_l)|| < \varepsilon.$$

Consequently, there exists a function $\tilde{x}(t)$, such that $x(t+s_i) - \tilde{x}(t) \to \infty$ for $i \to \infty$. Since γ is arbitrary, it follows that $\tilde{x}(t)$ is defined uniformly on $t \in \mathbb{R}$.

Next, we shall show that $\tilde{x}(t)$ is a solution of (3.53).

Since $x(t;t_0,\varphi_0)$ is a solution of (3.52) with appropriate initial data, we have

$$||{}^c_{t_0}D^\alpha_t x(t+s_i) - {}^c_{t_0}D^\alpha_t x(t+s_l)|| \le ||f(t+s_i,x_{t+s_i}) - f(t+s_l,x_{t+s_i})||$$

$$+ ||f(t+s_l,x_{t+s_i}) - f(t+s_l,x_{t+s_l})||,$$

for $t + s_\kappa \ne t_k$, $\kappa = i,l$ and $k = \pm 1, \pm 2, \ldots$.

As $x(t+s_i) \in \bar{B}_{q_1}$ for large s_i for each compact subset of $\mathbb{R}$ there exists an $n_1(\varepsilon) > 0$ such that if $l \ge i \ge n_1(\varepsilon)$, then

$$||f(t+s_i,x_{t+s_i}) - f(t+s_l,x_{t+s_i})|| < \frac{\varepsilon}{2}.$$

Since $x(t+s_\kappa) \in B_{\gamma(q)}$, $\kappa = i,l$, then from Lemma 3.5 it follows that there exists $n_2(\varepsilon) > 0$ such that if $l \ge i \ge n_2(\varepsilon)$, then

$$||f(t+s_l,x(t+s_i)) - f(t+s_l,x(t+s_l))|| < \frac{\varepsilon}{2}.$$

For $l \ge i \ge n(\varepsilon)$, $n(\varepsilon) = \max\{n_1(\varepsilon), n_2(\varepsilon)\}$, we obtain

$$||{}^c_{t_0}D^\alpha_t x(t+s_i) - {}^c_{t_0}D^\alpha_t x(t+s_l)|| \le \varepsilon,$$

where $t + s_\kappa \ne t^s_k$ which shows that $\lim_{i\to\infty} {}^c_{t_0}D^\alpha_t x(t+s_i)$ exists uniformly on all compact subsets of $\mathbb{R}$.

Let now $\lim_{i\to\infty} {}^c_{t_0}D^\alpha_t x(t+s_i) = {}^c_{t_0}D^\alpha_t \tilde{x}(t)$, and

$${}^c_{t_0}D^\alpha_t \tilde{x}(t) = \lim_{i\to\infty} \left[f(t+s_i,x(t+s_i)) - f(t+s_i,\tilde{x}(t)) + f(t+s_i,\tilde{x}(t)) \right]$$

$$= f^s(t,\tilde{x}(t)), \tag{3.61}$$

where $t \ne t^s_k$, $t^s_k = \lim_{i\to\infty} t_{k+i(s)}$.

On the other hand, for $t + s_i = t^s_k$ it follows

$$\tilde{x}(t^{s+}_k) - \tilde{x}(t^{s-}_k) = \lim_{i\to\infty}(x(t^s_k + s_i + 0) - x(t^s_k + s_i - 0))$$

$$= \lim_{i\to\infty} I^s_k(x(t^s_k + s_i)) = I^s_k(\tilde{x}(t^s_k)). \tag{3.62}$$

From (3.61) and (3.62), we get that $\tilde{x}(t)$ is a solution of (3.53).

We shall show that $\tilde{x}(t)$ is an almost periodic function.

Let the sequence $\{s_i\}$ moves the system (3.52) to $H(f, I_k, t_k)$. For any $\varepsilon > 0$ there exists $m_0(\varepsilon) > 0$ such that if $l \geq i \geq m_0(\varepsilon)$, then

$$E_\alpha(-w(s_i)^\alpha)w_2(2q_1) < \frac{w_1(\varepsilon)}{4},$$

and

$$\|f(\sigma + s_i, x_{\sigma + s_i}) - f(\sigma + s_l, x_{\sigma + s_l})\| < \frac{w_1(\varepsilon)}{4L_V},$$

where $x \in \mathscr{PC}$, $w > 0$.

For each fixed $t \in \mathbb{R}$ let $\tau_\varepsilon = \frac{w_1(\varepsilon)}{4L_V}$ be a translation number of f such that $t + \tau_\varepsilon \geq 0$.

Consider the function

$$V(\tau_\varepsilon + \sigma, \varphi(0), \varphi(s_i - s_l)),$$

where $t \leq \sigma \leq t + s_i$.

Then, in the same way as above, we get

$$^cD_+^\alpha V(\tau_\varepsilon + \sigma, \varphi(0), \varphi(s_i - s_l))$$

$$\leq -wV(\tau_\varepsilon + \sigma, \varphi(0), \varphi(s_i - s_l)) + L_V\|f^s(\sigma, \tilde{x}(\sigma)) - f^s(\tau_\varepsilon + \sigma, \tilde{x}(\sigma))\|$$

$$+ L_V\|f^s(\sigma + s_l - s_i, \tilde{x}(\sigma + s_l - s_i)) - f^s(\tau_\varepsilon + \sigma, \tilde{x}(\sigma + s_l - s_i))\|$$

$$\leq -wV(\tau_\varepsilon + \sigma, \varphi(0), \varphi(s_i - s_l)) + \frac{3w_1(\varepsilon)}{4}. \tag{3.63}$$

On the other hand,

$$V(\tau_\varepsilon + t_k^s, \varphi(0) + I_k^s(\varphi(t_k^s)), \varphi(t_k^s + s_l - s_i) + I_k^s(\varphi(t_k^s + s_l - s_i)))$$

$$\leq V(\tau_\varepsilon + t_k^s, \varphi(0), \varphi(s_l - s_i)). \tag{3.64}$$

From (3.63), (3.64) and Corollary 3.2 it follows

$$V(\tau_\varepsilon \mid t \mid s_i, \tilde{x}(t \mid s_l), \tilde{x}(t \mid s_l))$$

$$\leq E_\alpha(-w(s_i)^\alpha) \sup_{-r \leq \theta \leq 0} V(\tau_\varepsilon + t, \tilde{x}(t + \theta), \tilde{x}(t + \theta + s_i - s_l)) + \frac{3w_1(\varepsilon)}{4}$$

$$< w_1(\varepsilon). \tag{3.65}$$

Then from (3.65) for $l \geq i \geq m_0(\varepsilon)$, we have

$$\|\tilde{x}(t + s_i) - \tilde{x}(t + s_l)\| < \varepsilon. \tag{3.66}$$

Now, from definitions of the sequence $\{s_i\}$ and for $l \geq i \geq m_0(\varepsilon)$ it follows that $\rho(t_k + s_i, t_k + s_l) < \varepsilon$.

Then from (3.66) and the last inequality we obtain that the sequence $\tilde{x}(t + s_i)$ is convergent uniformly to the function $\tilde{x}(t)$.

The assertions 1 and 3 of the theorem follow immediately. We shall prove the assertion 2.

Let $v(t)$ be an arbitrary solution of (3.53). Set

$$u(t) = v(t) - \tilde{x}(t),$$

$$g^s(t, u(t)) = f^s(t, u(t) + \tilde{x}(t)) - f^s(t, \tilde{x}(t)),$$

$$B_k^s(u) = I_k^s(u + \tilde{x}) - I_k^s(\tilde{x}).$$

Now we consider the system

$$\begin{cases} {}_{t_0}^c D^\alpha u(t) = g^s(t, u(t)), \ t \neq t_k^s, \\ \Delta u(t_k^s) = B_k^s(u(t_k^s)), \ k = \pm 1, \pm 2, ..., \end{cases} \quad (3.67)$$

and let $W(t, u(t)) = V(t, \tilde{x}(t), \tilde{x}(t) + u(t))$. Then, from Lemma 3.7 it follows that the zero solution $u(t) = 0$ of (3.67) is globally perfectly uniform-asymptotically stable, and consequently $\tilde{x}(t)$ is globally perfectly uniform-asymptotically stable.

3.3 Uncertain Fractional Differential Systems

Uncertain differential equations, proposed by Liu (see [Liu 2007]), are differential equations driven by uncertain processes. Indeed, due to inaccuracy in the model parameter measurements, data input, and any kind of unpredictability, a real system always involves uncertainties. Quantifying the impact of uncertainties in physical systems has received considerable attention during the last decades. There are many publications devoted to different aspects of treatments of uncertain differential systems of integer order, including qualitative theory (see [Liu 2012], [Liu, Liu and Liao 2004], [Peng and Yao 2011], [Stamov 2009c], [Stamov and Alzabut 2011], [Stamov and Stamov 2013]). The goal of this section is to extend the theory to the fractional-order systems under uncertainty.

Let $t_0 \in \mathbb{R}$. Consider the following system of impulsive fractional differential equations

$$\begin{cases} {}_{t_0}^c D_t^\alpha x(t) = f(t, x(t)) + g(t, x(t)), \ t \neq t_k, \\ \Delta x(t_k) = I_k(x(t_k)) + J_k(x(t_k)), \ k = \pm 1, \pm 2, ..., \end{cases} \quad (3.68)$$

where $f, g : \mathbb{R} \times \mathbb{R}^n \to \mathbb{R}^n$, $0 < \alpha < 1$, $\{t_k\} \in \mathcal{B}$, $\Delta x(t_k) = x(t_k^+) - x(t_k)$, $I_k, J_k : \mathbb{R}^n \to \mathbb{R}^n$, $k = \pm 1, \pm 2, ...$.

The function $g(t, x)$ represents a structural uncertainty and functions $J_k(x)$, $k = \pm 1, \pm 2, ...$ represent uncertain perturbations in the system (3.68), and are characterized by

$$g \in U_g = \left\{ g : g(t, x) = e_g(t, x) . \delta_g(t, x), \ ||\delta_g(t, x)|| \leq ||m_g(t, x)|| \right\},$$

and

$$J_k \in U_J = \left\{ J_k : J_k(x) = e_k(x).\delta_k(x), \ ||\delta_k(x)|| \leq ||m_k(x)|| \right\}, \ k = \pm 1, \pm 2, ...,$$

where $e_g : \mathbb{R} \times \mathbb{R}^n \to \mathbb{R}^{n \times m}$, and $e_k : \mathbb{R}^n \to \mathbb{R}^{n \times m}$ are known matrix functions, whose entries are smooth functions, and δ_g, δ_k are unknown vector-valued functions, whose norms are bounded by the norms of the vector-valued functions $m_g(t, x)$, $m_k(x)$, respectively. Here $m_g : \mathbb{R} \times \mathbb{R}^n \to \mathbb{R}^m$, $m_k : \mathbb{R}^n \to \mathbb{R}^m$, $k = \pm 1, \pm 2, ...$ are given functions.

Let $x_0 \in \mathbb{R}^n$. Denote by $x(t) = x(t; t_0, x_0)$ the solution of system (3.68), satisfying the initial condition

$$x(t_0^+; t_0, x_0) = x_0. \tag{3.69}$$

Note that the system (3.22) is the nominal system for system (3.68).

Introduce the following conditions:

H3.15 The functions $f(t, x)$ and $e_g(t, x)$ are almost periodic in t uniformly with respect to $x \in \mathbb{R}^n$.

H3.16 The sequences $\{I_k(x)\}$ and $\{e_k(x)\}$, $k = \pm 1, \pm 2, ...$ are almost periodic uniformly with respect to $x \in \mathbb{R}^n$.

Definition 3.8 The almost periodic solution of the uncertain impulsive dynamical system (3.68) is said to be *globally perfectly robustly uniform-asymptotically stable*, if for any $g \in U_g$, $J_k \in U_J$, $k = \pm 1, \pm 2, ...$ the almost periodic solution of (3.68) is globally perfectly uniform-asymptotically stable.

Definition 3.9 ([Liu, Liu and Liao 2004]) The matrix function $X : \mathbb{R} \to \mathbb{R}^{n \times n}$ is said to be:

(a) *a positive definite matrix function*, if for any $t \in \mathbb{R}$, $X(t)$ is a positive definite matrix;

(b) *a positive definite matrix function bounded above*, if it is a positive definite matrix function, and there exists a positive real number $M > 0$ such that

$$\lambda_{\max}(X(t)) \leq M, \ t \in \mathbb{R},$$

where $\lambda_{\max}(X(t))$ is the maximum eigenvalue;

(c) *a uniformly positive definite matrix function*, if it is a positive definite matrix function, and there exists a positive real number $m > 0$ such that

$$\lambda_{\min}(X(t)) \geq m, \ t \in \mathbb{R},$$

where $\lambda_{\min}(X(t))$ is the minimum eigenvalue.

The proof of the following lemma is obvious.

Lemma 3.8 Let $X(t)$ be a positive definite matrix function and $Y(t)$ be a symmetric matrix.

Then for any $x \in \mathbb{R}^n$ and $t \in \mathbb{R}$ the following inequality holds

$$x^T Y(t)x \le \lambda_{\max}(X^{-1}(t)Y(t))x^T X(t)x.$$

We shall use the next lemma.

Lemma 3.9 ([Liu, Liu and Liao 2004]) Let $\Sigma(t)$ be a diagonal matrix function.

Then for any positive scalar function $\lambda(t)$ and for any $\xi, \eta \in \mathbb{R}^n$, the following inequality holds

$$2\xi^T \Sigma(t)\eta \le \lambda^{-1}(t)\xi^T \xi + \lambda(t)\eta^T \eta.$$

Now we shall prove the main theorem in this section.

Theorem 3.6 Assume that:

1. Conditions H3.1, H3.15 and H3.16 are met.
2. For system (3.68) there exists a function $V \in V_2$ such that H3.10, (3.25) hold, and the following relations are satisfied:

   (i) there exist $G_{1k} : \mathbb{R} \times \mathbb{R}^n \times \mathbb{R}^n \to \mathbb{R}^{1 \times m}$, $G_{2k} : \mathbb{R} \times \mathbb{R}^n \times \mathbb{R}^n \to \mathbb{R}^{m \times m}$, where G_{2k}, $k = \pm 1, \pm 2, \dots$ are positive definite matrix functions and for $x, y \in PC[\mathbb{R}, \mathbb{R}^n]$, $z \in \mathbb{R}^m$ it follows

   $$V(t, x(t) + I_k(x(t)) + e_k(x(t))z, y(t) + I_k(y(t)) + e_k(y(t))z)$$

   $$\le V(t, x(t) + I_k(x(t)), y(t) + I_k(y(t)))$$

   $$+ G_{1k}(t, x(t), y(t))z + z^T G_{2k}(t, x(t), y(t))z; \qquad (3.70)$$

   (ii) there exist positive constants χ_k, $k = \pm 1, \pm 2, \dots$ such that

   $$V(t_k^+, x(t_k) + I_k(x(t_k)), y(t_k) + I_k(y(t_k)))$$

   $$+ \chi_k^{-1} G_{1k} G_{1k}^T + (\chi_k + \lambda_{max}(G_{2k}))m_k^T m_k$$

   $$\le V(t_k, x(t_k), y(t_k)), \qquad (3.71)$$

   where $G_{1k} = G_{1k}(t_k, x(t_k), y(t_k))$, $G_{2k} = G_{2k}(t_k, x(t_k), y(t_k))$, $m_k = m_k(x(t_k))$;

   (iii) there exists a positive constant $\gamma < w$ such that for $x, y \in \mathbb{R}^n$ it follows

   $$L_V \left(||e_g(t, x)|| \, ||m_g(t, x)|| + ||e_g(t, y)|| \, ||m_g(t, y)|| \right) \le (w - \gamma)V(t, x, y). \qquad (3.72)$$

3. There exists a solution $x(t;t_0,x_0)$ of (3.68) such that

$$||x(t;t_0,x_0)|| < q,$$

where $t \geq t_0$, $q > 0$.

Then for the system (3.68) there exists a unique almost periodic solution $\tilde{x}(t)$ such that:

1. $||\tilde{x}(t)|| \leq q$, $q > 0$;
2. $\tilde{x}(t)$ globally perfectly robustly uniform-asymptotically stable.

Proof. From (3.70), (3.71), Lemma 3.8 and Lemma 3.9 for $k = \pm1,\pm2,...$, we have

$$V(t_k^+,x(t_k) + I_k(x(t_k)) + J_k(x(t_k)),y(t_k) + I_k(y(t_k)) + J_k(y(t_k)))$$

$$\leq V(t_k^+,x(t_k) + I_k(x(t_k)),y(t_k) + I_k(y(t_k))) + G_{1k}\delta(x(t_k)) + \delta(x(t_k))^T G_{2k}\delta(x(t_k))$$

$$\leq V(t_k^+,x(t_k) + I_k(x(t_k)),y(t_k) + I_k(y(t_k))) + \chi_k^{-1}G_{1k}G_{1k}^T + (\chi_k + \lambda_{max}(G_{2k}))m_k^T m_k$$

$$\leq V(t_k,x(t_k),y(t_k)). \tag{3.73}$$

On the other hand, for $t \neq t_k$, $k = \pm1,\pm2,...$, and $x,y \in \mathbb{R}^n$ in view of Theorem 3.3 and (3.72), we get

$$^cD_+^\alpha V(t,x,y) = \lim_{\chi \to 0^+} \sup \frac{1}{\chi^\alpha}\left[V(t,x,y)\right.$$

$$-V(t - \chi,x - \chi^\alpha(f(t,x) + g(t,x)),y - \chi^\beta(f(t,y) + g(t,y)))$$

$$-V(t - \chi,x - \chi^\alpha f(t,x),y - \chi^\beta f(t,y))$$

$$\left.+V(t - \chi,x - \chi^\alpha f(t,x),y - \chi^\beta f(t,y))\right]$$

$$\leq -wV(t,x,y) + L_V(||g(t,x)|| + ||g(t,y)||)$$

$$\leq -wV(t,x,y) + L_V\left(||e_g(t,x)||||m_g(t,x)|| + ||e_g(t,y)||||m_g(t,y)||\right)$$

$$\leq -\gamma V(t,x,y). \tag{3.74}$$

Therefore, in the light of inequalities (3.73) and (3.74) and condition 1 of the theorem it follows that the conditions of Theorem 3.3 are satisfied for the system (3.68). Hence, the conclusions are true.

Next, we shall consider impulsive functional differential systems of fractional order.

Let $r > 0$ and for $t \in \mathbb{R}$, $x_t(\theta) = x(t + \theta)$, $-r \leq \theta \leq 0$. Consider the following system of impulsive fractional functional differential equations

$$\begin{cases} {}_{t_0}^c D_t^\alpha x(t) = f(t,x_t) + g(t,x_t), \ t \neq t_k, \\ \Delta x(t_k) = I_k(x(t_k)) + J_k(x(t_k)), \ k = \pm1,\pm2,..., \end{cases} \tag{3.75}$$

where $f, g : \mathbb{R} \times \mathscr{PC} \to \mathbb{R}^n$, $0 < \alpha < 1$, $\{t_k\} \in \mathscr{B}$, $\Delta x(t_k) = x(t_k^+) - x(t_k)$, $I_k, J_k :$ $\mathbb{R}^n \to \mathbb{R}^n$, $k = \pm 1, \pm 2, \dots$.

The functions $g(t, \varphi)$, $\varphi \in \mathscr{PC}$ and $J_k(x)$, $x \in \mathbb{R}^n$, $k = \pm 1, \pm 2, \dots$ are characterized by

$$g \in U_g = \left\{ g : g(t, \varphi) = e_g(t, \varphi).\delta_g(t, \varphi), \ ||\delta_g(t, \varphi)|| \le ||m_g(t, \varphi)|| \right\},$$

and

$$J_k \in U_J = \left\{ J_k : J_k(x) = e_k(x).\delta_k(x), \ ||\delta_k(x)|| \le ||m_k(x)|| \right\}, \ k = \pm 1, \pm 2, \dots,$$

where $e_g : \mathbb{R} \times \mathscr{PC} \to \mathbb{R}^{n \times m}$, and $e_k : \mathbb{R}^n \to \mathbb{R}^{n \times m}$ are known matrix functions, whose entries are smooth functions of the state, and δ_g, δ_k are unknown vector-valued functions, whose norms are bounded by the norms of the vector-valued functions $m_g(t, \varphi), m_k(x)$, respectively. Here, again, $m_g : \mathbb{R} \times \mathscr{PC} \to \mathbb{R}^m$, $m_k : \mathbb{R}^n \to \mathbb{R}^m$, $k = \pm 1, \pm 2, \dots$ are given functions.

Let $\varphi_0 \in \mathscr{PC}$. Denote by $x(t) = x(t; t_0, \varphi_0)$ the solution of system (3.75), satisfying the initial conditions

$$\begin{cases} x(t) = \varphi_0(t - t_0), \ t_0 - r \le t \le t_0, \\ x(t_0^+) = \varphi_0(0), \end{cases}$$

System (3.52) is the nominal system of system (3.75).

We introduce the following condition:

H3.17 The functions $f(t, \varphi)$ and $e_g(t, \varphi)$ are almost periodic in t uniformly with respect to $\varphi \in \mathscr{PC}$ and the function $\varphi_0 \in \mathscr{PC}$ is almost periodic.

The proof of the next theorem is similar to the proof of Theorem 3.6. The Razumikhin technique is used.

Theorem 3.7 Assume that:

1. Conditions H3.1, H3.16 and H3.17 are met.
2. For system (3.75) there exists a function $V \in V_2$ such that H3.10, (3.54) hold, and the following relations are satisfied:
   (i) there exist $G_{1k} : \mathbb{R} \times \mathbb{R}^n \times \mathbb{R}^n \to \mathbb{R}^{1 \times m}$, $G_{2k} : \mathbb{R} \times \mathbb{R}^n \times \mathbb{R}^n \to \mathbb{R}^{m \times m}$, where G_{2k}, $k = \pm 1, \pm 2, \dots$ are positive definite matrix functions and for $x, y \in PC[\mathbb{R}, \mathbb{R}^n]$, $z \in \mathbb{R}^m$ it follows

$$V(t, x(t) + I_k(x(t)) + e_k(x(t))z, y(t) + I_k(y(t)) + e_k(y(t))z)$$

$$\le V(t, x(t) + I_k(x(t)), y(t) + I_k(y(t)))$$

$$+ G_{1k}(t, x(t), y(t))z + z^T G_{2k}(t, x(t), y(t))z;$$

(ii) there exist positive constants χ_k, $k = \pm 1, \pm 2, \ldots$ such that for $\varphi, \psi \in \mathscr{PC}$ it follows

$$V(t^+, \varphi(0) + I_k(\varphi), \psi(0) + I_k(\psi))$$

$$+ \chi_k^{-1} G_{1k} G_{1k}^T + (\chi_k + \lambda_{max}(G_{2k})) m_k^T m_k$$

$$\leq V(t, \varphi(0), \psi(0)), \ t = t_k, \ k = \pm 1, \pm 2, \ldots,$$

where $G_{1k} = G_{1k}(t_k, \varphi(t_k), \psi(t_k))$, $G_{2k} = G_{2k}(t_k, \varphi(t_k), \psi(t_k))$, $m_k = m_k(\varphi(t_k))$;

(iii) there exist a positive constant $\gamma < w$ such that for $\varphi, \psi \in \mathscr{PC}$ it follows

$$L_V \left(\|e_g(t, \varphi)\| \|m_g(t, \varphi)\| + \|e_g(t, \psi)\| \|m_g(t, \psi)\| \right)$$

$$\leq (cw - \gamma) V(t, \varphi(0), \psi(0))$$

is valid whenever $V(t + \theta, \varphi(\theta), \psi(\theta)) \leq V(t, \varphi(0), \psi(0))$ for $-r \leq \theta \leq 0$, $t \in \mathbb{R}$.

3. There exists a solution $x(t; t_0, \varphi_0)$ of (3.75) such that

$$\|x(t; t_0, \varphi_0)\| < q,$$

where $t \geq t_0$, $q > 0$.

Then for the system (3.75) there exists a unique almost periodic solution $\tilde{x}(t)$ such that:

1. $\|\tilde{x}(t)\| \leq q$, $q > 0$;
2. $\tilde{x}(t)$ globally perfectly robustly uniform-asymptotically stable.

3.4 Notes and Comments

The results in Section 3.1 for impulsive differential equations of Caputo fractional order in a Banach space are adapted from Stamov and Stamova (see [Stamov and Stamova 2014a]). Similar techniques have been applied in [Agarwal, Cuevas and Soto 2011], [Debbouche and El-Borai 2009], [El-Borai and Debbouche 2009] to prove the existence, uniqueness and stability of almost periodic and pseudo-almost periodic solutions of fractional differential equations without delays and impulses.

One of the universal methods for investigating qualitative properties of dynamical systems from different types is the second method of Lyapunov. Since the Lyapunov method is very important in the study of the almost periodic behavior of differential systems as well as in their applications, for integer order systems the method has emerged in succession and has been applied by many authors. The use of the method for existence and stability of almost periodic solutions is well established for functional differential systems (see, for example, [Zhou, Zhou and Wang 2011]) and for impulsive differential equations (see [He, Chen and Li 2010], [Stamov 2012]). For some analogous results for impulsive functional differential systems of integer order see [Liu, Huang and Chen 2012], [Stamov 2010b], [Stamov 2012], [Stamov and Alzabut 2011], [Stamov, Alzabut, Atanasov and Stamov 2011]. In Section 3.2, the Lyapunov method is extended and applied in the investigation of the existence and stability of almost periodic solutions for impulsive fractional-order systems with Caputo fractional derivatives. Theorem 3.3 for fractional impulsive differential systems is adapted from [Stamov and Stamova 2015b]. Theorem 3.4 for fractional impulsive integro-differential systems is due to Stamov (see [Stamov 2015]). Theorem 3.5 for fractional impulsive functional differential systems is taken from [Stamov and Stamova 2015a]. For results on the perfect uniform-asymptotic stability for integer order systems see [Lakshmikantham and Leela 1969] and [Yoshizawa 1966].

In Section 3.3, some extensions of the Lyapunov theory for almost periodic solutions of fractional order impulsive systems under uncertainty are discussed. All results of this section are new. For corresponding results for differential equations of integer order we refer to [Liu, Liu and Liao 2004], [Stamov 2012], [Stamov and Alzabut 2011], [Stamov and Stamov 2013]. Some fractional-order uncertain chaotic systems are studied in [Wang, Ding and Qi 2015].

Chapter 4

Applications

Fractional-order systems are valuable tools in modelling of many processes and phenomena in various fields of engineering, physics, biology and economics. Since a fractional-order derivative is nonlocal and has weakly singular kernels, it provides an excellent instrument for the description of memory and hereditary states of dynamical processes, which allows a greater flexibility in the system and is the main advantage of fractional models in comparison with the classical integer-order counterparts (see [Wu, Hei and Chen 2013]).

This chapter will treat fractional-order models with delays and impulses such as neural network models, biological models, models in the population dynamics and models in economics using the theoretical qualitative results for fractional–order systems given in the first three chapters.

Section 4.1 will deal with qualitative results of some fractional neural networks. First, a class of impulsive Caputo fractional-order cellular neural networks with time-varying delays will be considered. The problems of impulsive effects on the global Mittag–Leffler stability will be studied. Our results will provide a design method of impulsive control law which globally asymptotically stabilizes the impulse free fractional-order neural network time-delay model. In addition, the synchronization of fractional chaotic networks via non-impulsive linear controller is considered. Global Mittag–Leffler stability properties of the equilibriums of delayed Bidirectional Associative Memory (BAM) neural networks with Riemann–Liouville fractional operators and impulsive Caputo fractional neural networks with dynamical thresholds with finite and infinite delays will be also investigated. Finally, the existence and global perfect uniform-Mittag–Leffler stability of almost periodic solutions of Caputo impulsive fractional neural networks with time-varying delays will be studied. The uncertain case will be also discussed.

Many illustrative examples will be given to demonstrate the effectiveness of the obtained results.

In *Section* 4.2 different classes of fractional impulsive biological models will be considered. We shall state some existence theorems for almost periodic solutions and some stability results for a fractional impulsive Hutchinson's model, fractional impulsive Lasota–Wazewska models and fractional impulsive Lotka–Volterra models. Kolmogorov-type impulsive systems are also generalized to the fractional-order case. In addition, some models under uncertainties are investigated. The main results in this section are derived by using the fractional Lyapunov method. For the models with delays, the method is combined by the Razumikhin technique.

Finally, in *Section* 4.3 we shall formulate two fractional-order time-delay models: a model for price fluctuations in commodity markets and a Solow-type model. We shall study some qualitative properties of the solutions for such models that are subject to short-term perturbations during their development. The existence of almost periodic solutions as well as their uniform asymptotic stability and Mittag–Leffler stability for fractional impulsive price fluctuations models will be investigated. Stability properties will be discussed for the fractional-order impulsive Solow-type models.

4.1 Fractional Impulsive Neural Networks

Neural networks have received increasing interest due to their impressive applications in many areas of science and engineering (see [Arbib 1987] and [Haykin 1998]). Cellular Neural Networks (CNNs) were first introduced by Chua and Yang in 1988 (see [Chua and Yang 1988a], [Chua and Yang 1988b]) as a novel class of information-processing systems.

One cannot avoid time delay while working with different real phenomena. In many CNN applications, such as associative memories, the network is designed so that stable solutions represent stored information (see [Hopfield 1984]). The delay could be caused by finite switching speed of amplier circuits. Therefore, it is necessary to solve some adaptive control theory, optimization, linear and nonlinear programming, associative memory, pattern recognition and computer vision problems by using Delayed Cellular Neural Networks (DCNNs). Such applications heavily depend on the dynamic behavior of networks; therefore, the qualitative analysis of DCNNs is a necessary step for their practical design (see [Arik and Tavsanoglu 2000], [Cao and Wang 2005], [Zhang, Suda and Iwasa 2004]). Though delays arise frequently in practical applications, it is difficult to measure them precisely. In most situations, delays are variable, and therefore, the model of DCNNs with time-varying delays is put forward, which is naturally of better realistic significance (see [Chen 2002], [Huang and Cao 2003], [Zhang 2003]).

Besides delay effects, impulsive effects are also likely to exist in the neural networks. Indeed, the state of electronic networks is often subject to instantaneous perturbations and experience abrupt changes at certain instants, which may be caused by switching phenomenon, frequency change or other sudden noise, that exhibit impulsive effects. Accordingly, the study of impulsive integer-order DCNNs is of a great importance and has gained considerable popularity in recent years (see [Ahmad and Stamova 2008], [Liu, Huang and Chen 2012], [Liu, Teo and Hu 2005], [Long and Xu 2008], [Song, Xin and Huang 2012], [Stamov and Stamova 2007], [Stamova, Ilarionov and Vaneva 2010], [Stamova and Stamov 2011], [Stamova, Stamov and Li 2014], [Stamova, Stamov and Simeonova 2013], [Stamova, Stamov and Simeonova 2014], [Wang and Liu 2007a], [Zhou and Wan 2009]).

The rapid development in the field of fractional-order equations during the last two decades invoked an increasing interest of scientists in the area of fractional-order neural networks (see [Chen, Qu, Chai, Wu and Qi 2013], [Huang, Zhao, Wang and Li 2012], [Kaslik and Sivasundaram 2012a], [Zhang, Yu and Hu 2014], [Zhou, Li and Zhua 2008]). The researchers pointed out that fractional order neural networks are expected to be very effective in applications such as parameter estimations due to the fact they are characterized by infinite memory. Recently, some interesting results are obtained for fractional neural networks with delays (see [Chen, Chai, Wu, Ma and Zhai 2013], [Wang, Yu and Wen 2014], [Wu, Hei and Chen 2013]).

4.1.1 Stability and Synchronization

First, we shall investigate the global Mittag–Leffler stability of the following Caputo fractional-order impulsive DCNNs with time-varying delays

$$
\begin{cases}
{}^{c}_{t_0}D^{\alpha}_{t}x_i(t) = -c_ix_i(t) + \sum_{j=1}^{n} a_{ij}f_j(x_j(t)) \\
\qquad\qquad + \sum_{j=1}^{n} b_{ij}f_j(x_j(t - \tau_j(t))) + I_i, \ t \neq t_k, \\
\Delta x_i(t_k) = x_i(t_k^+) - x_i(t_k) = P_{ik}(x_i(t_k)), \ k - 1,2,...,
\end{cases}
\tag{4.1}
$$

$i = 1,2,...,n$, $0 < \alpha < 1$, where n corresponds to the number of units in the neural network, $x_i(t)$ corresponds to the state of the ith unit at time t, $f_j(x_j(t))$ denotes the output of the jth unit at time t. Further, a_{ij}, b_{ij}, I_i, c_i are constants, a_{ij} denotes the strength of the jth unit on the ith unit at time t, b_{ij} denotes the strength of the jth unit on the ith unit at time $t - \tau_j(t)$, I_i denotes the external bias on the ith unit, $\tau_j(t)$ corresponds to the transmission delay along the axon of the jth unit and satisfies $0 \leq \tau_j(t) \leq \tau$ ($\tau = const.$), c_i represents the rate with which the ith unit will reset its potential to the resting state in isolation when disconnected from the network and external inputs, t_k, $k = 1,2,...$ are the moments of impulsive perturbations and satisfy $t_0 < t_1 < t_2 < ...$ and $\lim\limits_{k\to\infty} t_k = \infty$. The numbers $x_i(t_k) = x_i(t_k^-)$ and $x_i(t_k^+)$ are, respectively, the states of the ith unit before and after an impulsive perturbation at the moment t_k and the functions P_{ik} represent the abrupt changes of the states $x_i(t)$ at the impulsive moments t_k.

Let $J \subseteq \mathbb{R}$ be an interval. Define the following class of functions:

$PCB[J, \mathbb{R}^n] = \{\sigma \in PC[J, \mathbb{R}^n] : \sigma(t) \text{ is bounded on } J\}$.

Let $\varphi_0 \in PCB[[-\tau, 0], \mathbb{R}^n]$ and $t_0 \in \mathbb{R}$. Denote by $x(t) = x(t; t_0, \varphi_0)$, $x \in \mathbb{R}^n$ the solution of system (4.1) that satisfies the initial conditions:

$$
\begin{cases}
x(t; t_0, \varphi_0) = \varphi_0(t - t_0), \, t_0 - \tau \leq t \leq t_0, \\
x(t_0^+; t_0, \varphi_0) = \varphi_0(0).
\end{cases}
\tag{4.2}
$$

The solution $x(t) = x(t; t_0, \varphi_0) = (x_1(t; t_0, \varphi_0), ..., x_n(t; t_0, \varphi_0))^T$ of problem (4.1), (4.2) is a piecewise continuous function (see [Stamova 2014b] and [Stamova and Stamov G 2014b]) with points of discontinuity of the first kind t_k, $k = 1, 2, ...$, where it is continuous from the left, i.e. the following relations are valid

$$x_i(t_k^-) = x_i(t_k), \, x_i(t_k^+) = x_i(t_k) + P_{ik}(x_i(t_k)).$$

Especially, a constant point $x^* \in \mathbb{R}^n$, $x^* = (x_1^*, x_2^*, ..., x_n^*)^T$ is called an equilibrium of (4.1), if $x^* = x^*(t; t_0, x^*)$ is a solution of (4.1).

We introduce the following conditions:

H4.1 There exist constants $L_i > 0$ such that

$$|f_i(u) - f_i(v)| \leq L_i|u - v|$$

for all $u, v \in \mathbb{R}$, $u \neq v$, and $f_i(0) = 0$, $i = 1, 2, ..., n$.

H4.2 The functions P_{ik} are continuous on $\mathbb{R}$, $i = 1, 2, ..., n$, $k = 1, 2, ...$

H4.3 $t_k < t_{k+1}$, $k = 1, 2, ...$ and $t_k \to \infty$ as $k \to \infty$.

H4.4 There exists a unique equilibrium

$$x^* = (x_1^*, x_2^*, ..., x_n^*)^T$$

of the system (4.1) such that

$$c_i x_i^* = \sum_{j=1}^n a_{ij} f_j(x_j^*) + \sum_{j=1}^n b_{ij} f_j(x_j^*) + I_i,$$

$$P_{ik}(x_i^*) = 0, \quad i = 1, 2, ..., n, \, k = 1, 2, ...$$

Remark 4.1 The problems of existence and uniqueness of equilibrium states of fractional-order neural networks without impulses have been investigated by several authors. For some efficient sufficient conditions which guarantee the existence and uniqueness of solutions for fractional DCNNs with constant delays see [Chen, Chai, Wu, Ma and Zhai 2013]. We will note that, if the system

$$
{}^c_{t_0}D^\alpha_t x_i(t) = -c_i x_i(t) + \sum_{j=1}^n a_{ij} f_j(x_j(t)) + \sum_{j=1}^n b_{ij} f_j(x_j(t - \tau_j(t))) + I_i,
$$

has a unique solution $x(t) = x(t; t_0, \varphi_0)$ with $\varphi_0 \in C[[-\tau, 0], \mathbb{R}^n]$ on the interval $[t_0, \infty)$, that means that the solution $x(t) = x(t; t_0, \varphi_0)$ of the problem (4.1), (4.2) is defined on each of the intervals $(t_{k-1}, t_k]$, $k = 1, 2, \ldots$. From the hypotheses H4.2 and H4.3 we conclude that it is continuable for $t \geq t_0$. For some recent existence results for fractional-order impulsive delay systems see [Chen, Chen and Wang 2009] and [Wang 2012].

For $\varphi \in \mathscr{PC}$, introduce the norm $||.||_\tau$ defined by $||\varphi||_\tau = \sup\limits_{-\tau \leq \theta \leq 0} ||\varphi(\theta)||$, where $||x|| = \sum\limits_{i=1}^n |x_i|$ is the norm of $x \in \mathbb{R}^n$. In the case $\tau = \infty$ we have $||\varphi||_\tau = ||\varphi||_\infty = \sup\limits_{\theta \in (-\infty, 0]} ||\varphi(\theta)||$.

Definition 4.1 The equilibrium $x^* = (x_1^*, x_2^*, \ldots, x_n^*)^T$ of system (4.1) is said to be *globally Mittag–Leffler stable*, if for $\varphi_0 \in PCB[[-\tau, 0], \mathbb{R}^n]$ there exist constants $w > 0$ and $d > 0$ such that

$$
||x(t; t_0, \varphi_0) - x^*|| \leq \{m(\varphi_0 - x^*)E_\alpha(-w(t - t_0)^\alpha)\}^d, \, t \geq t_0,
$$

where $m(0) = 0$, $m(\varphi) \geq 0$, and $m(\varphi)$ is Lipschitz with respect to $\varphi \in \mathscr{PC}$.

If $x^* = (x_1^*, \ldots, x_n^*)^T$ is an equilibrium of (4.1), one can derive from (4.1) that the error $e \in \mathscr{PC}$, $e = (e_1, e_2, \ldots, e_n)^T$, $e_i(t) = x_i(t) - x_i^*$ satisfies

$$
\begin{cases}
{}^c_{t_0}D^\alpha_t e_i(t) = -c_i e_i(t) + \sum_{j=1}^n a_{ij}\left(f_j\left(x_j^* + e_j(t)\right) - f_j(x_j^*)\right) \\
\qquad\qquad + \sum_{j=1}^n b_{ij}\left(f_j\left(x_j^* + e_j(t - \tau_j(t))\right) - f_j(x_j^*)\right), \, t \neq t_k, \\
\Delta e_i(t_k) = Q_{ik}(e_i(t_k)), \, k = \pm 1, \pm 2, \ldots,
\end{cases} \tag{4.3}
$$

where $Q_{ik}(e_i(t_k)) = P_{ik}(e_i(t_k) + x_i^*)$, $i = 1, 2, \ldots, n$, $k = 1, 2, \ldots$, and

$$
e(\theta) = \varphi_0(\theta) - x^*, \, \theta \in [-\tau, 0], \, e(t_0^+) = \varphi_0(0) - x^*.
$$

Further on we shall use piecewise continuous Lyapunov functions $V : [t_0, \infty) \times \mathbb{R}^n \to \mathbb{R}_+, V \in V_0$.

Theorem 4.1 Assume that:

1. Conditions H4.1–H4.4 are fulfilled.
2. The system parameters $a_{ij}, b_{ij}, c_i, (i, j = 1, 2, ..., n)$ satisfy

$$\min_{1 \le i \le n} \left(c_i - L_i \sum_{j=1}^{n} |a_{ji}| \right) > \max_{1 \le i \le n} \left(L_i \sum_{j=1}^{n} |b_{ji}| \right) > 0.$$

3. The functions P_{ik} are such that

$$P_{ik}(x_i(t_k)) = -\sigma_{ik}(x_i(t_k) - x_i^*), \quad 0 < \sigma_{ik} < 2,$$

$i = 1, 2, ..., n, k = 1, 2, ...$

Then the equilibrium x^* of (4.1) is globally Mittag–Leffler stable.

Proof. We define a Lyapunov function

$$V(t, e) = \sum_{i=1}^{n} |e_i(t)|.$$

Then, for $t \ge t_0$ and $t = t_k$, from condition 3 of the theorem we obtain

$$V(t_k^+, e(t_k^+)) = \sum_{i=1}^{n} |e_i(t_k) + Q_{ik}(e_i(t_k))|$$

$$= \sum_{i=1}^{n} |x_i(t_k) - x_i^* - \sigma_{ik}(x_i(t_k) - x_i^*)| = \sum_{i=1}^{n} |1 - \sigma_{ik}||x_i(t_k) - x_i^*|$$

$$< \sum_{i=1}^{n} |x_i(t_k) - x_i^*| = V(t_k, e(t_k)), k = 1, 2, ... \tag{4.4}$$

Let $t \ge t_0$ and $t \in [t_{k-1}, t_k)$. If $e_i(t) = 0$, $i = 1, 2, ..., n$, then $^cD_+^\alpha V(t, e(t)) = 0$. If $e_i(t) > 0$, $i = 1, 2, ..., n$, then

$$_{t_0}^{c}D_t^\alpha |e_i(t)| = \frac{1}{\Gamma(1 - \alpha)} \int_{t_0}^{t} \frac{|e_i(\sigma)|'}{(t - \sigma)^\alpha} d\sigma$$

$$= \frac{1}{\Gamma(1 - \alpha)} \int_{t_0}^{t} \frac{e_i'(\sigma)}{(t - \sigma)^\alpha} d\sigma = {}_{t_0}^{c}D_t^\alpha e_i(t).$$

If $e_i(t) < 0$, $i = 1, 2, ..., n$, then

$$_{t_0}^{c}D_t^\alpha |e_i(t)| = \frac{1}{\Gamma(1 - \alpha)} \int_{t_0}^{t} \frac{|e_i(\sigma)|'}{(t - \sigma)^\alpha} d\sigma$$

$$= -\frac{1}{\Gamma(1 - \alpha)} \int_{t_0}^{t} \frac{e_i'(\sigma)}{(t - \sigma)^\alpha} d\sigma = -{}_{t_0}^{c}D_t^\alpha e_i(t).$$

Therefore,

$$_{t_0}^{c}D_t^\alpha |e_i(t)| = \text{sgn}(e_i(t)) {}_{t_0}^{c}D_t^\alpha e_i(t).$$

Then for $t \geq t_0$ and $t \in [t_{k-1}, t_k)$ for the upper right derivative ${}^c D_+^\alpha V(t, e(t))$ along the solutions of system (4.3), we get

$$
{}^c D_+^\alpha V(t, e(t)) \leq \sum_{i=1}^n \left[-c_i |e_i(t)| + \sum_{j=1}^n L_j |a_{ij}| |e_j(t)| + \sum_{j=1}^n L_j |b_{ij}| |e_j(t - \tau_j(t))| \right]
$$

$$
= -\sum_{i=1}^n \left[c_i - L_i \sum_{j=1}^n |a_{ji}| \right] |e_i(t)| + \sum_{j=1}^n \sum_{i=1}^n L_j |b_{ij}| |e_j(t - \tau_j(t))|
$$

$$
\leq -\min_{1 \leq i \leq n} \left(c_i - L_i \sum_{j=1}^n |a_{ji}| \right) \sum_{i=1}^n |e_i(t)| + \max_{1 \leq i \leq n} \left(L_i \sum_{j=1}^n |b_{ji}| \right) \sum_{i=1}^n |e_j(t - \tau_j(t))|
$$

$$
\leq -k_1 V(t, e(t)) + k_2 \sup_{t - \tau \leq \theta \leq t} V(\theta, e(\theta)),
$$

where

$$
k_1 = \min_{1 \leq i \leq n} \left(c_i - L_i \sum_{j=1}^n |a_{ji}| \right) > 0,
$$

$$
k_2 = \max_{1 \leq i \leq n} \left(L_i \sum_{j=1}^n |b_{ji}| \right) > 0.
$$

From the above estimate, for any solution $e(t)$ of (4.3) which satisfies the Razumikhin condition

$$
V(\theta, e(\theta)) \leq V(t, e(t)), \quad t - \tau \leq \theta \leq t,
$$

we have

$$
{}^c D_+^\alpha V(t, e(t)) \leq -(k_1 - k_2) V(t, e(t)).
$$

By virtue of condition 2 of Theorem 4.1, there exits a real number $w > 0$ such that

$$
k_1 - k_2 \geq w,
$$

and it follows that

$$
{}^c D_+^\alpha V(t, e(t)) \leq -w V(t, e(t)), \tag{4.5}
$$

$t \neq t_k, t > t_0$.

Then using (4.4), (4.5) and Corollary 1.5, we get

$$
V(t, e(t)) \leq \sup_{-\tau \leq \theta \leq 0} V(t_0^+, \varphi_0(\theta) - x^*) E_\alpha(-w(t - t_0)^\alpha), \, t \in [t_0, \infty).
$$

So,

$$
\|x(t) - x^*\| = \sum_{i=1}^n |x_i(t) - x_i^*|
$$

$$
\leq \|\varphi_0 - x^*\|_\tau E_\alpha(-w(t - t_0)^\alpha), \, t \geq t_0.
$$

Let $m = ||\varphi_0 - x^*||_\tau$. Then

$$||x(t) - x^*|| \leq mE_\alpha(-w(t - t_0)^\alpha), \ t \geq t_0,$$

where $m \geq 0$ and $m = 0$ holds only if $\varphi_0(\theta) = x^*$ for $\theta \in [-\tau, 0]$, which implies that the equilibrium x^* of (4.1) is globally Mittag–Leffler stable.

Remark 4.2 Since the equilibrium of (4.1) is globally Mittag–Leffler stable, then it is globally asymptotically stable.

Remark 4.3 The impulsive fractional-order DCNN (4.1) is the corresponding closed-loop system to the control system

$$_{t_0}^c D_t^\alpha x_i(t) = -c_i x_i(t) + \sum_{j=1}^n a_{ij} f_j(x_j(t))$$

$$+ \sum_{j=1}^n b_{ij} f_j(x_j(t - \tau_j(t))) + I_i + u_i(t), \ t > t_0,$$

where

$$u_i(t) = \sum_{k=1}^\infty P_{ik}(x_i(t))\delta(t - t_k), \ i = 1, 2, ..., n \qquad (4.6)$$

is the control input, $\delta(t)$ is the Dirac impulsive function. The controller $u(t) = (u_1(t), ..., u_n(t))^T$ has an effect on suddenly change of the states of (4.1) at the time instants t_k due to which the states $x_i(t)$ of units change from the position $x_i(t_k)$ into the position $x_i(t_k^+)$, P_{ik} are the functions, which characterize the magnitudes of the impulse effects on the units x_i at the moments t_k, i.e., $u(t)$ is an impulsive control of the fractional-order DCNN

$$_{t_0}^c D_t^\alpha x_i(t) = -c_i x_i(t) + \sum_{j=1}^n a_{ij} f_j(x_j(t))$$

$$+ \sum_{j=1}^n b_{ij} f_j(x_j(t - \tau_j(t))) + I_i, \ t > t_0. \qquad (4.7)$$

Therefore, Theorem 4.1 presents a general design method of impulsive control law (4.6) for the impulse free fractional-order DCNN model (4.7). The constants σ_{ik} in condition 3 of Theorem 4.1 characterize the control gains of synchronizing impulses.

Hence, our results can be used to design impulsive control law under which the controlled neural networks (4.1) are globally asymptotically synchronized onto system (4.7).

In the next, we shall investigate drive-response impulsive fractional-order chaotic neural networks achieve synchronization under linear control. We will consider neural network system (4.1) as the master system, and the slave system is given by

$$
\begin{cases}
{}^{c}_{t_0}D^{\alpha}_t y_i(t) = -c_i y_i(t) + \sum_{j=1}^{n} a_{ij} f_j(y_j(t)) \\
\qquad + \sum_{j=1}^{n} b_{ij} f_j(y_j(t - \tau_j(t))) + I_i + u_i(t), \ t \neq t_k, \ t > t_0, \\
\Delta y_i(t_k) = y_i(t_k^+) - y_i(t_k) = P_{ik}(y_i(t_k)), \ k = 1, 2, ...,
\end{cases}
\tag{4.8}
$$

where $u_i(t)$ is a suitable non-impulsive controller.

We shall use a linear feedback scheme to realize synchronization between impulsive DCNNs (4.1) and (4.8), i.e., the controller $u_i(t)$ is defined by

$$
u_i(t) = -d_i(y_i(t) - x_i(t)),
$$

where $d_i > 0$, $i = 1, 2, ..., n$ represents the control gain.

Let $e_i(t) = y_i(t) - x_i(t)$, $i = 1, 2, ..., n$. Then from (4.1) and (4.8) the error system is given by

$$
\begin{cases}
{}^{c}_{t_0}D^{\alpha}_t e_i(t) = -c_i e_i(t) + \sum_{j=1}^{n} a_{ij} g_j(e_j(t)) \\
\qquad + \sum_{j=1}^{n} b_{ij} g_j(e_j(t - \tau_j(t))) - d_i e_i(t), \ t \neq t_k, \\
\Delta e_i(t_k) = P_{ik}(e_i(t_k)), \ k = 1, 2, ...,
\end{cases}
$$

where

$$
g_j(e_j(t)) = f_j(y_j(t)) - f_j(x_j(t))
$$

and

$$
g_j(e_j(t - \tau_j(t))) = f_j(y_j(t - \tau_j(t))) - f_j(x_j(t - \tau_j(t)))
$$

Introduce the following definition.

Definition 4.2 Master system (4.1) and slave system (4.8) are said to be *globally Mittag–Leffler synchronized*, if for $\varphi_0, \phi_0 \in PCB[[-\tau, 0], \mathbb{R}^n]$ there exist constants $w > 0$ and $d > 0$ such that

$$
\|x(t; t_0, \varphi_0) - y(t; t_0, \phi_0)\| \leq \{m(\varphi_0 - \phi_0) E_\alpha(-w(t - t_0)^\alpha)\}^d, \ t \geq t_0,
$$

where $m(0) = 0$, $m(\varphi) \geq 0$, and $m(\varphi)$ is Lipschitz with respect to $\varphi \in \mathscr{PC}$.

The proof of the next theorem is similar to the proof of Theorem 4.1.

Theorem 4.2 Assume that:

1. Conditions H4.1–H4.4 are fulfilled.
2. The system parameters a_{ij}, b_{ij}, c_i and the control gain d_i, $i = 1, 2, ..., n$ satisfy

$$\min_{1 \leq i \leq n} \left(c_i + d_i - L_i \sum_{j=1}^{n} |a_{ji}| \right) > \max_{1 \leq i \leq n} \left(L_i \sum_{j=1}^{n} |b_{ji}| \right) > 0.$$

3. The functions P_{ik} are such that

$$P_{ik}(e_i(t_k)) = -\sigma_{ik} e_i(t_k), \quad 0 < \sigma_{ik} < 2,$$

$i = 1, 2, ..., n, k = 1, 2, ...$

Then the master system (4.1) and slave system (4.8) are globally Mittag–Leffler (globally asymptotically) synchronized.

In the following, we shall give examples to illustrate the presented results.

Example 4.1 Consider the impulsive fractional-order DCNN with time-varying delays

$$_0^c D_t^\alpha x_i(t) = -c_i x_i(t)$$

$$+ \sum_{j=1}^{n} a_{ij} f_j(x_j(t)) + \sum_{j=1}^{n} b_{ij} f_j(x_j(t - \tau_j(t))) + I_i, \ t \neq t_k, \ t > 0, \quad (4.9)$$

where $0 < \alpha < 1$, $n = 2$, $I_1 = 0.66$, $I_2 = 0.69$, $c_1 = c_2 = 1$, $f_i(x_i) = \frac{1}{2}(|x_i + 1| - |x_i - 1|)$, $i = 1, 2, 0 \leq \tau_i(t) \leq \tau$ $(\tau = 1)$,

$$(a_{ij})_{2 \times 2} = \begin{pmatrix} a_{11} & a_{12} \\ a_{21} & a_{22} \end{pmatrix} = \begin{pmatrix} 0.2 & 0.2 \\ -0.2 & 0.2 \end{pmatrix},$$

$$(b_{ij})_{2 \times 2} = \begin{pmatrix} b_{11} & b_{12} \\ b_{21} & b_{22} \end{pmatrix} = \begin{pmatrix} 0.3 & -0.1 \\ 0.1 & 0.2 \end{pmatrix},$$

with

$$\begin{cases} x_1(t_k^+) = \dfrac{1.5 + x_1(t_k)}{4}, \ k = 1, 2, ..., \\[3mm] x_2(t_k^+) = \dfrac{1.8 + x_2(t_k)}{3}, \ k = 1, 2, ..., \end{cases} \quad (4.10)$$

where the impulsive moments are such that $0 < t_1 < t_2 < ...$, and $\lim_{k \to \infty} t_k = \infty$.

It is easy to verify that the condition 2 of Theorem 4.1 is satisfied for $L_1 = L_2 = 1$, $k_1 = 0.6$, $k_2 = 0.4$. Also we have that

$$0 < \sigma_{1k} = \frac{3}{4} < 2, \quad 0 < \sigma_{2k} = \frac{2}{3} < 2.$$

According to Theorem 4.1 the unique equilibrium

$$x^* = (x_1^*, x_2^*)^T = (0.5, 0.9)^T \qquad (4.11)$$

of (4.9), (4.10) is globally Mittag–Leffler (globally asymptotically) stable.

If we consider again the system (4.9) but with impulsive perturbations of the form

$$\begin{cases} x_1(t_k^+) = 1.5 - 2x_1(t_k), \, k = 1, 2, ..., \\ \\ x_2(t_k^+) = \dfrac{1.8 + x_2(t_k)}{3}, \, k = 1, 2, ..., \end{cases} \qquad (4.12)$$

then the point (4.11) will be again an equilibrium of (4.9), (4.12), but there is nothing we can say about its global asymptotic stability, because $\sigma_{1k} = 3 > 2$.

Remark 4.4 The example shows that by means of appropriate impulsive perturbations we can control the neural network system's dynamics.

Example 4.2 Consider the master impulsive fractional-order DCNN with time-varying delays (4.9) where $\alpha = 0.98$, $n = 2$, $I_1 = I_2 = 1$, $c_1 = c_2 = 2$, $f_i(x_i) = \dfrac{1}{2}(|x_i + 1| - |x_i - 1|)$, $i = 1, 2$, $0 \le \tau_i(t) \le \tau$ ($\tau = 1$),

$$(a_{ij})_{2 \times 2} = \begin{pmatrix} a_{11} & a_{12} \\ a_{21} & a_{22} \end{pmatrix} = \begin{pmatrix} 0.3 & 0.5 \\ -0.7 & 0.8 \end{pmatrix},$$

$$(b_{ij})_{2 \times 2} = \begin{pmatrix} b_{11} & b_{12} \\ b_{21} & b_{22} \end{pmatrix} = \begin{pmatrix} 0.2 & -0.6 \\ 0.5 & 0.3 \end{pmatrix},$$

with

$$\begin{cases} x_1(t_k^+) = \dfrac{2x_1(t_k)}{3}, \, k = 1, 2, ..., \\ \\ x_2(t_k^+) = \dfrac{2x_2(t_k)}{5}, \, k = 1, 2, ..., \end{cases} \qquad (4.13)$$

where the impulsive moments are such that $0 < t_1 < t_2 < ...$, $\lim\limits_{k \to \infty} t_k = \infty$, and the response system

$$_0^C D_t^\alpha y_i(t) = -c_i y_i(t) + \sum_{j=1}^n a_{ij} f_j(y_j(t))$$

$$+ \sum_{j=1}^n b_{ij} f_j(y_j(t - \tau_j(t))) + I_i - d_i(y_i(t) - x_i(t)), \, t \ne t_k, \, t > 0 \qquad (4.14)$$

with the feedback gains $d_1 = d_2 = 1.2$ and impulsive perturbations of the form

$$y_1(t_k^+) = \dfrac{2y_1(t_k)}{3}, \, y_2(t_k^+) = \dfrac{2y_2(t_k)}{5}, \, k = 1, 2, ... \qquad (4.15)$$

It is easy to verify that the condition 2 of Theorem 4.2 is satisfied for $L_1 = L_2 = 1$,

$$\min_{1 \leq i \leq n} \left(c_i + d_i - L_i \sum_{j=1}^{n} |a_{ji}| \right) = 1.9,$$

$$\max_{1 \leq i \leq n} \left(L_i \sum_{j=1}^{n} |b_{ji}| \right) = 0.9.$$

Also, we have that

$$0 < \sigma_{1k} = \frac{1}{3} < 2, \quad 0 < \sigma_{2k} = \frac{3}{5} < 2.$$

According to Theorem 4.2 the master system (4.9), (4.13) and response system (4.14), (4.15) are globally Mittag–Leffler (globally asymptotically) synchronized.

Remark 4.5 For $d_1 = d_2 = 0$ the condition 2 of Theorem 4.2 is not satisfied. Example 4.2 demonstrates that the linear controller $-1.2(y_i(t) - x_i(t))$ can realize the synchronization goal, which verifies the Theorem 4.2.

In the next section, we shall investigate stability properties of a fractional-order Bidirectional Associative Memory (BAM) neural network with time delays.

Note that, a series of integer-order BAM neural networks have been proposed by Kosko (see [Kosko 1987], [Kosko 1988], [Kosko 1992]). Due to their wide range of applications, in recent years the class of such neural networks with their various generalizations have attracted the attention of many mathematicians, physicists, and computer scientists. It is a special class of recurrent neural networks that can store bipolar vector pairs. The BAM neural network is composed of neurons arranged in two layers, the X-layer and Y-layer. The neurons in one layer are fully interconnected to the neurons in the other layer. Through iterations of forward and backward information ows between the two layer, it performs a two-way associative search for stored bipolar vector pairs and generalize the single-layer autoassociative Hebbian correlation to a two-layer pattern-matched heteroassociative circuits.

On the other hand, the memory is one of the main features in BAM neural networks. Hence, the study of integer-order BAM neural networks with time delays is of a great importance and has received an increasing interest. It is also important to note that the use of constant fixed delays in models of delayed feedback provides a good approximation in simple circuits consisting of only a small number of cells, neural networks usually have a spatial extent due to presence of a multitude of parallel pathways with a variety of axon sizes and lengths. Thus it is common to have a distribution of propagation delays. Recently, some authors have investigated the stability of integer-order BAM neural networks with distributed delays (see, for example, [Chen, Cao and Huang 2004], [Song and Cao 2007], [Zhao 2002], [Zhou and Wan 2009] and references therein).

However, few studies focused of fractional-order BAM neural networks with time-delays. Using the theory of fractional calculus, the generalized Gronwall inequality and estimates of Mittag–Leffer functions, the finite time stability of Caputo fractional-order BAM neural networks with distributed delay is investigated in [Cao and Bai 2014]. In [Yang, Song, Liu and Zhao 2014] a class of Caputo fractional-order BAM neural networks with delays in the leakage terms is considered. Instead of the Lyapunov approach, the above authors used inequality techniques and analysis methods to study the existence, uniqueness, and stability of the equilibrium point.

Here, we shall apply the Lyapunov–Razumikhin technique, and we shall extend the stability results for BAM neural networks with finite delays to the fractional-order case.

Consider the following fractional-order BAM neural network system involving Riemann–Liouville differential operators of order α:

$$
\begin{cases}
{}_0D_t^\alpha x_i(t) = -c_i x_i(t) + \sum_{j=1}^{n} a_{ji} f_j(y_j(t)) \\
\qquad + \sum_{j=1}^{n} w_{ji} f_j(y_j(t - \tau_{ji}(t))) + I_i, \\
{}_0D_t^\alpha y_j(t) = -d_j y_j(t) + \sum_{i=1}^{m} b_{ij} g_i(x_i(t)) \\
\qquad + \sum_{i=1}^{m} h_{ij} g_i(x_i(t - \sigma_{ij}(t))) + J_j,
\end{cases}
\tag{4.16}
$$

where $0 < \alpha < 1$, ${}_0D_t^\alpha$ is the Riemann–Liouville fractional derivative of order α, $t \geq 0$, $i = 1, 2, ..., m$, $j = 1, 2, ..., n$, $x_i(t)$ and $y_j(t)$ correspond to the states of the ith unit and jth unit, respectively, at time t, c_i, d_j are positive constants, time delays $\tau_{ji}(t)$ $(0 \leq \tau_{ji}(t) \leq \tau_{ji})$, $\sigma_{ij}(t)$ $(0 \leq \sigma_{ij}(t) \leq \sigma_{ij})$ correspond to the transmission delays at time t, a_{ji}, b_{ij} are the connection weights, f_j, g_i are the activation functions, I_i, J_j denote external inputs.

Let $\varphi \in PCB[[-\sigma, 0], \mathbb{R}^m]$, $\varphi = (\varphi_1, \varphi_2, ..., \varphi_m)^T$, $\phi \in PCB[[-\tau, 0], \mathbb{R}^n]$, $\phi = (\phi_1, \phi_2, ..., \phi_n)^T$. Denote by $(x(t), y(t))^T = (x(t; 0, \varphi), y(t; 0, \phi))^T \in \mathbb{R}^{m+n}$,

$$(x(t; 0, \varphi), y(t; 0, \phi))^T = (x_1(t; 0, \varphi), ..., x_m(t; 0, \varphi), y_1(t; 0, \phi), ..., y_n(t; 0, \phi))^T$$

the solution of system (4.16), satisfying the initial conditions

$$
\begin{cases}
x_i(t; 0, \varphi) = \varphi_i(t), \ -\sigma \leq t \leq 0, \ i = 1, 2, ..., m, \\
y_j(t; 0, \phi) = \phi_j(t), \ -\tau \leq t \leq 0, \ j = 1, 2, ..., n, \\
{}_0D_t^{\alpha-1} x_i(0; 0, \varphi) = \varphi_i(0), \ i = 1, 2, ..., m, \\
{}_0D_t^{\alpha-1} y_j(0; 0, \phi) = \phi_j(0), \ j = 1, 2, ..., n,
\end{cases}
\tag{4.17}
$$

where $\tau = \max\limits_{1 \leq i \leq m, 1 \leq j \leq n} \tau_{ji}$, $\sigma = \max\limits_{1 \leq i \leq m, 1 \leq j \leq n} \sigma_{ij}$.

Let $||z|| = \left(\sum\limits_{l=1}^{m+n} z_l^2 \right)^{1/2}$ define the norm of $z \in \mathbb{R}^{m+n}$.

We introduce the following conditions:

H4.5 There exist constants $L_j > 0$ such that

$$|f_j(u) - f_j(v)| \leq L_j |u - v|, \ f_j(0) = 0$$

for all $u, v \in \mathbb{R}$, $j = 1, 2, ..., n$.

H4.6 There exist constants $M_i > 0$ such that

$$|g_i(u) - g_i(v)| \leq M_i |u - v|, \ g_i(0) = 0$$

for all $u, v \in \mathbb{R}$, $i = 1, 2, ..., m$.

H4.7 There exists a unique equilibrium

$$(x^*, y^*)^T = (x_1^*, x_2^*, ..., x_m^*, y_1^*, y_2^*, ..., y_n^*)^T$$

of the system (4.16) such that

$$_0D_t^\alpha x_i^* = -c_i x_i^* + \sum_{j=1}^n (a_{ji} + w_{ji}) f_j(y_j^*) + I_i,$$

$$_0D_t^\alpha y_j^* = -d_j y_j^* + \sum_{i=1}^m (b_{ij} + h_{ij}) g_i(x_i^*) + J_j,$$

$$i = 1, 2, ..., m, \ j = 1, 2, ..., n.$$

Remark 4.6 The problems of existence and uniqueness of solutions of fractional functional differential equations involving Riemann–Liouville differential operators have been investigated in [Lakshmikantham 2008]. Efficient sufficient conditions for the existence and uniqueness of an equilibrium of Caputo fractional-order BAM neural networks with delays in the leakage terms are given in [Yang, Song, Liu and Zhao 2014].

Further on we shall use continuous Lyapunov functions $V : [0, \infty) \times \mathbb{R}^{m+n} \to \mathbb{R}_+$ such that $V \in C_0$.

We introduce the following notations:

$$A = (a_{ji})_{n \times m}, \ B = (b_{ij})_{m \times n}, \ W = (w_{ji})_{n \times m}, \ H = (h_{ij})_{m \times n},$$

$$C = diag(c_1, c_2, ..., c_m), \ D = diag(d_1, d_2, ..., d_n),$$

$$I = (I_1, I_2, ..., I_m)^T, \ J = (J_1, J_2, ..., J_n)^T,$$

$$L = (L_1, L_2, ..., L_n)^T, \ M = (M_1, M_2, ..., M_m)^T,$$

$\lambda_{\min}(P)$ is the smallest eigenvalue of a matrix P,

$\lambda_{\max}(P)$ is the greatest eigenvalue of a matrix P,

$$||P|| = \left[\lambda_{\max}(P^T P)\right]^{\frac{1}{2}} \text{ is the norm of matrix } P.$$

In the proof of the next result we shall use lemmas 1.7 and 1.8.

Theorem 4.3 Assume that:

1. Conditions H4.5, H4.6 and H4.7 hold.
2. There exist symmetric, positive definite matrices $P_{m \times m}$ and $Q_{n \times n}$ such that

$$\lambda_{\max}(P)\left(-2||C|| + ||A|| \, ||L|| + ||W|| \, ||L|| \frac{\lambda_{\max}(P) + \lambda_{\min}(Q)}{\lambda_{\min}(Q)}\right)$$

$$+ \lambda_{\max}(Q)\left(||B|| \, ||M|| + ||H|| \, ||M|| \frac{\lambda_{\max}(P)}{\lambda_{\min}(P)}\right) < 0,$$

$$\lambda_{\max}(Q)\left(-2||D|| + ||B|| \, ||M|| + ||H|| \, ||M|| \frac{\lambda_{\max}(Q) + \lambda_{\min}(P)}{\lambda_{\min}(P)}\right)$$

$$+ \lambda_{\max}(P)\left(||A|| \, ||L|| + ||W|| \, ||L|| \frac{\lambda_{\max}(Q)}{\lambda_{\min}(Q)}\right) < 0.$$

Then the equilibrium $col(x^*, y^*)$ of (4.16) is globally Mittag–Leffler stable (globally asymptotically stable).

Proof. Set $u(t) = x(t) - x^*$, $v(t) = y(t) - y^*$ and consider the following system

$$
\begin{cases}
{}_0D_t^\alpha u_i(t) = -c_i u_i(t) + \sum_{j=1}^{n} a_{ji}[f_j(y_j^* + v_j(t)) - f_j(y_j^*)] \\
\qquad + \sum_{j=1}^{n} w_{ji}[f_j(y_j^* + v_j(t - \tau_{ji}(t))) - f_j(y_j^*)], \\
{}_0D_t^\alpha v_j(t) = -d_j v_j(t) + \sum_{i=1}^{m} b_{ij}[g_i(x_i^* + u_i(t)) - g_i(x_i^*)] \\
\qquad + \sum_{i=1}^{m} h_{ij} g_i(x_i^* + u_i(t - \sigma_{ij}(t))) - g_i(x_i^*)],
\end{cases}
\tag{4.18}
$$

where $i = 1, 2, ..., m$, $j = 1, 2, ..., n$.

We define a Lyapunov function

$$
V(t, u, v) = u^T P u + v^T Q v. \tag{4.19}
$$

Since for the function $V(t, u, v)$, we have

$$
u^T P u + v^T Q v \leq \lambda_{\max}(P)||u||^2 + \lambda_{\max}(Q)||v||^2, \tag{4.20}
$$

then for any continuously differentiable functions $u(t)$ and $v(t)$ by Lemma 1.8 for the upper right-hand derivative ${}^cD_+^\alpha V(t, u(t), v(t))$, we get

$$
{}^cD_+^\alpha V(t, u(t), v(t))
$$

$$
\leq 2\lambda_{\max}(P)||u(t)||\,{}^cD_+^\alpha||u(t)|| + 2\lambda_{\max}(Q)||v(t)||\,{}^cD_+^\alpha||v(t)||,
$$

and using Lemma 1.7, we have

$$
{}^cD_+^\alpha V(t, u(t), v(t))
$$

$$
\leq 2\lambda_{\max}(P)||u(t)||D_+^\alpha||u(t)|| + 2\lambda_{\max}(Q)||v(t)||D_+^\alpha||v(t)||.
$$

Let $u(t)$ and $v(t)$ be determined by the system (4.18). Then for $t \geq 0$ using H4.5 and H4.6, for the upper right-hand derivative ${}^cD_+^\alpha V(t, u(t), v(t))$ of the function $V(t, u(t), v(t))$, we have

$$
{}^cD_+^\alpha V(t, u(t), v(t))
$$

$$
\leq 2\lambda_{\max}(P)||u(t)||\,sgn(u(t))D_+^\alpha u(t) + 2\lambda_{\max}(Q)||v(t)||\,sgn(v(t))D_+^\alpha v(t)
$$

$$
\leq 2\lambda_{\max}(P)||u(t)||\,sgn(u(t))(-Cu(t) + ALv(t) + WLv(t - \tau))
$$

$$
+ 2\lambda_{\max}(Q)||v(t)||\,sgn(v(t))(-Dv(t) + BMu(t) + HMu(t - \sigma))
$$

$$
\leq -2\lambda_{\max}(P)||C||\,||u(t)||^2 - 2\lambda_{\max}(Q)||D||\,||v(t)||^2
$$

$$+2\lambda_{\max}(P)\,||A||\,||L||\,||v(t)||\,||u(t)|| + 2\lambda_{\max}(Q)\,||B||\,||M||\,||u(t)||\,||v(t)||$$

$$+2\lambda_{\max}(P)\,||W||\,||L||\,\sup_{t-\tau\le\theta\le t}||v(\theta)||\,||u(t)||$$

$$+2\lambda_{\max}(Q)||H||\,||M||\,\sup_{t-\sigma\le\theta\le t}||u(\theta)||\,||v(t)||$$

for $t \ge 0$.

Using the inequality $2|a||b| \le a^2 + b^2$, we get for $t \ge 0$,

$$^cD_+^\alpha V(t,u(t),v(t)) \le -2\lambda_{\max}(P)||C||\,||u(t)||^2 - 2\lambda_{\max}(Q)||D||\,||v(t)||^2$$

$$+\Big(\lambda_{\max}(P)\,||A||\,||L|| + \lambda_{\max}(Q)\,||B||\,||M||\Big)\Big(||u(t)||^2 + ||v(t)||^2\Big)$$

$$+\lambda_{\max}(P)\,||W||\,||L||\Big(\sup_{t-\tau\le\theta\le t}||v(\theta)||^2 + ||u(t)||^2\Big)$$

$$+\lambda_{\max}(Q)\,||H||\,||M||\Big(\sup_{t-\sigma\le\theta\le t}||u(\theta)||^2 + ||v(t)||^2\Big). \qquad (4.21)$$

Since for the function $V(t,u(t),v(t))$, we have

$$\lambda_{\min}(P)||u(t)||^2 + \lambda_{\min}(Q)||v(t)||^2 \le u^T(t)Pu(t) + v^T(t)Qv(t)$$

$$\le \lambda_{\max}(P)||u(t)||^2 + \lambda_{\max}(Q)||v(t)||^2,\ t \ge 0,$$

then for $u(t)$ and $v(t)$ such that satisfy the Razumikhin condition

$$V(\theta,u(\theta),v(\theta)) \le V(t,u(t),v(t)),\ \ t - \max\{\tau,\sigma\} \le \theta \le t,$$

we obtain

$$\lambda_{\min}(P)||u(\theta)||^2 + \lambda_{\min}(Q)||v(\theta)||^2 \le u^T(\theta)Pu(\theta) + v^T(\theta)Qv(\theta)$$

$$\le u^T(t)Pu(t) + v^T(t)Qv(t) \le \lambda_{\max}(P)||u(t)||^2 + \lambda_{\max}(Q)||v(t)||^2, \qquad (4.22)$$

and hence

$$\begin{cases} ||u(\theta)||^2 \le \dfrac{\lambda_{\max}(P)||u(t)||^2 + \lambda_{\max}(Q)||v(t)||^2}{\lambda_{\min}(P)}, \\[4mm] ||v(\theta)||^2 \le \dfrac{\lambda_{\max}(P)||u(t)||^2 + \lambda_{\max}(Q)||v(t)||^2}{\lambda_{\min}(Q)}, \end{cases} \qquad (4.23)$$

for $t - \max\{\tau,\sigma\} \le \theta \le t, t \ge 0$.

From (4.21) and (4.23) for $t \ge 0$, we obtain

$$^cD_+^\alpha V(t,u(t),v(t)) \le -2\lambda_{\max}(P)||C||\,||u(t)||^2 - 2\lambda_{\max}(Q)||D||\,||v(t)||^2$$

$$+\Big(\lambda_{\max}(P)\,||A||\,||L|| + \lambda_{\max}(Q)\,||B||\,||M||\Big)\Big(||u(t)||^2 + ||v(t)||^2\Big)$$

$$+\lambda_{\max}(P)||W||\,||L||\left(\frac{\lambda_{\max}(P)||u(t)||^2+\lambda_{\max}(Q)||v(t)||^2}{\lambda_{\min}(Q)}+||u(t)||^2\right)$$

$$+\lambda_{\max}(Q)||H||\,||M||\left(\frac{\lambda_{\max}(P)||u(t)||^2+\lambda_{\max}(Q)||v(t)||^2}{\lambda_{\min}(P)}+||v(t)||^2\right)$$

$$=\left[\lambda_{\max}(P)\left(-2||C||+||A||\,||L||+||W||\,||L||\frac{\lambda_{\max}(P)+\lambda_{\min}(Q)}{\lambda_{\min}(Q)}\right)\right.$$

$$\left.+\lambda_{\max}(Q)\left(||B||\,||M||+||H||\,||M||\frac{\lambda_{\max}(P)}{\lambda_{\min}(P)}\right)\right]||u(t)||^2$$

$$+\left[\lambda_{\max}(Q)\left(-2||D||+||B||\,||M||+||H||\,||M||\frac{\lambda_{\max}(Q)+\lambda_{\min}(P)}{\lambda_{\min}(P)}\right)\right.$$

$$\left.+\lambda_{\max}(P)\left(||A||\,||L||+||W||\,||L||\frac{\lambda_{\max}(Q)}{\lambda_{\min}(Q)}\right)\right]||v(t)||^2.$$

From the condition 2 of Theorem 4.3, we derive for $t \geq 0$,

$$^cD_+^\alpha V(t,u(t),v(t)) \leq -p||u(t)||^2 - q||v(t)||^2$$

$$\leq -k_1\left(||u(t)||^2+||v(t)||^2\right), \tag{4.24}$$

where $p,q = const > 0$ and $k_1 = \min\{p,q\} > 0$.

Using (4.20), we get

$$V(t,u(t),v(t)) \leq w\left(||u(t)||^2+||v(t)||^2\right), \tag{4.25}$$

where $t \geq 0$, $w = \max\left\{\lambda_{\max}(P),\lambda_{\max}(Q)\right\}$.

Then, from the inequalities (4.24) and (4.25), we obtain

$$^cD_+^\alpha V(t,u(t),v(t)) \leq -\frac{k_1}{w}V(t,u(t),v(t))$$

for all $t \geq 0$ whenever $V(\theta,u(\theta),v(\theta)) \leq V(t,u(t),v(t))$, $t-\max\{\tau,\sigma\} \leq \theta \leq t$.

Since all conditions of Theorem 2.23 are met, then the zero solution of system (4.18) is globally Mittag–Leffler stable, i.e. the equilibrium $col(x^*,y^*)$ of (4.16) is globally Mittag–Leffler stable.

Example 4.3 Let $t \geq 0$. Consider the BAM neural network system of fractional order α, $0 < \alpha < 1$,

$$
\begin{cases}
{}_0D_t^\alpha x_i(t) = -c_i x_i(t) + \sum_{j=1}^{2} a_{ji} f_j(y_j(t)) \\
\qquad + \sum_{j=1}^{2} w_{ji} f_j(y_j(t - \tau_{ji}(t))) + I_i, \\
{}_0D_t^\alpha y_j(t) = -d_j y_j(t) + \sum_{i=1}^{2} b_{ij} g_i(x_i(t)) \\
\qquad + \sum_{i=1}^{2} h_{ij} g_i(x_i(t - \sigma_{ij}(t))) + J_j
\end{cases}
\tag{4.26}
$$

where $0 \leq \tau_{ji}(t) \leq 0.2$, $0 \leq \sigma_{ij}(t) \leq 0.1$, $i, j = 1, 2$,

$$
f_j(u) = g_i(u) = \frac{1}{2}(|u+1| - |u-1|), \quad i, j = 1, 2, u \in \mathbb{R},
$$

$$
x(t) = \begin{pmatrix} x_1(t) \\ x_2(t) \end{pmatrix}, y(t) = \begin{pmatrix} y_1(t) \\ y_2(t) \end{pmatrix}, C = \begin{pmatrix} 9 & 0 \\ 0 & 9 \end{pmatrix}, D = \begin{pmatrix} 6 & 0 \\ 0 & 6 \end{pmatrix},
$$

$$
A = \begin{pmatrix} 1/2 & 1/2 \\ 1/2 & -1/2 \end{pmatrix}, B = \begin{pmatrix} 1/3 & 1/3 \\ -1/3 & 1/3 \end{pmatrix},
$$

$$
W = \begin{pmatrix} 1/2 & -1/2 \\ 1/2 & 1/2 \end{pmatrix}, H = \begin{pmatrix} -1/3 & 1/3 \\ 1/3 & 1/3 \end{pmatrix},
$$

and

$$
I = \begin{pmatrix} I_1 \\ I_2 \end{pmatrix}, J = \begin{pmatrix} J_1 \\ J_2 \end{pmatrix}.
$$

and initial conditions

$$
\begin{cases}
x_i(t; 0, \varphi) = \varphi_i(t), t \in [-0.1, 0], i = 1, 2, \\
y_j(t; 0, \phi) = \phi_j(t), t \in [-0.2, 0], j = 1, 2.
\end{cases}
$$

Upon substituting $I_1 = I_2 = 0.875$ and $J_1 = J_2 = 1.416667$, we find that system (4.26) has an equilibrium $x_1^* = x_2^* = 0.125$, $y_1^* = y_2^* = 0.25$.

Let $P = \begin{pmatrix} 2 & 0 \\ 0 & 2 \end{pmatrix}$ and $Q = \begin{pmatrix} 1 & 0 \\ 0 & 1 \end{pmatrix}$. Since $L = M = \begin{pmatrix} 1 & 0 \\ 0 & 1 \end{pmatrix}$, we have

$$
\lambda_{\max}(P)\left(-2||C|| + ||A|| \, ||L|| + ||W|| \, ||L|| \frac{\lambda_{\max}(P) + \lambda_{\min}(Q)}{\lambda_{\min}(Q)}\right)
$$

$$
+ \lambda_{\max}(Q)\left(||B|| \, ||M|| + ||H|| \, ||M|| \frac{\lambda_{\max}(P)}{\lambda_{\min}(P)}\right) = -36 + \frac{14}{3}\sqrt{2} < 0,
$$

and

$$\lambda_{\max}(Q)\left(-2||D||+||B||\,||M||+||H||\,||M||\frac{\lambda_{\max}(Q)+\lambda_{\min}(P)}{\lambda_{\min}(P)}\right)$$

$$+\lambda_{\max}(P)\left(||A||\,||L||+||W||\,||L||\frac{\lambda_{\max}(Q)}{\lambda_{\min}(Q)}\right)=-12+\frac{17}{6}\sqrt{2}<0.$$

Thus, all conditions of Theorem 4.3 are satisfied and the equilibrium $x_1^* = x_2^* = 0.125$, $y_1^* = y_2^* = 0.25$ of (4.26) is globally Mittag–Leffler stable.

Next, we shall study the global Mittag–Leffler stability (global asymptotic stability) of some fractional neural networks with dynamical thresholds.

Consider the following Caputo fractional impulsive delayed cellular neural network with dynamical thresholds

$$\begin{cases} {}_0^C D_t^\alpha x(t) = -x(t) + af\left(x(t) - bx(t-\tau(t)) - c\right), \ t \neq t_k, \ t \geq 0, \\ \Delta x(t_k) = x(t_k^+) - x(t_k) = I_k(x(t_k)), \ k = 1, 2, ..., \end{cases} \tag{4.27}$$

where $0 < \alpha < 1$, $a > 0$, b and c are nonnegative constants, $f : \mathbb{R} \to \mathbb{R}$, $\tau(t)$ corresponds to the transmission delay and satisfies $0 \leq \tau(t) \leq \tau$ ($\tau = const$), $t - \tau(t) \to \infty$ as $t \to \infty$, $I_k : \mathbb{R} \to \mathbb{R}$, $k = 1, 2, ..., t_k < t_{k+1} < ..., \lim_{k \to \infty} t_k = \infty$.

Note that for $\alpha = 1$ the equation (4.27) is investigated in [Stamova 2009], and generalizes numerous non-impulsive neural network models.

Let $\varphi \in PCB[[-\tau, 0], \mathbb{R}]$. Denote by $x(t) = x(t; 0, \varphi)$, $x \in \mathbb{R}$ the solution of (4.27), satisfying the initial conditions

$$\begin{cases} x(t; 0, \varphi) = \varphi(t), \ -\tau \leq t \leq 0, \\ x(0^+; 0, \varphi) = \varphi(0). \end{cases} \tag{4.28}$$

Introduce the following notation:
$|\varphi|_\tau = \sup_{\theta \in [-\tau, 0]} |\varphi(\theta)|$ is the norm of the function $\varphi \in PCB[[-\tau, 0], \mathbb{R}]$.

We introduce the following conditions:

H4.8 There exists a constant $L > 0$ such that

$$|f(u) - f(v)| \leq L|u - v|$$

for all $u, v \in \mathbb{R}$.

H4.9 There exists a constant $M > 0$ such that for all $u \in \mathbb{R}$

$$|f(u)| \leq M < \infty.$$

H4.10 $a > 0, b \geq 0, a(1-b) < 1$.

H4.11 For any $k = 1, 2, \ldots$ the functions I_k are continuous in $\mathbb{R}$.

H4.12 $0 < t_1 < t_2 < \ldots < t_k < t_{k+1} < \ldots$ and $t_k \to \infty$ as $k \to \infty$.

H4.13 There exists a unique equilibrium x^* of the equation (4.27) defined on $[0, \infty)$ such that

$$x^* = af(x^* - bx^* - c),$$

$$I_k(x_i^*) = 0, \ \ k = 1, 2, \ldots.$$

The main results here are obtained by means of piecewise continuous Lyapunov functions $V : [0, \infty) \times \mathbb{R} \to \mathbb{R}_+$ such that $V \in V_0$.

Let $y(t) = x(t) - bx(t - \tau(t)) - c$. Using the fact that the Caputo fractional derivative of a constant is zero, we transform (4.27) to the form

$$\begin{cases} {}_0^c D_t^\alpha y(t) = -y(t) - c + af(y(t)) - abf(y(t - \tau(t))), \ t \neq t_k, \ t \geq 0, \\ \Delta y(t_k) = J_k(y(t_k)), \ k = 1, 2, \ldots, \end{cases} \tag{4.29}$$

where $J_k(y(t_k)) = I_k(y(t_k) + bx(t_k - \tau(t_k)) + c) - I_k(bx(t_k - \tau(t_k)) + c), k = 1, 2, \ldots$.

Set $u(t) = y(t) - y^*$, where y^* is the equilibrium of (4.29), and consider the following equation

$$\begin{cases} {}_0^c D_t^\alpha u(t) = -u(t) + a[f(u(t) + y^*) - f(y^*)] \\ \qquad\qquad - ab[f(u(t - \tau(t)) + y^*) - f(y^*)], \ t \neq t_k, \ t \geq 0, \\ \Delta u(t_k) = P_k(u(t_k)), \ k = 1, 2, \ldots, \end{cases} \tag{4.30}$$

where $P_k(u) = J_k(u + y^*) - J_k(y^*) = J_k(u + y^*), k = 1, 2, \ldots$.

Theorem 4.4 Assume that:
1. Conditions H4.8–H4.13 hold.
2. There exists a constant $d > 0$ such that

$$0 < d \leq 1 - La(1 + b).$$

3. The functions P_k are such that

$$P_k(u(t_k)) = -\sigma_k u(t_k), \ \ 0 < \sigma_k < 2, k = 1, 2, \ldots.$$

Then the equilibrium x^* *of* (4.27) is globally Mittag–Leffler stable.

Proof. We define a Lyapunov function

$$V(t, u) = \frac{1}{2} u^2.$$

Then for $t = t_k$, from the condition 3 of Theorem 4.4, we obtain

$$V(t_k^+, u(t_k) + P_k(u(t_k))) = \frac{1}{2}(u(t_k) + P_k(u(t_k)))^2$$

$$= \frac{1}{2}(1 - \sigma_k)^2 u^2(t_k) < V(t_k, u(t_k)), \ k = 1, 2, ... \qquad (4.31)$$

Let $t \geq 0$ and $t \neq t_k$. Then, by Lemma 1.8 for the Caputo upper right-hand derivative ${}^cD_+^\alpha V(t, u(t))$ of V with respect to system (4.30) we get

$$^cD_+^\alpha V(t, u(t)) = \frac{1}{2}{}^cD_+^\alpha \left(u^2(t) \right) \leq u(t)^c D_+^\alpha u(t)$$

$$\leq u(t)\left(-u(t) + a\left[f(u(t) + y^*) - f(y^*) \right] \right.$$

$$\left. -ab\left[f(u(t - \tau(t)) + y^*) - f(y^*) \right] \right)$$

$$= -u^2(t) + au(t)\left[f(u(t) + y^*) - f(y^*) \right] - abu(t)\left[f(u(t - \tau(t)) + y^*) - f(y^*) \right].$$

Since, for the function f assumption H4.8 is true, then

$$^cD_+^\alpha V(t, u(t)) \leq -u^2(t) + aLu^2(t) + abLu(t)u(t - \tau(t)), \ t \neq t_k, \ k = 1, 2, ...$$

From the above estimate for any solution $u(t)$ of (4.30) such that $V(\theta, u(\theta)) \leq V(t, u(t))$, $t - \tau \leq \theta \leq t$ for $t \neq t_k$, $k = 1, 2, ...$, we have

$$^cD_+^\alpha V(t, u(t)) \leq (-1 + aL(1 + b))2V(t, u(t)) \leq -2dV(t, u(t)), \qquad (4.32)$$

where, by condition 2 of Theorem 4.4, $d > 0$.

From (4.31) and (4.32) by Corollary 1.3, we have

$$V(t, u(t)) \leq V(0^+, u(0^+))E_\alpha(-2d\,t^\alpha), \ t \geq 0,$$

or

$$|u(t)| \leq |u(0^+)|\{E_\alpha(-2d\,t^\alpha)\}^{1/2} \leq |u|_\tau \{E_\alpha(-2d\,t^\alpha)\}^{1/2}, \ t \geq 0.$$

Thus, the zero solution of system (4.30) is globally Mittag–Leffler stable, and therefore the equilibrium x^* of (4.27) is globally Mittag–Leffler stable (and hence, globally asymptotically stable).

We consider the following Caputo fractional impulsive delayed cellular neural network model with dynamical thresholds

$$\begin{cases} {}_0^cD_t^\alpha x(t) = -x(t) + af\left(x(t) - c - b\int_0^\infty m(\theta)x(t - \theta)\,d\theta \right), \ t \neq t_k, \\ \Delta x(t_k) = x(t_k^+) - x(t_k) = I_k(x(t_k)), \ k = 1, 2, ..., \end{cases}$$

$$(4.33)$$

where $0 < \alpha < 1$, $t \geq 0$, $x : \mathbb{R}_+ \to \mathbb{R}$, $m : \mathbb{R}_+ \to \mathbb{R}_+$ is delayed-ker-function, $a > 0$, b and c are nonnegative constants, $f : \mathbb{R} \to \mathbb{R}$, $I_k : \mathbb{R} \to \mathbb{R}$, $k = 1, 2, ...$, $0 < t_1 < t_2 < ... < t_k < t_{k+1} < ...$, $\lim_{k \to \infty} t_k = \infty$.

Let $\varphi \in PCB[(-\infty, 0], \mathbb{R}]$. Denote by $x(t) = x(t; 0, \varphi)$, $x \in \mathbb{R}$ the solution of equation (4.33), satisfying the initial conditions

$$\begin{cases} x(t; 0, \varphi) = \varphi(t), & -\infty < t \leq 0, \\ x(0^+; 0, \varphi) = \varphi(0). \end{cases} \tag{4.34}$$

Let $|\varphi|_\infty = \sup_{\theta \in (-\infty, 0]} |\varphi(\theta)|$ be the norm of the function $\varphi \in PCB[(-\infty, 0], \mathbb{R}]$.

We introduce the following conditions:

H4.14 $\displaystyle\int_0^\infty m(\theta) \, d\theta = 1$.

H4.15 $\displaystyle\int_0^\infty \theta m(\theta) \, d\theta < \infty$.

H4.16 There exists a unique equilibrium x^* of the equation (4.33) defined on $[0, \infty)$ such that

$$x^* = f\left(x^* - c - b \int_0^\infty m(\theta) x^* \, d\theta\right),$$

$$I_k(x_i^*) = 0, \quad k = 1, 2, ...$$

Further on we shall use piecewise continuous Lyapunov functions $V : \mathbb{R}_+ \times \mathbb{R} \to \mathbb{R}_+$ such that $V \in V_0$.

Set $y(t) = x(t) - x^*$ and consider the following system:

$$\begin{cases} {}_0^c D_t^\alpha y(t) = -y(t) + af\left(y(t) - b \int_0^\infty m(s)(y(t-s) + x^*) \, ds \right. \\ \qquad\qquad \left. + x^* - c\right) - x^*, \; t \neq t_k, \; t \geq 0, \\ \Delta y(t_k) = J_k(y(t_k)), \; k = 1, 2, ..., \end{cases} \tag{4.35}$$

where $J_k(y) = I_k(y + x^*)$, $k = 1, 2, ...$

Theorem 4.5 Assume that:
1. Conditions H4.8–H4.12, H4.14–H4.16 hold.
2. $La(1 + b) < 1$.
3. The functions J_k are such that

$$|y(t_k) + J_k(y(t_k))| \leq |y(t_k)|, \; y \in \mathbb{R}, \; t_k > 0.$$

Then the equilibrium x^* of (4.33) is globally Mittag–Leffler stable.

Proof. We define a Lyapunov function

$$V(t, y) = |y|.$$

Then, for $t = t_k$, from the condition 3 of Theorem 4.5, we obtain

$$V(t_k^+, y(t_k) + J_k(y(t_k))) = |y(t_k) + J_k(y(t_k))|$$
$$\leq |y(t_k)| = V(t_k, y(t_k)), \, k = 1, 2, ... \quad (4.36)$$

Let $t \geq 0$ and $t \neq t_k$. Then, since

$${}_0^c D_t^\alpha |u(t)| = sgn(u(t)) {}_0^c D_t^\alpha u(t),$$

for the upper right-hand derivative ${}^c D_+^\alpha V(t, y(t))$ of $V(t, y(t))$ in the Caputo sense with respect to (4.35) we get

$${}^c D_+^\alpha V(t, y(t))$$

$$\leq sgn(y(t)) \left(-y(t) + af \left(y(t) - b \int_0^\infty m(\theta)(y(t - \theta) + x^*) d\theta + x^* - c \right) - x^* \right).$$

Since x^* is the equilibrium of (4.33), then for $t \geq 0$ and $t \neq t_k$ in view of H4.16 it satisfies the following equation

$$-x^* + af \left(x^* - b \int_0^\infty m(\theta) x^* d\theta - c \right) = 0. \quad (4.37)$$

From (4.37) and condition H4.8, it follows

$${}^c D_+^\alpha V(t, y(t)) \leq -|y(t)| + aL \left| y(t) - b \int_0^\infty m(\theta) y(t - \theta) d\theta \right|$$

$$\leq -|y(t)| + aL|y(t)| + abL \int_0^\infty m(\theta) |y(t - \theta)| d\theta.$$

From the above estimate for any solution $y(t)$ of (4.35) such that $V(\theta, y(\theta)) \leq V(t, y(t))$, $t - \tau \leq \theta \leq t$ for $t \neq t_k$, $k = 1, 2, ...$, we have

$${}^c D_+^\alpha V(t, y(t)) \leq (-1 + aL(1 + b)) V(t, y(t)),$$

and from condition 2 of Theorem 4.5 it follows that there exists $w > 0$ such that

$${}^c D_+^\alpha V(t, y(t)) \leq -wV(t, y(t)). \quad (4.38)$$

From (4.36) and (4.38) and Corollary 1.3, we have

$$V(t, y(t)) \leq V(0^+, y(0^+)) E_\alpha(-w t^\alpha), \, t \geq 0,$$

or

$$|y(t)| \leq |y(0^+)| E_\alpha(-w t^\alpha) \leq |y|_\infty E_\alpha(-w t^\alpha), \, t \geq 0.$$

Thus, the zero solution of system (4.35) is globally Mittag–Leffler stable, and hence the equilibrium x^* of (4.33) is globally Mittag–Leffler stable (globally asymptotically stable).

4.1.2   Almost Periodic Solutions

In the next section, we shall study the existence of almost periodic solutions for a Caputo fractional impulsive DCCN of the type (4.1), and we shall investigate the effect of the impulses on the global asymptotic stability behavior of such solutions.

Let $0 < \alpha < 1$. Consider the fractional impulsive DCCN

$$\begin{cases} {}_{t_0}^{c}D_t^{\alpha}x_i(t) = -c_i x_i(t) + \displaystyle\sum_{j=1}^{n} a_{ij}(t)f_j(x_j(t)) \\ \qquad\qquad + \displaystyle\sum_{j=1}^{n} b_{ij}(t)f_j(x_j(t-\tau_j(t))) + I_i,\ t \neq t_k, \\ \Delta x_i(t_k) = x_i(t_k^+) - x_i(t_k) = P_{ik}x_i(t_k),\ k = \pm 1, \pm 2, ..., \end{cases} \tag{4.39}$$

$i = 1, 2, ..., n$, where $\{t_k\} \in \mathscr{B}$, $a_{ij}(t)$ denotes the strength of the jth unit on the ith unit at time t, $b_{ij}(t)$ denotes the strength of the jth unit on the ith unit at time $t - \tau_j(t)$, P_{ik}, $i = 1, 2, ..., n$, $k = \pm 1, \pm 2, ...$ are constants.

Let, for simplicity denote by $X(t,x) = (X_1(t,x), X_2(t,x), ..., X_n(t,x))$,

$$X_i(t,x) = -c_i x_i(t) + \sum_{j=1}^{n} a_{ij}(t)f_j(x_j(t)) + \sum_{j=1}^{n} b_{ij}(t)f_j(x_j(t-\tau_j(t))) + I_i,$$

where $i = 1, 2, ..., n$ and let $P_k = diag[P_{1k}, P_{2k}, ..., P_{nk}]$ be a diagonal $n \times n$ matrix.

We introduce the following conditions:

H4.17 The functions $a_{ij}(t)$, $b_{ij}(t)$ and $\tau_j(t)$, $i, j = 1, 2, ..., n$ are almost periodic on t.

H4.18 There exist constants $H_i > 0$ such that

$$|f_i(u)| \leq H_i$$

for all $u \in \mathbb{R}$, $i = 1, 2, ..., n$.

H4.19 The sequences of constants $\{P_{ik}\}$ are almost periodic uniformly on $i = 1, 2, ..., n$ and

$$-1 < P_{ik} \leq 0$$

for $i = 1, 2, ..., n$, $k = \pm 1, \pm 2, ...$

H4.20 The function $\varphi_0 \in \mathscr{PC}$ is almost periodic.

Let conditions H3.1, H4.1, H4.17–H4.20 hold, and let $\{s'_m\}$ be an arbitrary sequence of real numbers. Then, there exists a subsequence $\{s_p\}$, $s_p = s'_{m_p}$ such that the system (4.39) moves to the system

$$\begin{cases} {}^{c}_{t_0}D^{\alpha}_t x_i(t) = -c_i x_i(t) + \sum_{j=1}^{n} a^s_{ij}(t) f_j(x_j(t)) \\ \qquad + \sum_{j=1}^{n} b^s_{ij}(t) f_j(x_j(t - \tau^s_j(t))) + I_i, \ t \neq t^s_k, \\ \Delta x_i(t^s_k) = x_i(t^{s+}_k) - x_i(t^s_k) = P^s_{ik} x_i(t^s_k), \ k = \pm 1, \pm 2, ... \end{cases} \tag{4.40}$$

and the set of systems in the form (4.40) we shall denote by $H(X, P_k, t_k)$.

Theorem 4.6 Assume that:
1. Conditions H3.1, H4.1, H4.17–H4.20 are fulfilled.
2. There exists a positive number A such that for ant $t \geq t_0$ the system parameters $a_{ij}(t)$, $b_{ij}(t)$, c_i $(i, j = 1, 2, ..., n)$ satisfy

$$\min_{1 \leq i \leq n} \left(c_i - L_i \sum_{j=1}^{n} |a_{ji}(t)| \right) - \max_{1 \leq i \leq n} \left(L_i \sum_{j=1}^{n} |b_{ji}(t)| \right) = A > 0.$$

3. There exists a solution $x(t; t_0, \varphi_0)$ of (4.39) such that

$$||x(t; t_0, \varphi_0)|| < q,$$

where $t \geq t_0$, $q > 0$.

Then for the system (4.39) there exists a unique almost periodic solution $\tilde{x}(t)$ such that:
(a) $||\tilde{x}(t)|| \leq q_1$, $q_1 < q$;
(b) $H(\tilde{x}, t_k) \subset H(X, P_k, t_k)$.

Proof. Let $t_0 \in \mathbb{R}$. We define a Lyapunov function

$$V(x, y) = \sum_{i=1}^{n} |x_i - y_i|.$$

Let $\{s_p\}$ be any sequence of real numbers such that $s_p \to \infty$ as $p \to \infty$ and $\{s_p\}$ moves the system (4.39) to a system at $H(X, P_k, t_k)$.

For any real number γ, let $p_0 = p_0(\gamma)$ be the smallest value of p, such that $s_{p_0} + \gamma \geq t_0$. Since $||x(t; t_0, \varphi_0)|| \leq q_1$, $q_1 < q$ for all $t \geq t_0$, then $||x(t + s_p; t_0, \varphi_0)|| \leq q_1$ for $t \geq \gamma$, $p \geq p_0$.

Let U, $U \subset (\gamma, \infty)$ be compact. Then, for any $\varepsilon > 0$, choose an integer $n_0(\varepsilon, \gamma) \geq p_0(\gamma)$, so large that for $l \geq p \geq n_0(\varepsilon, \gamma)$ and $t \in \mathbb{R}$, $t \neq t_k$, $k = \pm 1, \pm 2, ...$, we have $V(\varphi(\theta + s_p), \varphi(\theta + s_l)) \leq V(\varphi(s_p), \varphi(s_l))$ for $-\tau \leq \theta \leq 0$, $\varphi \in \mathscr{PC}$,

$$4\gamma E_\alpha(-A(t + s_p - s_l)^\alpha) < \varepsilon, \tag{4.41}$$

and

$$|a_{ij}(t + s_p) - a_{ij}(t + s_l)| \leq A\varepsilon \left(4 \sum_{j=1}^{n} H_j \right)^{-1}. \tag{4.42}$$

Then for $t = t_k$ by the condition H4.19 we get

$$V(x(t_k^+ + s_p), x(t_k^+ + s_l)) = \sum_{i=1}^{n} |x_i(t_k^+ + s_p) - x_i(t_k^+ + s_l)|$$

$$= \sum_{i=1}^{n} |x_i(t_k + s_p) + P_{ik}x_i(t_k + s_p) - x_i(t_k + s_l) - P_{ik}x_i(t_k + s_l)|$$

$$\leq \sum_{i=1}^{n} (1 + P_{ik})|x_i(t_k + s_p) - x_i(t_k + s_l)|$$

$$\leq V(x(t_k + s_p), x(t_k + s_l))|. \tag{4.43}$$

Let $t \geq t_0$. As in Theorem 4.1, we can prove that for $t + s_\kappa \neq t_k$, $\kappa = p, l$, $k = \pm 1, \pm 2, ...$, we have

$$_{t_0}^{c}D_t^{\alpha}|x_i(t + s_p) - x_i(t + s_l)|$$

$$= sgn(x_i(t + s_p) - x_i(t + s_l)) _{t_0}^{c}D_t^{\alpha}(x_i(t + s_p) - x_i(t + s_l)).$$

Then for $t \geq t_0$, $t \in [t_{k-1}, t_k)$, $\varphi \in \mathscr{PC}$ for the upper right-hand derivative

$$^{c}D_+^{\alpha}V(\varphi(s_p), \varphi(s_l))$$

with respect to system (4.39), we have

$$^{c}D_+^{\alpha}V(\varphi(s_p), \varphi(s_l)) \leq \sum_{i=1}^{n} \left[-c_i|\varphi_i(s_p) - \varphi_i(s_l)| \right.$$

$$+ \sum_{j=1}^{n} \left[L_j|a_{ij}(t + s_p)||\varphi_j(s_p) - \varphi_j(s_l)| + \varepsilon H_j \right]$$

$$+ \sum_{j=1}^{n} L_j \left[|b_{ij}(t + s_p)||\varphi_j(s_p - \tau_j(s_p)) - \varphi_j(s_l - \tau_j(s_l))| + \varepsilon H_j \right] \right]$$

$$= -\sum_{i=1}^{n} \left[c_i - L_i \sum_{j=1}^{n} |a_{ji}(t + s_p)| \right] |\varphi_i(s_p) - \varphi_i(s_l)|$$

$$+ \sum_{j=1}^{n} \sum_{i=1}^{n} L_j|b_{ij}(t + s_p)||\varphi_j(s_p - \tau_j(s_p)) - \varphi_j(s_l - \tau_j(s_l))|$$

$$\leq -\min_{1 \leq i \leq n} \left(c_i - L_i \sum_{j=1}^{n} |a_{ji}(t + s_p)| \right) \sum_{i=1}^{n} |\varphi_i(s_p) - \varphi_i(s_l)|$$

$$+ \max_{1 \leq i \leq n} \left(L_i \sum_{j=1}^{n} |b_{ji}(t + s_p)| \right) \sum_{i=1}^{n} |\varphi_i(s_p - \tau_j(s_p)) - \varphi_i(s_l - \tau_j(s_l))| + \frac{A\varepsilon}{2}$$

$$\leq -k_1 V(\varphi(s_p), \varphi(s_l)) + k_2 \sup_{-\tau \leq \theta \leq 0} V(\varphi(s_p + \theta), \varphi(s_l + \theta)) + \frac{A\varepsilon}{2},$$

where

$$k_1 = \min_{1 \leq i \leq n} \left(c_i - L_i \sum_{j=1}^{n} |a_{ji}(t + s_p)| \right) > 0,$$

$$k_2 = \max_{1 \le i \le n} \left(L_i \sum_{j=1}^{n} |b_{ji}(t + s_p)| \right) > 0.$$

Now for $\varphi(t + s_p)$, $\varphi(t + s_l) \in \mathscr{PC}$ which satisfy the Razumikhin condition

$$V(\varphi(s_p + \theta), \varphi(s_l + \theta)) \le V(x(s_p), x(s_l)), \quad -\tau \le \theta \le 0,$$

by virtue of condition 2 of Theorem 4.6, it follows that

$$^{c}D_{+}^{\alpha}V(\varphi(s_p), \varphi(s_l)) \le -A\left(V(\varphi(s_p), \varphi(s_l)) - \frac{\varepsilon}{2} \right), \tag{4.44}$$

$t \ne t_k$, $t > t_0$.

So, for $t \in U$ from (4.41), using (4.43), (4.44) and Corollary 3.2, for $t \in [t_0, \infty)$ we get

$$||x(t + s_p) - x(t + s_l)|| = \sum_{i=1}^{n} |x_i(t + s_p) - x_i(t + s_l)|$$

$$\le E_\alpha(-A(t + s_p - s_l)^\alpha) \sup_{-\tau \le \theta \le 0} ||\varphi_0(s_p + \theta) - \varphi_0(s_l + \theta)|| + \frac{\varepsilon}{2}$$

$$= ||\varphi_0(s_p) - \varphi_0(s_l)||_\tau E_\alpha(-A(t + s_p - s_l)^\alpha) + \frac{\varepsilon}{2} < \varepsilon. \tag{4.45}$$

Consequently, there exists a function $\tilde{x}(t) = (\tilde{x}_1(t), \tilde{x}_2(t), ..., \tilde{x}_n(t))$, such that $x(t + s_p) - \tilde{x}(t) \to \infty$ for $p \to \infty$. Since γ is arbitrary, it follows that $\tilde{x}(t)$ is defined uniformly on $t \in \mathbb{R}$.

On the other hand, since $x(t; t_0, \varphi_0)$ is a solution of (4.39), (4.2), for $t + s_\kappa \ne t_k$, $\kappa = p, l$; $k = \pm 1, \pm 2, ...$, by similar arguments as above we can prove that

$$||^{c}_{t_0}D_t^{\alpha}\varphi(s_p) - ^{c}_{t_0}D_t^{\alpha}\varphi((s_l))|| < \varepsilon$$

which shows that $\lim_{p \to \infty} ^{c}_{t_0}D_t^{\alpha}\varphi(s_p)$ exists uniformly on all compact subsets of $\mathbb{R}$.

Let now $\lim_{p \to \infty} ^{c}_{t_0}D_t^{\alpha}\varphi(s_p) = ^{c}_{t_0}D_t^{\alpha}\tilde{x}(t)$, and

$$^{c}_{t_0}D_t^{\alpha}\tilde{x}(t) = \lim_{p \to \infty} \left[X(t + s_p, x(t + s_p)) - X(t + s_p, \tilde{x}(t)) + X(t + s_p, \tilde{x}(t)) \right]$$

$$= X^s(t, \tilde{x}(t)), \tag{4.46}$$

where $t \ne t_k^s$, $t_k^s = \lim_{p \to \infty} t_{k+p(s)}$, and

$$X^s(t, x) = (X_1^s(t, x), X_2^s(t, x), ..., X_n^s(t, x)),$$

$$X_i^s(t, x) = -c_i x_i(t) + \sum_{j=1}^{n} a_{ij}^s(t) f_j(x_j(t)) + \sum_{j=1}^{n} b_{ij}^s(t) f_j(x_j(t - \tau_j^s(t))) + I_i,$$

$i = 1, 2, ..., n.$

On the other hand, for $t + s_p = t_k^s$ it follows

$$\tilde{x}(t_k^{s+}) - \tilde{x}(t_k^{s-}) = \lim_{p \to \infty} (x(t_k^s + s_p + 0) - x(t_k^s + s_p - 0))$$

$$= \lim_{p \to \infty} P_k^s x(t_k^s + s_p) = P_k^s \tilde{x}(t_k^s), \tag{4.47}$$

where $P_k^s = diag[P_{1k}^s, P_{2k}^s, ..., P_{nk}^s]$.

From (4.46) and (4.47), we get that $\tilde{x}(t)$ is a solution of (4.40).

To show that $\tilde{x}(t)$ is an almost periodic function, let we consider the sequence $\{s_p\}$ that moves the system (4.39) to $H(X, P_k, t_k)$ and the Lyapunov function

$$V(\varphi(0), \varphi(s_p - s_l)).$$

Then by (4.41) and (4.42), we have

$$^c D_+^\alpha V(\varphi(0), \varphi(s_p - s_l)) = ||\varphi(0) - \varphi(s_p - s_l)||$$

$$\leq -AV(\varphi(0), \varphi(s_p - s_l)) + \frac{A\varepsilon}{2}. \tag{4.48}$$

Again,

$$V(\varphi(0) + P_k^s \varphi(t_k^s), \varphi(t_k^s + s_l - s_p) + P_k^s \varphi(t_k^s + s_l - s_p))$$

$$\leq V(\varphi(0), \varphi(s_l - s_p)). \tag{4.49}$$

From (4.48), (4.49) and Corollary 1.5 it follows

$$V(\tilde{x}(t + s_p), \tilde{x}(t + s_l))$$

$$\leq E_\alpha(-A(s_p)^\alpha) V(\tilde{x}(t), \tilde{x}(t + s_p - s_l)) + \frac{\varepsilon}{2} < \varepsilon.$$

From the last inequality for $l \geq p \geq m_0(\varepsilon)$, we have

$$||\tilde{x}(t + s_p) - \tilde{x}(t + s_l)|| < \varepsilon. \tag{4.50}$$

Now, from the definitions of the sequence $\{s_p\}$ and for $l \geq p \geq m_0(\varepsilon)$ it follows that $\rho(t_k + s_p, t_k + s_l) < \varepsilon$.

Then from (4.50) and the last inequality we obtain that the sequence $\tilde{x}(t + s_p)$ converges uniformly to the function $\tilde{x}(t)$.

The assertions (a) and (b) of the theorem follow immediately.

Next, we shall consider the asymptotic stability of almost periodic solutions of (4.39).

Let $\varphi_1 \in \mathscr{PC}$. Denote by $x_1(t) = x_1(t; t_0, \varphi_1)$ the solution of (4.39), satisfying the initial conditions

$$\begin{cases} x_1(t; t_0, \varphi_1) = \varphi_1(t - t_0), t_0 - \tau \leq t \leq t_0, \\ x_1(t_0^+; t_0, \varphi_1) = \varphi_1(0). \end{cases}$$

By the next definition we shall introduce the notion of global perfect uniform-Mittag–Leffler stability.

Definition 4.3 The solution $x_1(t)$ of (4.39) is said to be *globally perfectly uniform-Mittag–Leffler stable*, if it is uniformly stable, globally Mittag–Leffler stable, and the solutions of (4.39) are uniformly bounded.

Theorem 4.7 Assume that conditions of Theorem 4.6 are met. Then the almost periodic solution of (4.39) is globally perfectly uniform-Mittag–Leffler stable.

Proof. Let $\overline{\omega}(t)$ be an arbitrary solution of (4.40). Set

$$u(t) = \overline{\omega}(t) - \tilde{x}(t),$$

$$Y^s(t, u(t)) = X^s(t, u(t) + \tilde{x}(t)) - X^s(t, \tilde{x}(t)).$$

Now we consider the system

$$\begin{cases} {}^c_{t_0}D^\alpha_t u = Y^s(t, u(t)), \ t \neq t^s_k, \\ \Delta u(t^s_k) = P^s_k u(t^s_k), \ k = \pm 1, \pm 2, ..., \end{cases} \tag{4.51}$$

and let $W(t, u(t)) = V(t, \tilde{x}(t), \tilde{x}(t) + u(t))$. Then, from Corollary 3.2 it follows that the zero solution $u(t) = 0$ of (4.51) is globally perfectly uniform-Mittag–Leffler stable, and consequently $\tilde{x}(t)$ is globally perfectly uniform-Mittag–Leffler stable.

Remark 4.7 Since the Mittag–Leffler stability always implies asymptotic stability, then from Theorem 4.7 it follows that the almost periodic solution of the fractional impulsive DCNN (4.39) is globally perfectly uniform-asymptotically stable.

Next, as a special case of the fractional CNN (4.1) we shall consider the following fractional impulsive CNN without delays

$$\begin{cases} {}^c_{t_0}D^\alpha_t x_i(t) = -c_i x_i(t) + \sum_{j=1}^{n} a_{ij}(t) f_j(x_j(t)) + I_i, \ t \neq t_k, \\ \Delta x_i(t_k) = x_i(t^+_k) - x_i(t_k) = P_{ik} x_i(t_k), \ k = \pm 1, \pm 2, ..., \end{cases} \tag{4.52}$$

$i = 1, 2, ..., n, \{t_k\} \in \mathscr{B}.$

The next theorem follows directly from theorems 4.6 and 4.7.

Theorem 4.8 Assume that:

1. Conditions H3.1, H4.1, H4.18 and H4.19 are fulfilled.

2. For $t \geq t_0$ the functions $a_{ij}(t)$ are almost periodic on t and $a_{ij}(t)$, c_i $(i,j = 1,2,...,n)$ satisfy

$$\min_{1 \leq i \leq n} \left(c_i - L_i \sum_{j=1}^{n} |a_{ji}(t)| \right) > 0.$$

3. There exists a solution $x(t;t_0,x_0)$, $x_0 \in \mathbb{R}^n$, of (4.52) such that

$$||x(t;t_0,x_0)|| < q,$$

where $t \geq t_0$, $q > 0$.

Then for the system (4.52) there exists a unique globally perfectly uniform-Mittag–Leffler stable (globally perfectly uniform-asymptotically stable) almost periodic solution $\tilde{x}(t)$ and $||\tilde{x}(t)|| \leq q_1$, $q_1 < q$.

4.1.3 The Uncertain Case

We shall consider the following uncertain impulsive fractional-order CNN

$$\begin{cases} {}^c_{t_0}D^\alpha_t x_i(t) = -(c_i + \tilde{c}_i)x_i(t) \\ \qquad + \sum_{j=1}^{n} \left(a_{ij}(t)f_j(x_j(t)) + \tilde{a}_{ij}(t)\tilde{f}_j(x_j(t)) \right) + I_i + \tilde{I}_i, \ t \neq t_k, \\ \Delta x_i(t_k) = (P_{ik} + \tilde{P}_{ik})x_i(t_k), \ k = \pm 1, \pm 2, ..., \end{cases} \quad (4.53)$$

$i = 1,2,...,n$, $\{t_k\} \in \mathscr{B}$.

The functions $\tilde{f}_j(u)$, $\tilde{a}_{ij}(t)$ and constants $\tilde{I}_i$, $\tilde{c}_i$ $\tilde{P}_{ik}$ represent uncertain terms of system (4.52), and let

$$\min_{u \in \mathbb{R}} m_j(u) \leq |\tilde{f}_j(u)| \leq \max_{u \in \mathbb{R}} m_j(u), \ |\tilde{a}_{ij}(t)| \leq A_{ij}(t),$$

where the functions $m_j(u) : \mathbb{R} \to \mathbb{R}_+$, $A_{ij}(t) : \mathbb{R} \to \mathbb{R}_+$, $i,j = 1,2,...,n$, and are bounded.

Note that, the system (4.52) is the nominal system of system (4.53).

We introduce the following conditions:

H4.21 The functions $\tilde{a}_{ij}(t)$, $i,j = 1,2,...,n$, and the sequences of constants $\{P_{ik}\}$, $\{\tilde{P}_{ik}\}$ are almost periodic.

H4.22 There exist constants $L_{m_i} > 0$ such that

$$\max_{u \in \mathbb{R}} m_j(u) - \min_{u \in \mathbb{R}} m_j(u) \leq L_{m_i}|u - v|, \ u,v \in \mathbb{R}, \ i,j = 1,2,...,n.$$

Theorem 4.9 Assume that:

1. Conditions H4.21 and H4.22 are met.
2. Conditions of Theorem 4.8 are satisfied, and there exists a solution $x(t;t_0,x_0)$ of (4.53) such that $x(t;t_0,x_0) < q$, where $t \geq t_0$, $q > 0$.
3. For $t \geq 't_0$ and $i = 1, 2, ..., n$, $k = \pm 1, \pm 2, ...$ the following inequalities are met

$$\tilde{P}_{ik} \in (-1 - P_{ik}, -P_{ik}], \tag{4.54}$$

$$0 < \max_{1 \leq i \leq n} \left(L_{m_i} \sum_{j=1}^{n} |\tilde{a}_{ji}(t)| - \tilde{c}_i \right) < \min_{1 \leq i \leq n} \left(c_i - L_i \sum_{j=1}^{n} |a_{ji}(t)| \right). \tag{4.55}$$

Then for the system (4.53) there exists a unique almost periodic solution $\tilde{x}(t)$ such that:

1. $\|\tilde{x}(t)\| \leq q$;
2. $\tilde{x}(t)$ is globally perfectly robustly uniform-asymptotically stable.

Proof. Let $t_0 \in \mathbb{R}$, and we define the Lyapunov function

$$V(x,y) = \sum_{i=1}^{n} |x_i - y_i|.$$

Then, for $t = t_k, t \geq t_0$ by (4.54) we get

$$V(x(t_k^+), y(t_k^+)) = \sum_{i=1}^{n} |x_i(t_k^+) - y_i(t_k^+)|$$

$$= \sum_{i=1}^{n} |x_i(t_k) + (P_{ik} + \tilde{P}_{ik})x_i(t_k) - (y_i(t_k) + (P_{ik} + \tilde{P}_{ik})y_i(t_k))|$$

$$\leq \sum_{i=1}^{n} (1 + P_{ik} + \tilde{P}_{ik})|x_i(t_k) - y_i(t_k)| \leq V(x(t_k), y(t_k)). \tag{4.56}$$

On the other hand, for $t \neq t_k$, $k = \pm 1, \pm 2, ..., t \geq t_0$ and $i = 1, 2, ..., n$, we have

$$_{t_0}^{c}D_t^{\alpha}|x_i(t) - y_i(t)| = sign(x_i(t) - y_i(t))_{t_0}^{c}D_t^{\alpha}(x_i(t) - y_i(t)).$$

Then for $t \geq t_0$, $t \in [t_{k-1}, t_k)$, using (4.55) for the upper right derivative $^cD_+^{\alpha}V(x(t),y(t))$ along the solutions of system (4.53) there exists a positive constant $\tilde{A}$ such that

$$^cD_+^{\alpha}V(x(t),y(t)) \leq \sum_{i=1}^{n} \Big[-(c_i + \tilde{c}_i)|x_i(t) - y_i(t)|$$

$$+ \sum_{j=1}^{n} (|a_{ij}(t)|L_j + |\tilde{a}_{ij}(t)|L_{m_j})|x_j(t) - y_j(t)| \Big]$$

$$\leq \sum_{i=1}^{n} \left[-(c_i + \tilde{c}_i) + L_i \sum_{j=1}^{n} |a_{ji}(t)| + L_{m_i} \sum_{j=1}^{n} |\tilde{a}_{ji}(t)| \right] |x_i(t) - y_i(t)|$$

$$\leq -\tilde{A}V(x(t), y(t)). \tag{4.57}$$

Then, in view of (4.56) and (4.57), it follows that all conditions of Theorem 3.6 hold for system (4.53), and the proof of Theorem 4.9 is complete.

4.2 Fractional Impulsive Biological Models

Various applications of the theory of ordinary differential equations and dynamical systems have been found in population dynamics, biology, biotechnologies, chemistry, etc. Indeed, integer-order biological models due to their theoretical and practical significance, have been studied extensively in the literature.

On the other hand, time delays have been incorporated into biological models by many researchers to represent resource regeneration times, maturation periods, feeding times, reaction times, etc. We refer to the monographs of Gopalsamy ([Gopalsamy 1992]), Kuang ([Kuang 1993]) and Takeuchi ([Takeuchi 1996]) and the references cited therein for discussions of general delayed biological systems.

Also, impulsive biological systems have gained increasing popularity during the last few decades. Many results on the dynamical behaviors for impulsive Lotka–Volterra and related systems, such as the permanence, extinction, stability and asymptotic behavior have been found in [Ahmad and Stamova 2007a], [Ahmad and Stamova 2007b], [Ahmad and Stamova 2012], [Ahmad and Stamova 2013], [Ballinger and Liu, 1997], [Dong, Chen and Sun 2006], [Dou, Chen and Li 2004], [Hu 2013], [Jiang and Lu 2007], [Kou, Adimy and Ducrot 2009], [Li and Fan 2007], [Li, Wang, Zhang and Yang 2009], [Liu and Rohlf 1998], [Stamov 2009a], [Stamov 2010a], [Stamov 2012], [Stamova 2009], [Stamova 2010], [Stamova 2011b], [Zhang and Teng 2011].

In the past decade, fractional calculus has been incorporated into some biological models (see [Abbas, Banerjee and Momani 2011], [Agrawal, Srivastava and Das 2012], [Ahmed, El-Sayed and El-Saka 2007], [Das and Gupta 2011], [El-Saka, Ahmed, Shehata and El-Sayed 2009], [El-Sayed, El-Mesiry and El-Saka 2007], [El-Sayed, Rida and Arafa 2009]). Fractional biological models are generalizations of biological models to arbitrary non-integer orders. Indeed, recent investigations have shown that many physical systems can be represented more accurately through fractional derivative formulation (see [Das 2011], [Hilfer 2000], [Kilbas, Srivastava and Trujillo 2006], [Magin 2006], [Miller and Ross 1993], [Podlubny 1999]).

In this section, impulsive fractional biological systems involving Caputo fractional derivatives will be proposed. Qualitative properties of the solutions will be investigated.

4.2.1 Hutchinson's Model

The delay differential equation

$$\dot{x}(t) = rx(t)\left(1 - \frac{x(t-\tau)}{K}\right), t \geq 0, \tag{4.58}$$

called Hutchinson's equation (see [Hutchinson 1948]) is a single species population growth model, where $x(t)$ is the population number at time t, $K > 0$ is the carrying capacity for x, $\tau > 0$ is a constant delay and represents the maturation period. The constant r is called *intrinsic growth rate*, or the growth rate in the absence of any limiting factors.

The biological model (4.58) is a delayed generalization of the basic logistic equation, and has been studied by many authors; see for example [Cunningham 1954], [Gopalsamy 1992], [Kuang 1993].

The impulsive generalization of the model (4.58), i.e. the case where at certain moments of time $t_1, t_2, ..., 0 < t_1 < t_2 < ..., \lim_{k \to \infty} t_k = \infty$, impulsive factors act on the population momentarily, so that the population number $x(t)$ varies by jumps, and some related models have been investigated in [Ahmad and Stamova 2013], [Stamov 2012], [Stamova 2009]. Precisely, numerous authors investigated qualitative properties of the solutions of the impulsive Hutchinson's equation of the form

$$\begin{cases} \dot{x}(t) = rx(t)\left(1 - \frac{x(t-\tau)}{K}\right), t \neq t_k, t \geq 0, \\ \\ \Delta x(t_k) = x(t_k^+) - x(t_k) = d_k(x(t_k) - K), k = 1, 2, ..., \end{cases} \tag{4.59}$$

where the values $x(t_k)$ and $x(t_k^+)$ are the population numbers of the species before and after the impulsive effect at the time t_k, respectively, and $d_k > -1$ for all $k = 1, 2, ...$

In this section we shall investigate a Caputo fractional generalization of the model (4.59) in the form

$$\begin{cases} {}_0^C D_t^\alpha x(t) = rx(t)\left(1 - \frac{x(t-\tau)}{K}\right), t \neq t_k, t \geq 0, \\ \\ \Delta x(t_k) = x(t_k^+) - x(t_k) = d_k(x(t_k) - K), k = 1, 2, ..., \end{cases} \tag{4.60}$$

where $0 < \alpha < 1, -1 < d_k \leq 0, k = 1, 2, ...$

Due to the applicable point of view, we are interested in only those solutions of (4.60) corresponding to initial conditions of the form

$$x(\theta) \geq 0, \ \theta \in [-\tau, 0], \ x(0^+) = x_0 > 0.$$

It is easy to show that, as in the integer-order case, the point $x^* = K$ is an equilibrium of the model (4.60).

Choose

$$V(t,x) = (x - K)^2.$$

For $t \geq 0$ and $t \neq t_k$, for the upper right-hand derivative $^cD_+^\alpha V(t, x(t))$ along the solutions of (4.60) we have

$$^cD_+^\alpha V(t, x(t)) = {}^cD_+^\alpha \left((x(t) - K)^2 \right).$$

By Lemma 1.8 we obtain

$$^cD_+^\alpha V(t, x(t)) \leq 2(x(t) - K)^c D_+^\alpha \left(x(t) - K \right).$$

Using the fact that a Caputo fractional derivative of a constant is zero, for any solution $x(t)$ of (4.60) such that $V(\theta, x(\theta)) \leq V(t, x(t))$, $t - \tau \leq \theta \leq t$ for $t \neq t_k$, $k = 1, 2, ...$, we have

$$^cD_+^\alpha V(t, x(t)) \leq -\frac{2r}{K} x(t) V(t, x(t)). \tag{4.61}$$

Suppose that for any closed interval contained in $t \in (t_{k-1}, t_k]$, $k = 1, 2, ...$, there exist positive numbers r_* and r^* such that

$$r_* \leq x(t) \leq r^*. \tag{4.62}$$

Efficient sufficient conditions, which guarantee the validity of (4.62) for fractional logistic equations can be found in [Abbas, Banerjee and Momani 2011], [Das 2011], [Das and Gupta 2011], [El-Saka, Ahmed, Shehata and El-Sayed 2009], [El-Sayed, El-Mesiry and El-Saka 2007], [El-Sayed, Rida and Arafa 2009].

From (4.61) and (4.62), we have

$$^cD_+^\alpha V(t, x(t)) \leq -\frac{2rr_*}{K} V(t, x(t)), \ t \neq t_k, \ t \geq 0. \tag{4.63}$$

Also,

$$V(t_k^+, x(t_k^+)) = \left(x(t_k^+) - K \right)^2$$

$$= (1 + d_k)^2 \left(x(t_k) - K \right)^2 \leq V(t_k, x(t_k)), \ k = 1, 2, ... \tag{4.64}$$

Then using (4.63), (4.64) and Corollary 1.5, we get

$$V(t, x(t)) \leq \sup_{-\tau \leq \theta \leq 0} V(0^+, x(\theta)) E_\alpha(-wt^\alpha), \ t \geq 0,$$

where $w = \frac{2rr_*}{K}$.

So,

$$|x(t) - K|^2 \leq \sup_{-\tau \leq \theta \leq 0} |x(\theta) - K|^2 E_\alpha(-wt^\alpha), \ t \geq 0,$$

or

$$|x(t) - K| \leq \sup_{-\tau \leq \theta \leq 0} |x(\theta) - K| \left\{ E_\alpha(-wt^\alpha) \right\}^{1/2}, \ t \geq 0.$$

Therefore, the equilibrium K of the equation (4.60) is globally Mittag–Leffler stable (and hence, globally asymptotically stable).

4.2.2 Lasota–Wazewska Models

One of the important problems associated with the study of a model dynamics in an almost periodic environment is the existence of almost periodic solutions, and at the present time many results for integer-order biological models have been obtained (see [Ahmad and Stamov 2009a], [Ahmad and Stamov 2009b], [Ding and Nieto 2013], [He, Chen and Li 2010], [Li and Ye 2013], [Zhou, Wang and Zhou 2013]).

In this part of Section 4.2, we shall consider an impulsive Caputo fractional-order Lasota–Wazewska model with time-varying delays. Existence and uniqueness of almost periodic solutions will be investigated. Some sufficient conditions will be provided for the global perfect uniform-asymptotic stability of such solutions.

It is well known that the functional differential equation

$$\dot{x}(t) = -\mu x(t) + v e^{-\gamma x(t-\tau)} \tag{4.65}$$

was used by Wazewska–Czyzewska and Lasota (see [Wazewska–Czyzewska and Lasota 1976]) as a model for the survival of red blood cells in an animal. Here $x(t)$ denotes the number of red blood cells at time t, $\mu > 0$ is the probability of death of a red blood cell, v and γ are positive constants related to the production of red blood cells per unit time, and τ is the (constant) time delay between the production of immature red blood cells and their maturation for release in circulating bloodstream. The qualitative properties of the Lasota–Wazewska equation (4.65) have been extensively studied by many authors, see for example [Gyori and Ladas 1991], [Kulenovic, Ladas and Sficas 1989], [Qiuxiang and Rong 2006].

In the recent past, attention has been given to impulsive Lasota-Wazewska systems, and interesting results concerning the dynamical behavior of such systems have appeared: see, for instance [Liu and Takeuchi 2007], [Stamov 2012], [Wang, Yu and Niu 2012], [Yan 2003].

Consider the following impulsive Caputo fractional Lasota–Wazewska model with time-varying delays

$$
\begin{cases}
{}^{c}_{t_0}D^{\alpha}_t x(t) = -\mu(t)x(t) + \sum_{j=1}^{m} v_j(t)e^{-\gamma_j(t)x(t-\tau_j(t))}, \ t \neq t_k, \\[2mm]
\Delta x(t_k) = x(t_k^+) - x(t_k) = \mu_k x(t_k) + v_k, \ k = \pm 1, \pm 2, ...,
\end{cases}
\tag{4.66}
$$

where $0 < \alpha < 1$, $\mu(t)$, $v_j(t)$, $\gamma_j(t)$, $\tau_j(t) \in C[\mathbb{R}, \mathbb{R}_+]$, $0 \leq \tau_j(t) \leq \tau$, $j = 1, 2, ..., m$, $\tau = const$, $\mu_k, v_k \in \mathbb{R}$, $k = \pm 1, \pm 2, ...$, $\{t_k\} \in \mathscr{B}$.

Let for simplicity denote

$$
X(t,x) = -\mu(t)x(t) + \sum_{j=1}^{m} v_j(t)e^{-\gamma_j(t)x(t-\tau_j(t))}.
$$

Let $\varphi_0 \in PCB[[-\tau, 0], \mathbb{R}]$ and $t_0 \in \mathbb{R}_+$. Denote by $x(t) = x(t; t_0, \varphi_0)$, $x \in \mathbb{R}$ the solution of (4.66) that satisfies the initial conditions:

$$
\begin{cases}
x(t; t_0, \varphi_0) = \varphi_0(t - t_0), t_0 - \tau \leq t \leq t_0, \\[2mm]
x(t_0^+; t_0, \varphi_0) = \varphi_0(0).
\end{cases}
\tag{4.67}
$$

The solution $x(t) = x(t; t_0, \varphi_0)$ of problem (4.66), (4.67) is a piecewise continuous function with points of discontinuity of the first kind t_k, $k = \pm 1, \pm 2, ...$, at which it is continuous from the left, i.e. the following relations are valid

$$
x(t_k^-) = x(t_k), \ x(t_k^+) = x(t_k) + \mu_k x(t_k) + v_k, \ k = \pm 1, \pm 2, ...
$$

For the system (4.66) like a biological model, as well-known, we shall consider only positive solutions, and in the next, we shall suppose that

$$
x(t) = \varphi_0(t - t_0) \geq 0, \ \sup \varphi_0(\theta) < \infty, \ \varphi_0(0) > 0.
$$

We introduce the following conditions:

H4.23 The functions $\mu(t)$, $v_j(t)$, $\gamma_j(t)$, $\tau_j(t)$, $j = 1, 2, ..., m$ are almost periodic on t and there exist positive numbers μ^*, v^*, γ^* such that:

$$
\mu(t) \leq \mu^*, \ v_j(t) \leq v^*, \ \gamma_j(t) \leq \gamma^*, \ j = 1, 2, ..., m.
$$

H4.24 The sequences of constants $\{\mu_k\}$ and $\{v_k\}$ are almost periodic and $-1 < \mu_k \leq 0$, $k = \pm 1, \pm 2, ...$

H4.25 The function $\varphi_0 \in PCB[[-\tau, 0], \mathbb{R}]$ is almost periodic.

Let conditions H3.1, H4.23–H4.25 hold, and let $\{s'_m\}$ be an arbitrary sequence of real numbers. Then it follows from Lemma 3.5 that there exists a subsequence $\{s_p\}$, $s_p = s'_{m_p}$, such that the system (4.66) moves to the system

$$
\begin{cases}
{}^c_{t_0}D^\alpha_t x(t) = -\mu^s(t)x(t) + \sum_{j=1}^{m} v^s_j(t)e^{-\gamma^s_j(t)x(t-\tau^s_j(t))}, \ t \neq t^s_k, \\
\Delta x(t^s_k) = x(t^{s+}_k) - x(t^s_k) = \mu^s_k x(t^s_k) + v^s_k, \ k = \pm 1, \pm 2, ...,
\end{cases}
\tag{4.68}
$$

and the set of systems in the form (4.68) we shall denote by $H(X, \mu_k, v_k, t_k)$.

Denote by

$$
X^s(t,x) = -\mu^s(t)x(t) + \sum_{j=1}^{m} v^s_j(t)e^{-\gamma^s_j(t)x(t-\tau^s_j(t))}.
$$

Theorem 4.10 Assume that:

1. Conditions H3.1, H4.23–H4.25 are fulfilled.
2. There exists a solution $x(t;t_0, \varphi_0)$ of (4.66) such that

$$
x(t;t_0, \varphi_0) < q,
$$

where $t \geq t_0$, $q > 0$.
3. The following inequality is valid

$$
\mu(t) - \sum_{j=1}^{m} v_j(t)\gamma_j(t) > v^*m + q(m+1).
\tag{4.69}
$$

Then for the system (4.66) there exists a unique almost periodic solution $\tilde{x}(t)$ such that:

(a) $\tilde{x}(t) \leq q_1$, $q_1 < q$;
(b) $H(\tilde{x}, t_k) \subset H(X, \mu_k, v_k, t_k)$.

Proof. Let $t_0 \in \mathbb{R}_+$. We define a Lyapunov function

$$
V(x,y) = |x - y|.
$$

Let $\{s_p\}$ be any sequence of real numbers such that $s_p \to \infty$ as $p \to \infty$ and $\{s_p\}$ moves the system (4.66) to a system at $H(X, \mu_k, v_k, t_k)$.

For any real number ξ, let $p_0 = p_0(\xi)$ be the smallest value of p, such that $s_{p_0} + \xi \geq t_0$. Since $x(t;t_0, \varphi_0) \leq q_1$, $q_1 < q$ for all $t \geq t_0$, then $x(t + s_p;t_0, \varphi_0) \leq q_1$ for $t \geq \xi$, $p \geq p_0$.

Let U, $U \subset (\xi, \infty)$ be compact. Then, for any $\varepsilon > 0$ choose an integer $n_0(\varepsilon, \xi) \geq p_0(\xi)$, so large that for $l \geq p \geq n_0(\varepsilon, \xi)$, $t \in \mathbb{R}$, $t \neq t_k$, $k = \pm 1, \pm 2, ...$, and any $w \in \left(v^* m + q_1(m+1), \mu(t) - \sum_{j=1}^{m} v_j(t)\gamma_j(t) \right)$ we have

$$\frac{2wq_1}{w - (v^* m + q_1(m+1))} E_\alpha(-w(t+s_p-s_l)^\alpha)|x(t_0+s_p) - x(t_0+s_l)| < \varepsilon,$$

(4.70)

where $E_\alpha(.)$ is the corresponding Mittag–Leffler function.

Then for $t = t_k$ by the condition H4.24, we get

$$V(x(t_k^+ + s_p), x(t_k^+ + s_l)) = |x(t_k^+ + s_p) - x(t_k^+ + s_l)|$$
$$= |x(t_k + s_p) + \mu_k x(t_k + s_p) - x(t_k + s_l) - \mu_k x(t_k + s_l)|$$
$$\leq (1+\mu_k)|x(t_k + s_p) - x(t_k + s_l)|$$
$$\leq V(x(t_k + s_p), x(t_k + s_l)).$$

(4.71)

For $t \geq t_0$, $t + s_\kappa \neq t_k$, $\kappa = p, l$; $k = \pm 1, \pm 2, ...$, we have

$$_{t_0}^{c}D_t^\alpha |x(t+s_p) - x(t+s_l)|$$
$$= sgn(x(t+s_p) - x(t+s_l))_{t_0}^{c}D_t^\alpha(x(t+s_p) - x(t+s_l)).$$

Then for $t \geq t_0$, $t \in [t_{k-1}, t_k)$ using (4.69) for the upper right-hand derivative $^cD_+^\alpha V(\varphi(t+s_p), \varphi(t+s_l))$ along the solutions of system (4.66), we have

$$^cD_+^\alpha V(\varphi(t+s_p), \varphi(t+s_l)) \leq |-\mu(t+s_p)\varphi(t+s_p) + \mu(t+s_l)\varphi(t+s_l)|$$

$$+ \sum_{j=1}^{m} \left| v_j(t+s_p)e^{-\gamma_j(t+s_p)\varphi(t+s_p-\tau_j(t+s_p))} - v_j(t+s_l)e^{-\gamma_j(t+s_l)\varphi(t+s_l-\tau_j(t+s_l))} \right|$$

$$\leq -\mu(t+s_p)|\varphi(t+s_p) - \varphi(t+s_l)| + |\mu(t+s_l) - \mu(t+s_p)|\varphi(t+s_l)$$

$$+ \sum_{j=1}^{m} \left(v_j(t+s_p) \left| e^{-\gamma_j(t+s_p)\varphi(t+s_p-\tau_j(t+s_p))} - e^{-\gamma_j(t+s_l)\varphi(t+s_l-\tau_j(t+s_l))} \right| \right.$$

$$\left. + \left| v_j(t+s_p) - v_j(t+s_l) \right| \left| e^{-\gamma_j(t+s_l)\varphi(t+s_l-\tau_j(t+s_l))} \right| \right)$$

$$\leq -\mu(t+s_p)|\varphi(t+s_p) - \varphi(t+s_l)| + \varepsilon q_1$$

$$+ \sum_{j=1}^{m} \left(v_j(t+s_p)\gamma_j(t+s_p)|\varphi(t+s_p - \tau_j(t+s_p)) - \varphi(t+s_l - \tau_j(t+s_l))| + \varepsilon q_1 \right) + \varepsilon \right)$$

$$\leq -\mu(t+s_p)V(\varphi(t+s_p), \varphi(t+s_l))$$

$$+ \sum_{j=1}^{m} v_j(t+s_p)(\gamma_j(t+s_p) \sup_{t-\tau \leq \sigma \leq t} V(\varphi(\sigma+s_p), \varphi(\sigma+s_l))$$

$$+ \varepsilon((v^* + q_1)m + q_1).$$

Now, for $\varphi(t+s_p)$, $\varphi(t+s_l) \in \mathscr{PC}$ which satisfy the Razumikhin condition

$$V(\varphi(\sigma+s_p), \varphi(\sigma+s_l)) \leq V(\varphi(t+s_p), \varphi(t+s_l)), \quad t-\tau \leq \sigma \leq t,$$

by virtue of condition 3 of Theorem 4.10, it follows that

$$^cD_+^\alpha V(x(t+s_p), x(t+s_l)) \leq -wV(x(t+s_p), x(t+s_l)) + \varepsilon((v^*+q_1)m+q_1),$$

where $t+s_\kappa \neq t_k$, $t+s_\kappa > t_0$, $\kappa = p, l$.

So, for $t \in U$ from (4.69), (4.70), (4.71) and Corollary 3.2, for $t \in [t_0, \infty)$ we get

$$|x(t+s_p) - x(t+s_l)|$$

$$\leq E_\alpha(-w(t+s_p-s_l)^\alpha)|x(t_0+s_p) - x(t_0+s_l)| + \frac{\varepsilon(v^*m+q_1(m+1))}{w} < \varepsilon.$$

Consequently, there exists a function $\tilde{x}(t)$, such that $x(t+s_p) - \tilde{x}(t) \to \infty$ for $p \to \infty$. Since ξ is arbitrary, it follows that $\tilde{x}(t)$ is defined uniformly on $t \in \mathbb{R}$.

On the other hand, since $x(t; t_0, \varphi_0)$ is a solution of (4.66), (4.67), for $t+s_\kappa \neq t_k$, $\kappa = p, l$; $k = \pm 1, \pm 2, \ldots$, by the same way as above we have that

$$|^c_{t_0}D_t^\alpha x(t+s_p) - ^c_{t_0}D_t^\alpha x(t+s_l)| < \varepsilon,$$

which shows that $\lim\limits_{p \to \infty} {}^c_{t_0}D_t^\alpha x(t+s_p)$ exists uniformly on all compact subsets of $\mathbb{R}$.

Let now $\lim\limits_{p \to \infty} {}^c_{t_0}D_t^\alpha x(t+s_p) = {}^c_{t_0}D_t^\alpha \tilde{x}(t)$, and

$$^c_{t_0}D_t^\alpha \tilde{x}(t) = \lim_{p \to \infty} \left[X(t+s_p, x(t+s_p)) \right.$$

$$\left. -X(t+s_p, \tilde{x}(t)) + X(t+s_p, \tilde{x}(t)) \right] = X^s(t, \tilde{x}(t)), \quad (4.72)$$

where $t \neq t_k^s$, $t_k^s = \lim\limits_{p \to \infty} t_{k+s_p}$.

On the other hand, for $t+s_p = t_k^s$ it follows

$$\tilde{x}(t_k^{s+}) - \tilde{x}(t_k^{s-}) = \lim_{p \to \infty} (x(t_k^s+s_p+0) - x(t_k^s+s_p-0))$$

$$= \lim_{p \to \infty} \mu_k^s x(t_k^s+s_p) = \mu_k^s \tilde{x}(t_k^s). \quad (4.73)$$

From (4.72) and (4.73), we get that $\tilde{x}(t)$ is a solution of (4.68).

To show that $\tilde{x}(t)$ is an almost periodic function, let we consider the sequence $\{s_p\}$ which moves the system (4.66) to $H(X, \mu_k, \nu_k, t_k)$ and consider the function

$$V(\varphi(0), \varphi(s_p-s_l)).$$

Then by (4.69) and (4.70), we have

$$^cD_+^\alpha V(\varphi(0), \varphi(s_p-s_l)) = |\varphi(0) - \varphi(s_p-s_l)|$$

$$\leq -wV(\varphi(0), \varphi(s_p-s_l)) + \varepsilon((v^*+q_1)m+q_1). \quad (4.74)$$

Again,

$$V(\varphi(0) + \mu_k^s \varphi(0), \varphi(t_k^s + s_l - s_p) + \mu_k^s \varphi(t_k^s + s_l - s_p))$$

$$\leq V(\varphi(0), \varphi(s_l - s_p)). \tag{4.75}$$

From (4.74), (4.75) and Lemma 3.7, it follows

$$V(\tilde{x}(t + s_p), \tilde{x}(t + s_l))$$

$$\leq E_\alpha(-w(s_p)^\alpha) \sup_{-\tau \leq \theta \leq 0} V(\tilde{x}(t + \theta), \tilde{x}(t + s_p - s_l + \theta))$$

$$+ \frac{\varepsilon(v^*m + q_1(m+1))}{w} < \varepsilon.$$

From the last inequality for $l \geq p \geq m_0(\varepsilon)$, we have

$$|\tilde{x}(t + s_p) - \tilde{x}(t + s_l)| < \varepsilon. \tag{4.76}$$

Now, from definitions of the sequence $\{s_p\}$ and for $l \geq p \geq m_0(\varepsilon)$ it follows that $\rho(t_k + s_p, t_k + s_l) < \varepsilon$.

Then from (4.76) and the last inequality we obtain that the sequence $\tilde{x}(t + s_p)$ is convergent uniformly to the function $\tilde{x}(t)$.

The assertion (b) of the theorem follows immediately.

In the next theorem we give stability criteria for the almost periodic solution.

Theorem 4.11 Assume that the conditions of Theorem 4.10 hold. Then the almost periodic solution $\tilde{x}(t)$ of (4.66) is globally perfectly uniform-asymptotically stable.

Proof. Let $\overline{\omega}(t)$ be an arbitrary solution of (4.68).

Set
$$u(t) = \overline{\omega}(t) - \tilde{x}(t),$$
$$Y^s(t, u(t)) = X^s(t, u(t) + \tilde{x}(t)) - X^s(t, \tilde{x}(t)).$$

Now, we consider the system

$$\begin{cases} {}_{t_0}^c D_t^\alpha u(t) = Y^s(t, u(t)), \ t \neq t_k^s, \\ \Delta u(t_k^s) = \mu_k^s u(t_k^s) + v_k^s, \ k = \pm 1, \pm 2, ..., \end{cases} \tag{4.77}$$

and let $W(t, u(t)) = V(t, \tilde{x}(t), \tilde{x}(t) + u(t))$. Then, from Lemma 3.7 it follows that the zero solution $u(t) = 0$ of (4.77) is globally perfectly uniform-asymptotically stable, and consequently $\tilde{x}(t)$ is globally perfectly uniform-asymptotically stable.

Now we shall extend the obtained qualitative results to the uncertain case. Consider the following uncertain impulsive fractional-order Lasota–Wazewska model with time-varying delays

$$
\begin{cases}
{}^{c}_{t_0}D^{\alpha}_t x(t) = -\big(\mu(t) + \tilde{\mu}(t)\big)x(t) \\
\qquad\qquad + \displaystyle\sum_{j=1}^{m} \big(v_j(t) + \tilde{v}_j(t)\big)e^{-(\gamma_j(t)+\tilde{\gamma}_j(t))x(t-\tau_j(t))}, \ t \neq t_k, \\
\Delta x(t_k) = \big(\mu_k + \tilde{\mu}_k\big)x(t_k) + v_k + \tilde{v}_k, \ k = \pm 1, \pm 2, ...,
\end{cases}
\tag{4.78}
$$

where $\tilde{\mu}$, $\tilde{v}_j$, $\tilde{\gamma}_j \in C[\mathbb{R}, \mathbb{R}_+]$, $j = 1, 2, ..., m$ are unknown bounded functions, $\tilde{\mu}_k$, $\tilde{v}_k \in \mathbb{R}$, $k = \pm 1, \pm 2, ...$ are unknown bounded vectors and represent the uncertainty of the system.

Note that, the Lasota–Wazewska system (4.66) is the nominal system of (4.78).

Theorem 4.12 Assume that:

1. The functions $\tilde{\mu}(t)$, $\tilde{v}_j(t)$, $j = 1, 2, ..., m$, are almost periodic on t and the sequence of constants $\{\tilde{v}_k\}$, $k = \pm 1, \pm 2, ...$, is almost periodic.

2. The conditions of Theorem 4.10 are fulfilled and there exists a solution $x(t; t_0, \varphi_0)$ of (4.78) such that $x(t; t_0, \varphi_0) < q$, where $t \geq t_0$, $q > 0$.

3. For $k = \pm 1, \pm 2, ...$, $t \geq t_0$ the following inequalities are met

$$
\tilde{\mu}_k \in (-1 - \mu_k, -\mu_k],
\tag{4.79}
$$

$$
\tilde{\mu}(t) < \sum_{j=1}^{m} \big(\tilde{v}_j(t)(\gamma_j(t) + \tilde{\gamma}_j(t)) + v_j(t)\tilde{\gamma}_j(t)\big) < \tilde{\mu}(t) + mv^*.
\tag{4.80}
$$

Then for the system (4.78) there exists a unique almost periodic solution $\tilde{x}(t)$ such that:

1. $|\tilde{x}(t)| \leq q$;

2. $\tilde{x}(t)$ is globally perfectly robustly uniform-asymptotically stable.

Proof. Let $t_0 \in \mathbb{R}_+$. We define a Lyapunov function

$$
V(x, y) = |x - y|.
$$

Then for $k = \pm 1, \pm 2, ...$, $\varphi, \psi \in \mathscr{PC}$, by (4.79) we get

$$
V\big(\varphi(0) + (\mu_k + \tilde{\mu}_k)\varphi(0) + v_k + \tilde{v}_k, \psi(0) + (\mu_k + \tilde{\mu}_k)\psi(0) + v_k + \tilde{v}_k\big)
$$

$$
= |\varphi(0) + (\mu_k + \tilde{\mu}_k)\varphi(0) - \psi(0) - (\mu_k + \tilde{\mu}_k)\psi(0)|
$$

$$\leq (1 + \mu_k + \tilde{\mu}_k)|\varphi(0) - \psi(0)|$$
$$\leq V(\varphi(0), \psi(0)). \tag{4.81}$$

For $t \in \mathbb{R}$ and $t \in [t_{k-1}, t_k)$, we have

$$_{t_0}^{c}D_t^{\alpha}|x(t) - y(t)| = sgn(x(t) - y(t))_{t_0}^{c}D_t^{\alpha}(x(t) - y(t)).$$

Then for $t \neq t_k$, $k = \pm 1, \pm 2, ...$, $\varphi, \psi \in \mathscr{P}C$ for the upper right-hand derivative $^cD_+^{\alpha}V(x(t), y(t))$ along the solutions of system (4.78), we have

$$^cD_+^{\alpha}V(\varphi(0), \psi(0)) \leq -(\mu(t) + \tilde{\mu}(t))|\varphi(0) - \psi(0)|$$

$$+ \sum_{j=1}^{m} \left(\nu_j(t) + \tilde{\nu}_j(t)\right)\left|e^{-(\gamma_j(t) + \tilde{\gamma}_j(t))\varphi} - e^{-(\gamma_j(t) + \tilde{\gamma}_j(t))\psi}\right|$$

$$\leq -(\mu(t) + \tilde{\mu}(t))V(\varphi(0), \psi(0))$$

$$+ \sum_{j=1}^{m} \left(\nu_j(t) + \tilde{\nu}_j(t)\right)(\gamma_j(t) + \tilde{\gamma}_j(t)) \sup_{t-\tau \leq \theta \leq t} V(\varphi(\theta), \psi(\theta)).$$

Now for any $\varphi(t)$ and $\psi(t)$ which satisfy the condition

$$V(\varphi(\theta), \psi(\theta)) \leq V(\varphi(0), \psi(0)), \quad t - \tau \leqslant \theta \leq t$$

and by the virtue of (4.79) and (4.80) it follows that there exists

$$w \in \left(\sum_{j=1}^{m} \left(\tilde{\nu}_j(t)(\gamma_j(t) + \tilde{\gamma}_j(t)) + \nu_j(t)\tilde{\gamma}_j(t)\right) - \tilde{\mu}(t), m\nu^* \right)$$

such that

$$^cD_+^{\alpha}V(\varphi(0), \psi(0)) \leq -wV(\varphi(0), \psi(0)), \tag{4.82}$$

where $t \neq t_k$.

Then in view of (4.81) and (4.82) and Theorem 3.7, the proof of Theorem 4.12 is complete.

4.2.3 *Lotka–Volterra Models*

In this part, we shall consider a fractional Lotka–Volterra type competitive system with impulsive perturbations at fixed moments of time. In general, Lotka–Volterra systems are very important in the models of multi-species population dynamics. These kinds of systems are of a great interest not only for population dynamics or in chemical kinetics, but they establish the basis of many models studied in optimal control, biology, medicine, bio-technologies, ecology, economics, neural networks, etc. (see [Ahmad and Stamova 2013], [Gopalsamy 1992], [Takeuchi 1996]).

Let $t_0 \in \mathbb{R}_+$ and $||u|| = \sum_{i=1}^{n} |u_i|$ define the norm of $u \in \mathbb{R}^n$. We consider the following impulsive n-dimensional fractional-order competitive system involving Caputo fractional derivatives

$$\begin{cases} {}_{t_0}^{c}D_t^{\alpha}u_i(t) = u_i(t)\left[r_i(t) - a_{ii}(t)u_i(t) - \sum_{j=1,j\neq i}^{n} a_{ij}(t)u_j(t)\right], t \neq t_k, \\ \Delta u_i(t_k) = u_i(t_k^+) - u_i(t_k) = p_{ik}u_i(t_k), \ k = \pm 1, \pm 2, ..., \end{cases} \quad (4.83)$$

where $n \geq 2$, $0 < \alpha < 1$, $r_i \in C[\mathbb{R}, \mathbb{R}_+]$, $a_{ij} \in C[\mathbb{R}, \mathbb{R}_+]$, $p_{ik} \in \mathbb{R}$, $1 \leq i, j \leq n$, $k = \pm 1, \pm 2, ...$ and $\{t_k\} \in \mathscr{B}$. The numbers $u_i(t_k)$ and $u_i(t_k^+)$ are respectively, the population densities of species i before and after an impulsive perturbation at the moment t_k. The constant p_{ik} characterizes the magnitude of the impulsive effect on the species i at the moment t_k.

Let $u_0 = col(u_{10}, u_{20}, ..., u_{n0})$, $u_{i0} \in \mathbb{R}$ for $1 \leq i \leq n$. We shall denote by $u(t) = u(t; t_0, u_0)$, $u(t) = col(u_1(t), u_2(t), ..., u_n(t))$ the solution of the system (4.83) with an initial condition

$$u(t_0^+; t_0, u_0) = u_0. \quad (4.84)$$

The solution $u(t) = u(t; t_0, u_0)$ of problem (4.83), (4.84) is a piecewise continuous function with points of discontinuity of the first kind at the moments t_k, $k = \pm 1, \pm 2, ...$, at which it is continuous from the left, i.e. the following relations are valid:

$$u_i(t_k^-) = u_i(t_k), \ u_i(t_k^+) = u_i(t_k) + p_{ik}u_i(t_k), \ k = \pm 1, \pm 2, ..., \ 1 \leq i \leq n.$$

Furthermore, we shall restrict our attention only to those solutions which evolve in the phase space $\mathbb{R}_+^n$, and we will assume that if $u_{i0} > 0$ for some i, then $u_i(t) > 0$ for all $t \geq t_0$. Note that this assumption is natural from the biological point of view.

Let, for simplicity, denote by $U(t, u) = (U_1(t, u), U_2(t, u), ..., U_n(t, u))$,

$$U_i(t, u) = u_i(t)\left[r_i(t) - \sum_{j=1}^{n} a_{ij}(t)u_j(t)\right],$$

where $1 \leq i \leq n$, and $P_k = diag[p_{1k}, p_{2k}, ..., p_{nk}]$ is a diagonal $n \times n$ matrix, $k = \pm 1, \pm 2, ...$

We introduce the following conditions:

H4.26 The functions $r_i(t), a_{ij}(t)$, $1 \leq i, j \leq n$ are continuous, nonnegative and almost periodic on t.

H4.27 The sequences of constants $\{p_{ik}\}$ are almost periodic uniformly with $1 \leq i \leq n$, and

$$-1 < p_{ik} \leq 0$$

for $k = \pm 1, \pm 2, ...$

Let the conditions H3.1, H4.26 and H4.27 hold, and let $\{s'_m\}$ be an arbitrary sequence of real numbers. Then, there exists a subsequence $\{s_p\}$, $s_p = s'_{m_p}$ such that the system (4.83) moves to the system

$$\begin{cases} {}^{c}_{t_0}D^{\alpha}_t u_i(t) = u_i(t)\left[r^s_i(t) - -a^s_{ii}(t)u_i(t) - \sum_{j=1}^{n} a^s_{ij}(t)u_j(t)\right], \ t \neq t^s_k, \\ \Delta u_i(t^s_k) = u_i(t^{s+}_k) - u_i(t^s_k) = p^s_{ik}u_i(t^s_k), \ k = \pm 1, \pm 2, ..., \end{cases} \tag{4.85}$$

and the set of systems in the form (4.85) we shall denote by $H(U, P_k, t_k)$.

Denote by $U^s(t,x) = (U^s_1(t,x), U^s_2(t,x), ..., U^s_n(t,x))$, where

$$U^s_i(t,x) = u_i(t)\left[r^s_i(t) - \sum_{j=1}^{n} a^s_{ij}(t)u_j(t)\right], \ 1 \leq i \leq n,$$

and $P^s_k = diag[p^s_{1k}, p^s_{2k}, ..., p^s_{nk}]$.

Theorem 4.13 Assume that:

1. Conditions H3.1, H4.26 and H4.27 are fulfilled.
2. There exists a solution $u(t; t_0, u_0)$ of (4.83) such that for $t \geq t_0$

$$0 < \underline{u} \leq u(t; t_0, u_0) < \bar{u},$$

where $\underline{u} = \max(u^I_1(t), u^I_2(t), ..., u^I_n(t))$, $\bar{u} = \min(u^S_1(t), u^S_2(t), ..., u^S_n(t))$,

$$u^I_i(t) = \inf_{t \in [t_0, \infty)} u_i(t), \ 1 \leq i \leq n,$$

$$u^S_i(t) = \sup_{t \in [t_0, \infty)} u_i(t), \ 1 \leq i \leq n.$$

3. There exists a positive number A such that the functions $r_i(t)$, $a_{ij}(t)$, $1 \leq i, j \leq n$ satisfy

$$2\sum_{j=1}^{n} a_{ij}(t)u^I_j(t) - r_i(t) = A > 0.$$

Then for the system (4.83) there exists a unique almost periodic solution $\tilde{u}(t)$ such that:

1. $\|\tilde{u}(t)\| \leq A_1$, $0 < A_1 < \bar{u}$;
2. $H(\tilde{u}, t_k) \subset H(U, P_k, t_k)$;
3. $\tilde{u}(t)$ is globally perfectly uniform-asymptotically stable.

Proof. Let $t_0 \in \mathbb{R}_+$. We define a Lyapunov function

$$V(u, v) = \sum_{i=1}^{n} |u_i - v_i|.$$

Let $\{s_p\}$ be any sequence of real numbers such that $s_p \to \infty$ as $p \to \infty$ and $\{s_p\}$ moves the system (4.83) to a system at $H(U, P_k, t_k)$.

For any real number ξ, let $p_0 = p_0(\xi)$ be the smallest value of p, such that $s_{p_0} + \xi \geq t_0$. Since $u(t; t_0, u_0) \leq A_1$, $A_1 < \bar{u}$ for all $t \geq t_0$, then $u(t + s_p; t_0, u_0) \leq A_1$ for $t \geq \xi$, $p \geq p_0$.

Let J, $J \subset (\xi, \infty)$ be compact. Then, for any $\varepsilon > 0$, choose an integer $n_0(\varepsilon, \xi) \geq p_0(\xi)$, so large that for $l \geq p \geq n_0(\varepsilon, \xi)$ and $t \in J$, $t \neq t_k$, $k = \pm 1, \pm 2, ...$, we have

$$4\bar{u}E_\alpha(-A(t + s_p - s_l)^\alpha) < \varepsilon. \tag{4.86}$$

Now, for $t = t_k$ by the condition H4.27, we get

$$V(u(t_k^+ + s_p), u(t_k^+ + s_l)) = \sum_{i=1}^{n} |u_i(t_k^+ + s_p) - u_i(t_k^+ + s_l)|$$

$$= \sum_{i=1}^{n} |u_i(t_k + s_p) + p_{ik} u_i(t_k + s_p) - u_i(t_k + s_l) - p_{ik} u_i(t_k + s_l)|$$

$$= \sum_{i=1}^{n} (1 + p_{ik}) |u_i(t_k + s_p) - u_i(t_k + s_l)| \leq V(u(t_k + s_p), u(t_k + s_l)). \tag{4.87}$$

Let $t \neq t_k$, $k = \pm 1, \pm 2, ...$ Then, we have

$${}_{t_0}^{c}D_t^\alpha |u_i(t + s_p) - u_i(t + s_l)|$$

$$= \text{sgn}(u_i(t + s_p) - u_i(t + s_l)) {}_{t_0}^{c}D_t^\alpha (u_i(t + s_p) - u_i(t + s_l)).$$

Then for $t \in [t_{k-1}, t_k)$, using (4.86) for the upper right-hand derivative ${}^cD_+^\alpha V(u(t + s_p), u(t + s_l))$ along the solutions of system (4.83), we have

$${}^cD_+^\alpha V(u(t + s_p), u(t + s_l)) \leq \sum_{i=1}^{n} \left| u_i(t + s_p) r_i(t + s_p) - a_{ii}(t + s_p) u_i^2(t + s_p) \right.$$

$$- \sum_{j=1, j\neq i}^{n} a_{ij}(t + s_p) u_j(t + s_p) u_i(t + s_p)$$

$$-u_i(t + s_l) r_i(t + s_l) + a_{ii}(t + s_l) u_i^2(t + s_l) + \sum_{j=1, j\neq i}^{n} a_{ij}(t + s_l) u_j(t + s_l) u_i(t + s_l) \right|$$

$$\leq \sum_{i=1}^{n} \left[r_i(t + s_l) - a_{ii}(t + s_p)(u_i(t + s_p) + u_i(t + s_l)) \right.$$

$$- \sum_{j=1, j\neq i}^{n} \left(a_{ij}(t + s_p) u_j(t + s_p) + a_{ji}(t + s_p) u_j(t + s_l) \right) \right] |u_i(t + s_p) - u_i(t + s_l)|$$

$$+ \sum_{i=1}^{n} \Big[u_i(t+s_p)(r_i(t+s_p) - r_i(t+s_l)) - (a_{ii}(t+s_p) - a_{ii}(t+s_l))u_i^2(t+s_l)$$

$$- \sum_{j=1, j \neq i}^{n} (a_{ij}(t+s_p) - a_{ij}(t+s_l))u_j(t+s_p)u_i(t+s_l) \Big]$$

$$\leq -A \sum_{i=1}^{n} |u_i(t+s_p) - u_i(t+s_l)| + (\bar{u}n + (\underline{u}n)^2)\varepsilon$$

$$\leq -A\Big(V(u(t+s_p), u(t+s_l)) - \frac{\varepsilon}{A}(\bar{u}n + (\underline{u}n)^2)\Big). \tag{4.88}$$

Now for $t \in J$, using (4.87), (4.88) and Corollary 3.1, we get

$$||u(t+s_p) - u(t+s_l)|| = \sum_{i=1}^{n} |u_i(t+s_p) - u_i(t+s_l)|$$

$$\leq E_\alpha(-A(t+s_p - s_l)^\alpha) \sum_{i=1}^{n} |u_i(t_0+s_p) - u_i(t_0+s_l)| + \frac{\varepsilon}{A}(\bar{u}n + (\underline{u}n)^2)$$

$$= E_\alpha(-A(t+s_p - s_l)^\alpha)||u(t_0+s_p) - u(t_0+s_l)|| + \frac{\varepsilon}{A}(\bar{u}n + (\underline{u}n)^2) = H\varepsilon,$$

where $H = \dfrac{1}{2} + \dfrac{1}{A}(\bar{u}n + (\underline{u}n)^2)$.

Consequently, there exists a function $\tilde{u}(t) = (\tilde{u}_1(t), \tilde{u}_2(t), ..., \tilde{u}_n(t))$, such that $u(t+s_p) - \tilde{u}(t) \to \infty$ for $p \to \infty$. Since ξ is arbitrary, it follows that $\tilde{u}(t)$ is defined uniformly on $t \in \mathbb{R}$.

On the other hand, since $u(t; t_0, u_0)$ is a solution of (4.83), (4.84), for $t + s_\kappa \neq t_k$, $\kappa = p, l$; $k = \pm 1, \pm 2, ...$, by the same way as above it follows that

$$||_{t_0}^c D_t^\alpha u(t+s_p) - _{t_0}^c D_t^\alpha u(t+s_l)|| < \varepsilon,$$

which shows that $\lim\limits_{p \to \infty} {}_{t_0}^c D_t^\alpha u(t+s_p)$ exists uniformly on all compact subsets of $\mathbb{R}$.

Let now $\lim\limits_{p \to \infty} {}_{t_0}^c D_t^\alpha u(t+s_p) = {}_{t_0}^c D_t^\alpha \tilde{u}(t)$ and

$$_{t_0}^c D_t^\alpha \tilde{u}(t) = \lim_{p \to \infty} \big[U(t+s_p, u(t+s_p)) - U(t+s_p, \tilde{u}(t)) + U(t+s_p, \tilde{u}(t)) \big]$$

$$= U^s(t, \tilde{u}(t)), \tag{4.89}$$

where $t \neq t_k^s$, $t_k^s = \lim\limits_{p \to \infty} t_{k+s_p}$.

On the other hand, for $t + s_p = t_k^s$ it follows

$$\tilde{u}(t_k^{s+}) - \tilde{u}(t_k^{s-}) = \lim_{p \to \infty} (u(t_k^s + s_p + 0) - u(t_k^s + s_p - 0))$$

$$= \lim_{p \to \infty} P_k^s u(t_k^s + s_p) = P_k^s \tilde{u}(t_k^s). \tag{4.90}$$

From (4.89) and (4.90), we get that $\tilde{u}(t)$ is a solution of (4.85).

To show that $\tilde{u}(t)$ is an almost periodic function, let we consider the sequence $\{s_p\}$ that moves the system (4.83) to $H(U, P_k, t_k)$ and consider the function

$$V(\tilde{u}(\sigma), \tilde{u}(\sigma + s_p - s_l)).$$

Then by (4.86) the same way like above, we have

$$^cD_+^\alpha V(\tilde{u}(\sigma), \tilde{u}(\sigma + s_p - s_l)) = ||\tilde{u}(\sigma) - \tilde{u}(\sigma + s_p - s_l))||$$

$$\le -AV(\tilde{u}(\sigma), \tilde{u}(\sigma + s_p - s_l)). \tag{4.91}$$

Again,

$$V(\tilde{u}(t_k^s) + P_k^s\tilde{u}(t_k^s), \tilde{u}(t_k^s + s_p - s_l) + P_k^s\tilde{u}(t_k^s))$$

$$\le V(\tilde{u}(t_k^s), \tilde{u}(t_k^s + s_p - s_l)). \tag{4.92}$$

On the other hand, from definitions of the sequences $\{s_p\}$ it follows that $\rho(t_k + s_p, t_k + s_l) < \varepsilon$. Then by (4.91), (4.92) and Corollary 3.1, it follows

$$V(\tilde{u}(t + s_p), \tilde{u}(t + s_l))) \le E_\alpha(-A(s_p)^\alpha)V(\tilde{u}(t), \tilde{u}(t + s_p - s_l)) < \varepsilon.$$

Now, for $l \ge p \ge m_0(\varepsilon)$, we have

$$||\tilde{u}(t + s_p) - \tilde{u}(t + s_l)|| < \varepsilon. \tag{4.93}$$

Then from (4.93) we obtain that the sequence $\tilde{u}(t + s_p)$ converges uniformly to the function $\tilde{u}(t)$.

The assertions 1 and 2 of the theorem follow immediately.

To prove the assertion 3, consider an arbitrary solution $u^*(t)$ of (4.85), and set

$$x(t) = u^*(t) - \tilde{u}(t),$$

$$X^s(t, x(t)) = U^s(t, x(t) + \tilde{u}(t)) - U^s(t, \tilde{u}(t)).$$

Now we consider the system

$$\begin{cases} {}_{t_0}^cD_t^\alpha x(t) = X^s(t, x(t)), & t \ne t_k^s, \\ \Delta x(t_k^s) = P_k^s x(t_k^s), & k = \pm 1, \pm 2, ..., \end{cases} \tag{4.94}$$

and let $W(t, x(t)) = V(t, \tilde{u}(t), \tilde{u}(t) + x(t))$. Then, from Lemma 3.6 it follows that the zero solution $x(t) = 0$ of (4.94) is globally perfectly uniform-asymptotically stable, and consequently $\tilde{u}(t)$ is globally perfectly uniform-asymptotically stable.

What follows is an uncertain fractional-order competitive system for which (4.83) is the "nominal system"

$$\begin{cases} {}^{c}_{t_0}D^{\alpha}_t u_i(t) = u_i(t)\left[r_i(t) - \sum_{j=1}^{n}\left(a_{ij}(t) - b_{ij}(t)\right)u_j(t)\right], \ t \neq t_k, \\ \Delta u_i(t_k) = (p_{ik} - q_{ik})u_i(t_k), \ k = \pm 1, \pm 2, ..., \end{cases} \qquad (4.95)$$

where the functions $b_{ij} \in C[\mathbb{R}, \mathbb{R}_+]$, and the constants $q_{ik} \in \mathbb{R}$, $1 \leq i, j \leq n$, $k = \pm 1, \pm 2, ...$

The functions $b_{ij}(t)$ and the sequences $\{q_{ik}\}$, $1 \leq i, j \leq n$, $k = \pm 1, \pm 2, ...,$ represent the structural uncertainty or uncertain perturbations, and are characterized by

$$b_{ij} \in U_b = \left\{ b_{ij} : b_{ij}(t) = e_{b_{ij}}(t).\delta_{b_{ij}}(t), \ m^L_{b_{ij}}(t) \leq \delta_{b_{ij}}(t) \leq m^R_{b_{ij}}(t) \right\},$$

and

$$q_{ik} \in U_q = \left\{ q_{ik} : q_{ik} = e_{ik}.\delta_{ik}, \ |\delta_{ik}| \leq |m_{ik}| \right\}, \ k = \pm 1, \pm 2, ...,$$

where $e_{b_{ij}} : \mathbb{R} \to \mathbb{R}_+$ are known smooth functions of the state, and $e_{ik} \in \mathbb{R}$ are known constants. The functions $\delta_{b_{ij}}(t)$ and the constants δ_{ik} are unknown bounded functions and unknown bounded constants respectively, by the functions $m^L_{ij}(t)$, $m^R_{ij}(t)$ and the norm of the constants $m_{ik} \in \mathbb{R}$, respectively. Here $m^L_{ij}, m^R_{ij} : \mathbb{R} \to \mathbb{R}_+$ are given smooth functions.

Theorem 4.14 Assume that:

1. Conditions H3.1, H4.26 and H4.27 are fulfilled.

2. The conditions of Theorem 4.13 are fulfilled and there exists a solution $u(t; t_0, u_0)$ of (4.95) such that for $t \geq t_0$

$$0 < \underline{u} \leq u(t; t_0, u_0) < \overline{u},$$

where $\underline{u} = \max(u^I_1(t), u^I_2(t), ..., u^I_n(t))$, $\overline{u} = \min(u^S_1(t), u^S_2(t), ..., u^S_n(t))$,

$$u^I_i(t) = \inf_{t \in [t_0, \infty)} u_i(t), \ 1 \leq i \leq n,$$

$$u^S_i(t) = \sup_{t \in [t_0, \infty)} u_i(t), \ 1 \leq i \leq n.$$

3. The sequence $\{q_{ik}\}$ is almost periodic and $p_{ik} \leq q_{ik} < 1 + p_{ik}$, $1 \leq i \leq n$, $k = \pm 1, \pm 2, ...$

4. The functions $b_{ij}(t)$ are almost periodic, $r^I_i > 0$, a^I_{ij}, $b^I_{ij} \geq 0$, a^S_{ij}, $b^S_{ij} < \infty$, $r^S_i < \infty$ for $1 \leq i, j \leq n$ and

$$0 < \sum_{j=1}^{n} e_{b_{ij}} m^R_{b_{ij}}(t)u^I_j(t) < \sum_{j=1}^{n} a_{ij}(t)u^I_j(t) - \frac{1}{2}r_i(t).$$

Then for the system (4.95) there exists a unique, strictly positive and globally perfectly robustly uniform-asymptotically stable almost periodic solution.

Proof. Let $t_0 \in \mathbb{R}_+$ and consider again the Lyapunov's function

$$V(u,v) = \sum_{i=1}^{n} |u_i - v_i|.$$

Then for $t = t_k$ and condition 3 of the theorem, we get

$$V(u(t_k^+), v(t_k^+)) = \sum_{i=1}^{n} |u_i(t_k^+) - v_i(t_k^+)|$$

$$= \sum_{i=1}^{n} \left| u_i(t_k) + (p_{ik} - q_{ik})u_i(t_k) - \left(v_i(t_k) + (p_{ik} - q_{ik})v_i(t_k)\right) \right|$$

$$\leq \sum_{i=1}^{n} (1 + p_{ik} - q_{ik})|u_i(t_k) - v_i(t_k)| \leq V(u(t_k), v(t_k)). \qquad (4.96)$$

On the other hand, for $t \neq t_k$, $k = \pm 1, \pm 2, \dots$, using conditions 2 and 4 of the theorem, by similar arguments as in the proof of Theorem 4.13, for the upper right-hand derivative $^cD_+^q V(u(t), v(t))$ along the solutions of system (4.95) we have

$$^cD_+^\alpha V(u(t), v(t)) \leq -\tilde{A}V(u(t), v(t)), \qquad (4.97)$$

where $0 < \tilde{A} = 2 \sum_{j=1}^{n} \left(a_{ij}(t) - e_{b_{ij}}m_{b_{ij}}^R(t)\right)u_j^I(t) - r_i(t)$.

Then in view of (4.96) and (4.97), Theorem 3.6 and Theorem 4.13, the proof of Theorem 4.14 is complete.

4.2.4 *Kolmogorov-type Models*

Finally, more general Kolmogorov-type systems of fractional order will be investigated. Kolmogorov systems are very important models of multi-species population dynamics, and such systems with integer-order derivatives have been studied extensively (see [Ahmad and Stamova 2012], [Ahmad and Stamova 2013], [Gopalsamy 1992], [Stamova and Stamov G 2014a].

Here, using fractional integrals we shall develop an impulsive fractional-order Kolmogorov model. The parametric stability of the proposed model will be investigated. The concept of parametric stability which addresses simultaneously the twin problem of existence and stability of a moving equilibrium has been introduced in [Ikeda, Ohta and Siljak 1991] and is applied by many authors. See, for example [Stamova 2009], [Zecevic and Siljak 2010].

Let $t_0 \in \mathbb{R}_+$, $\|.\|$ define the norm in $\mathbb{R}^n$, and Ω be an open set in the cone $\mathbb{R}_+^n = \{u \in \mathbb{R}^n, u \geq 0\}$ containing the origin.

Consider the following Kolmogorov system of fractional differential equations with variable impulsive perturbations

$$\begin{cases} {}_{t_0}^{c}D_t^{\alpha}u_i(t) = u_i(t)f_i(t,u(t),p), \ t \neq \tau_k(u(t)), \\ \Delta u_i(t) = u_i(t^+) - u_i(t^-) = I_{ik}(u_i(t)), \ t = \tau_k(u(t)), \ k = 1,2,..., \end{cases} \tag{4.98}$$

$i = 1, 2, ..., n$, where $p \in \mathbb{R}^m$ is a constant parameter vector, $u = (u_1, ..., u_n) \in \Omega$, $f = (f_1, f_2, ..., f_n)$, $f : [t_0, \infty) \times \Omega \times \mathbb{R}^m \to \mathbb{R}^n$, $0 < \alpha < 1$, $I_{ik} : \mathbb{R}_+ \to \mathbb{R}$, $\tau_k : \Omega \to \mathbb{R}^n$, $k = 1, 2, ...$, $\tau_k = (\tau_{1k}, \tau_{2k}, ..., \tau_{nk})$, and $I_k = (I_{1k}, I_{2k}, ..., I_{nk})$.

Let $u_0 = (u_{10}, ..., u_{n0}) \in \Omega$. Denote by $u(t) = u(t; t_0, u_0, p)$ the solution of system (4.98) that satisfies the initial condition:

$$u_i(t_0^+; t_0, u_0, p) = u_{i0}, \ i = 1, 2, ..., n. \tag{4.99}$$

The solutions $u(t)$ of system (4.98) are piecewise continuous functions with points of discontinuity of the first kind at which they are left continuous; i.e., at the moments t_{l_k} when the integral curve of the solution $u(t)$ meets the hypersurfaces

$$\sigma_k = \left\{ (t,u) \in [t_0, \infty) \times \Omega : t = \tau_k(u) \right\}$$

the following relations are satisfied:

$$u(t_{l_k}^-) = u(t_{l_k}), \ u(t_{l_k}^+) = u(t_{l_k}) + I_{l_k}(u(t_{l_k})).$$

The points $t_{l_1}, t_{l_2}, ... \ (t_0 < t_{l_1} < t_{l_2})$ are the impulsive moments. Let us note that, in general, $k \neq l_k$. Note, again, that it is possible that the integral curve of the problem under consideration does not meet the hypersurface σ_k at the moment t_k.

Let $\tau_0(u) \equiv t_0$ for $u \in \Omega$. We assume that the functions $\tau_k(u)$ are continuous and the following relations hold:

$$t_0 < \tau_1(u) < \tau_2(u) < ..., \ \tau_k(u) \to \infty \text{ as } k \to \infty$$

uniformly on $u \in \Omega$. We also suppose that the functions f, I_k and τ_k are smooth enough on $[t_0, \infty) \times \Omega$ and Ω, respectively, to guarantee the existence, uniqueness and continuability of the solution $u(t) = u(t; t_0, u_0, p)$ of the IVP (4.98), (4.99) on the interval $[t_0, \infty)$ for each $u_0 \in \Omega$, and $t_0 \in \mathbb{R}_+$ and absence of the phenomenon "beating" (see [Bainov and Dishliev 1997]). For more results about such systems, we refer the reader to [Stamov 2012], [Stamova 2009] and the references therein.

We assume that solutions of (4.98) with initial conditions (4.99) are nonnegative, and if $u_{i0} > 0$ for some i, then $u_i(t) > 0$ for all $t \geq t_0$. If, moreover, $(t_k, u_i) \in (t_0, \infty) \times (0, \infty)$, then $u_i(t_k) + I_{ik}(u_i(t_k)) > 0$ for all $i = 1, 2, ..., n$ and $k = 1, 2, ...$. Note that these assumptions are natural from the applicable point of view.

We also assume that for some nominal value p^* of the parameter vector p, there is an equilibrium state u^*, that is,

$$\begin{cases} f(t,u^*,p^*) = 0, \, t \geq t_0, \, t \neq \tau_k(u), \\ \Delta u^*(t) = u^*(t^+) - u^*(t) = 0, \, \tau_k(u) > t_0, \, k = 1,2,..., \end{cases} \tag{4.100}$$

and u^* is stable. Suppose that the parameter vector p is changed from p^* to another value, and that there exists an equilibrium $u^\varepsilon(p) \in \Omega$.

The question arises: Is the new equilibrium stable as u^* was, or is its stability destroyed by the change of p?

Consider the equilibrium $u^\varepsilon : \mathbb{R}^m \to \Omega$ as a function $u^\varepsilon(p)$ and introduce the following definitions of parametric stability.

Definition 4.4 The system (4.98) is said to be *parametrically stable* at $p^* \in R^m$, if there exists a neighborhood $N(p^*)$ such that for any $p \in N(p^*)$:

$$(\forall t_0 \in \mathbb{R}_+)(\forall \varepsilon > 0)(\exists \delta = \delta(t_0,\varepsilon,p) > 0)(\forall u_0 \in \Omega : ||u_0 - u^\varepsilon(p)|| < \delta)$$

$$(\forall t \geq t_0) : ||u(t;t_0,u_0,p) - u^\varepsilon(p)|| < \varepsilon.$$

Remark 4.8 If the system (4.98) is not stable in the above sense, we say it is *parametrically unstable* at p^*. This means that if for any neighborhood $N(p^*)$, there exists a $p \in N(p^*)$ for which either there is no equilibrium $u^\varepsilon(p)$ of (4.98), or there is an equilibrium $u^\varepsilon(p)$, which is unstable in the sense of Lyapunov.

Definition 4.5 The system (4.98) is said to be *parametrically uniformly stable* at $p^* \in \mathbb{R}^m$, if the number δ from Definition 4.4 is independent on $t_0 \in \mathbb{R}_+$.

Definition 4.6 The system (4.98) is said to be *parametrically uniformly asymptotically stable* at $p^* \in R^m$, if there exists a neighborhood $N(p^*)$ such that for any $p \in N(p^*)$:

(i) it is parametrically uniformly stable at p^*;

(ii) for all $p \in N(p^*)$, there exists a number $\mu = \mu(p) > 0$ such that $||u_0 - u^\varepsilon(p)|| < \mu$ implies

$$\lim_{t \to \infty} ||u(t;t_0,u_0,p) - u^\varepsilon(p)|| = 0.$$

We shall use Lyapunov functions $V : [t_0,\infty) \times \Omega \to \mathbb{R}_+$, which belong to the class V_0, and satisfy the following condition

H4.28 $V(t,u^\varepsilon(p)) = 0, \, t \in [t_0,\infty), \, p \in N(p^*)$.

Let $t_1,t_2,...$ ($t_0 < t_1 < t_2 < ...$) be the moments at which the integral curve $(t,u(t;t_0,u_0,p))$ of the IVP (4.98), (4.99) meets the hypersurfaces $\sigma_k, k = 1,2,...$, i.e., each of the points t_k is a solution of some of the equations $t = \tau_k(u(t))$, $k = 1,2,....$

In the proof of the main results, we shall use the following lemma.

Lemma 4.1 Assume that:

1. The function $F : [t_0, \infty) \times \mathbb{R}_+ \times \mathbb{R}_+ \to \mathbb{R}$ is continuous in each of the sets $(t_{k-1}, t_k] \times \mathbb{R}_+ \times \mathbb{R}_+$, $k = 1, 2, \ldots$ and $F(t, v, \mu)$ is non-decreasing in v for each $t \in [t_0, \infty)$ and $\mu \in \mathbb{R}_+$, where $\mu = \mu(p)$ is a parameter.

2. The functions $\psi_k \in C[\mathbb{R}_+ \times \mathbb{R}_+, \mathbb{R}_+]$, $\psi_k = \psi_k(v, \mu)$, $k = 1, 2, \ldots$ are non-decreasing with respect to v.

3. The maximal solution $R(t; t_0, v_0, \mu)$ of the problem

$$\begin{cases} {}^{c}_{t_0}D^{\alpha}_t v = F(t, v, \mu), \ t \geq t_0, \ t \neq t_k, \\ v(t_0^+) = v_0 \geq 0, \\ v(t_k^+) = \psi_k(v(t_k), \mu), \ t_k > t_0, \ k = 1, 2, \ldots \end{cases}$$

is defined in the interval $[t_0, \infty)$.

4. There exists a function $V \in V_0$ such that $V(t_0^+, u_0) \leq v_0$,

$$V(t^+, u + I_k(u, p)) \leq \psi_k(V(t, u), \mu), \ p \in \mathbb{R}^m, \ \mu \in \mathbb{R}_+, \ u \in \Omega, \ t = t_k, \ k = 1, 2, \ldots,$$

and the inequality

$$^{c}D^{\beta}_{+}V(t, u) \leq F(t, V(t, u), \mu), \ t \neq t_k, \ k = 1, 2, \ldots$$

is valid for $0 < \beta < 1$, $t \in [t_0, \infty)$, $u \in \Omega$, $\mu \in \mathbb{R}_+$.

Then

$$V(t, u(t; t_0, u_0, p)) \leq R(t; t_0, v_0, \mu), \ t \in [t_0, \infty).$$

The proof of Lemma 4.1 is similar to that of Lemma 1.5. We omit it here.

In the case when $F(t, v, \mu) \equiv 0$ for $t \in [t_0, \infty)$, $v \in \mathbb{R}_+$, $\mu \in \mathbb{R}_+$ and $\psi_k(v, \mu) \equiv v$ for $v \in \mathbb{R}_+$, $\mu \in \mathbb{R}_+$, $k = 1, 2, \ldots$, we deduce the following corollary from Lemma 4.1.

Corollary 4.1 Assume that there exists a function $V \in V_0$ such that

$$^{c}D^{\beta}_{+}V(t, u) \leq 0, \ u \in \Omega, \ t \geq t_0, \ t \neq t_k, \ k = 1, 2, \ldots,$$

$$V(t^+, u + I_k(u, p)) \leq V(t, u), \ p \in \mathbb{R}^m, \ u \in \Omega, \ t = t_k, \ k = 1, 2, \ldots.$$

Then

$$V(t, u(t; t_0, u_0, p)) \leq V(t_0^+, u_0), \ t \in [t_0, \infty).$$

Theorem 4.15 Assume that there exists a function $V \in V_0$ such that H4.28 holds,

$$w_1(\|u - u^{\varepsilon}(p)\|) \leq V(t, u), \ w_1 \in K, \ (t, u) \in [t_0, \infty) \times \Omega, \tag{4.101}$$

$$V(t^+, u + I_k(u,p)) \le V(t,u), \ u \in \Omega, \ p \in N(p^*), \ t = t_k, k = 1,2,..., \quad (4.102)$$

and the inequality

$$^cD_+^\beta V(t,u) \le 0, \ t \ne t_k, \ k = 1,2,...$$

is valid for $0 < \beta < 1, t \in [t_0, \infty), u \in \Omega$.

Then the system (4.98) is parametrically stable at p^*.

Proof. Let $\varepsilon > 0$. From the properties of the function V, it follows that there exists a constant $\delta = \delta(t_0, \varepsilon, p) > 0$ such that if $u \in \Omega : ||u - u^\varepsilon(p)|| < \delta$, then

$$\sup_{||u - u^\varepsilon(p)|| < \delta} V(t_0^+, u) < w_1(\varepsilon).$$

Let $u_0 \in \Omega : ||u_0 - u^\varepsilon(p)|| < \delta$. Then

$$V(t_0^+, u_0) < w_1(\varepsilon). \quad (4.103)$$

Let $u(t) = u(t; t_0, u_0, p)$ be the solution of problem (4.98), (4.99). Since the conditions of Corollary 4.1 are met, then

$$V(t, u(t; t_0, u_0, p)) \le V(t_0^+, u_0), \ t \in [t_0, \infty).$$

From the last inequality, (4.101), (4.102) and (4.103), there follow the inequalities

$$w_1(||u(t; t_0, u_0, p) - u^\varepsilon(p)||) \le V(t, u(t; t_0, u_0, p))$$

$$\le V(t_0^+, u_0) < w_1(\varepsilon),$$

which imply that $||u(t; t_0, u_0, p) - u^\varepsilon(p)|| < \varepsilon$ for $t \ge t_0$. This implies that the Kolmogorov system (4.98) is parametrically stable at p^*.

Theorem 4.16 Let the conditions of Theorem 4.15 hold, and a function $w_2 \in K$ exists such that

$$V(t,u) \le w_2(||u - u^\varepsilon(p)||), \ (t,u) \in [t_0, \infty) \times \Omega.$$

Then the system (4.98) is parametrically uniformly stable at p^*.

Theorem 4.17 Assume that there exists a function $V \in V_0$ such that H4.28 and (4.102) hold,

$$w_1(||u - u^\varepsilon(p)||) \le V(t,u) \le w_2(||u - u^\varepsilon(p)||), \ w_1, w_2 \in K, \ (t,u) \in [t_0, \infty) \times \Omega,$$

and the inequality

$$^cD_+^\beta V(t,u) \le -w_3(||(u(t) - u^\varepsilon(p)||), \ t \ne t_k, \ k = 1,2,... \quad (4.104)$$

is valid for $w_3 \in K, t \in [t_0, \infty)$ and $u \in \Omega$.

Then the system (4.98) is parametrically uniformly asymptotically stable at p^*.

Theorem 4.16 and Theorem 4.17 are similar to the Theorem 2.9 and Theorem 2.10, respectively.

Corollary 4.2 If in Theorem 4.17 condition (4.104) is replaced by the condition

$$^cD_+^\beta V(t,u) \leq -wV(t,u), t \in [t_0,\infty), t \neq t_k, k = 1,2,...,$$

where $u \in \Omega$, $w = const > 0$, then the system (4.98) is parametrically uniformly asymptotically stable.

Example 4.4 Let $\tau_0(u) \equiv 0$ for $u \in \mathbb{R}_+$, and $\tau_k(u) = u^2 + k$, $k = 1,2,....$ Then, we have $\tau_k \in C[\mathbb{R}_+,(0,\infty)]$, $k = 1,2,...$, $\tau_k(u) \to \infty$ as $k \to \infty$ uniformly on $u \in \mathbb{R}_+$, and also

$$0 \equiv \tau_0(u) < \tau_1(u) < \tau_2(u) < ..., u \in \mathbb{R}_+.$$

We consider the following impulsive α-fractional-order competitive equation

$$\begin{cases} {}_0^cD_t^\alpha u(t) = u(t)[a - bu(t)], t \neq \tau_k(u), \\ \Delta u(t) = u(t^+) - u(t) = p_k\left(u(t) - \frac{a}{b}\right), t = \tau_k(u), k = 1,2,..., \end{cases} \quad (4.105)$$

where $0 < \alpha < 1$, $t \in \mathbb{R}_+$, the systems parameters $a,b > 0$, $-1 < p_k \leq 0$, $k = 1,2,....$

Let $u(t;0,u_0)$ be the solution of (4.105) with initial condition

$$u(0^+) = u_0,$$

where $u_0 \in \mathbb{R}_+$. As for the general Kolmogorov system (4.98), we assume that the solutions $u(t;0,u_0)$ of (4.105) are non-negative, and if $u(t) > 0$, then $u(t^+) > 0$ for $t \geq 0$.

Obviously there exist two equilibriums: $u_1^* = 0$ and $u_2^* = \frac{a}{b} \in \mathbb{R}_+$ of the model (4.105).

Consider the non-zero equilibrium $u^\varepsilon = u_2^* = \frac{a}{b}$ and define a Lyapunov function

$$V(t,u) = |u - u^\varepsilon|.$$

Let $t \geq 0$ and $t \neq \tau_k(u(t))$, $k = 1,2,....$ Then for the upper right-hand derivative $^cD_+^\alpha V(t,u(t))$ of the function V along the solutions of (4.105), we get

$$^cD_+^\alpha V(t,u(t)) \leq sgn(u(t) - u^\varepsilon) {}_0^cD_t^\alpha(u(t) - u^\varepsilon)$$

$$= sgn(u(t) - u^\varepsilon) {}_0^cD_t^\alpha u(t) = sgn(u(t) - u^\varepsilon)u(t)(a - bu(t)).$$

Since u^ε is an equilibrium of (4.105), then

$$^cD_+^\alpha V(t,u(t))$$

$$\leq -b(u(t) - u^{\varepsilon})^2 - bu^{\varepsilon}|u(t) - u^{\varepsilon}| \leq 0, \tag{4.106}$$

$t \in \mathbb{R}_+, t \neq \tau_k(u(t)), k = 1, 2, ...$

Also, for $t = \tau_k(u(t))$, $k = 1, 2, ...$, we have

$$V(t^+, u(t) + \Delta u(t)) = V(t^+, u(t) + p_k u(t) - p_k u^{\varepsilon})$$

$$= |1 + p_k||u(t) - u^{\varepsilon}| = (1 + p_k)V(t, u(t)) \leq V(t, u(t)). \tag{4.107}$$

Thus:

(i) from (4.106), (4.107) and Theorem 4.16 it follows that the equilibrium $u_2^* = \frac{a}{b}$ is parametrically uniformly stable for all permissible values of the parameters a and b.

(ii) If there exists a constant $L > 0$ such that $u \geq L$ for all $u \in \mathbb{R}_+$, then (4.106) implies

$$^c D_+^{\alpha} V(t, u(t)) \leq -bL|u(t) - u^{\varepsilon}|, \ t \in \mathbb{R}_+, \ t \neq \tau_k(u(t)), \ k = 1, 2, ...$$

and from Theorem 4.17 it follows that the equilibrium $u_2^* = \frac{a}{b}$ is parametrically uniformly asymptotically stable for all permissible values of the parameters a and b.

Example 4.5 Consider the following fractional impulsive Lotka–Volterra model of two interacting species with fixed moments of impulsive perturbations

$$\begin{cases} {}_0^c D_t^{\alpha} x_1(t) = \dfrac{r_1}{K_1} x_1(t) \Big(K_1 - x_1(t) - e_{12}\alpha_{12}x_2(t - \tau_2(t)) \Big), t \neq t_k, \\[2mm] {}_0^c D_t^{\alpha} x_2(t) = \dfrac{r_2}{K_2} x_2(t) \Big(K_2 - x_2(t) - e_{21}\alpha_{21}x_1(t - \tau_1(t)) \Big), t \neq t_k, \\[2mm] x_1(t_k^+) = (d_{1k} + 1)x_1(t_k) - d_{1k}\dfrac{K_1 - K_2 e_{12}\alpha_{12}}{1 - e_{12}e_{21}\alpha_{12}\alpha_{21}}, \ k = 1, 2, ..., \\[2mm] x_2(t_k^+) = (d_{2k} + 1)x_2(t_k) - d_{2k}\dfrac{K_2 - K_1 e_{21}\alpha_{21}}{1 - e_{12}e_{21}\alpha_{12}\alpha_{21}}, \ k = 1, 2, ..., \end{cases} \tag{4.108}$$

where $0 < \alpha < 1$, $x_1(t)$ and $x_2(t)$ are populations of the two species at time t, r_1 and r_2 are intrinsic growth rates, K_1 and K_2 are the carrying capacities of the environment, α_{12} and α_{21} are inter-specific coefficients, and $0 \leq \tau_i(t) \leq \tau_0$, $i = 1, 2, t \geq 0$. All parameters r_1, r_2, K_1, K_2 and α_{12} and α_{21} are positive numbers. The uncertain parameters are e_{12} and e_{21}, which can take values from the interval $[0, 1]$ and represent the interaction strength between the species. The moments t_k are such that $t_k < t_{k+1}, k = 1, 2, ...$ and $\lim_{k \to \infty} t_k = \infty$. The values $x_i(t_k)$ and $x_i(t_k^+)$ are the population numbers of i–th species before and after the impulsive effect at the time t_k, respectively, and $d_{ik} > -1$ for all $i = 1, 2$ and $k = 1, 2, ...$

It is easy to show that for (4.108) there exists an equilibrium x^ε at

$$
\begin{cases}
x_1^\varepsilon = \dfrac{K_1 - K_2 e_{12}\alpha_{12}}{1 - e_{12}e_{21}\alpha_{12}\alpha_{21}}, \\[3mm]
x_2^\varepsilon = \dfrac{K_2 - K_1 e_{21}\alpha_{21}}{1 - e_{12}e_{21}\alpha_{12}\alpha_{21}},
\end{cases}
\tag{4.109}
$$

which is positive for all permissible values of e_{12} and e_{21} whenever the carrying capacity ratio K_1/K_2 satisfies the condition

$$
e_{12}\alpha_{12} < \frac{K_1}{K_2} < \frac{1}{e_{21}\alpha_{21}}.
\tag{4.110}
$$

Theorem 4.18 Assume that:

1. Condition (4.110) hold.

2. $0 < t_1 < t_2 < \dots$ and $\lim\limits_{k\to\infty} t_k = \infty$.

3. For any closed interval contained in $(t_{k-1}, t_k]$, $k = 1, 2, \dots$, there exist positive numbers r_* and r^* such that for $i = 1, 2$,

$$
r_* \le \frac{r_i x_i}{K_i} \le r^*.
\tag{4.111}
$$

4. $-1 < d_{1k} \le 0,\ -1 < d_{2k} \le 0,\ k = 1, 2, \dots$.

5. There exists a constant $c > 0$ such that

$$
r_* > c + (\alpha_{12} + \alpha_{21})r^*.
$$

Then the system (4.108) is parametrically uniformly asymptotically stable for all permissible values of e_{12} and e_{21}.

Proof. Choose

$$
V(t, x_1, x_2) = \left(x_1 - x_1^\varepsilon\right)^2 + \left(x_2 - x_2^\varepsilon\right)^2.
$$

For $t \ge 0$ and $t \ne t_k$, for the upper right-hand derivative ${}^cD_+^\alpha V(t, x_1(t), x_2(t))$ of the function V along the solutions of (4.108) using Lemma 1.8, we have

$$
{}^cD_+^\alpha V(t, x_1(t), x_2(t)) = {}^cD_+^\alpha \left(x_1 - x_1^\varepsilon\right)^2 + {}^cD_+^\alpha \left(x_2 - x_2^\varepsilon\right)^2
$$

$$
\le 2(x_1(t) - x_1^\varepsilon)\, {}^cD_+^\alpha \left(x_1 - x_1^\varepsilon\right) + 2(x_2(t) - x_2^\varepsilon)\, {}^cD_+^\alpha \left(x_2 - x_2^\varepsilon\right)
$$

$$
= 2(x_1(t) - x_1^\varepsilon)\frac{r_1 x_1(t)}{K_1}\left[K_1 - x_1(t) - e_{12}\alpha_{12}x_2(t - \tau_2(t))\right]
$$

$$
+ 2(x_2(t) - x_2^\varepsilon)\frac{r_2 x_2(t)}{K_2}\left[K_2 - x_2(t) - e_{21}\alpha_{21}x_1(t - \tau_1(t))\right].
$$

Since $(x_1^{\varepsilon}, x_2^{\varepsilon})$ is an equilibrium of (4.108), from (4.111) we obtain

$$^{c}D_{+}^{\alpha}V(t, x_1(t), x_2(t))$$

$$\leq -2r_*(x_1(t) - x_1^{\varepsilon})^2 + 2r^*|\alpha_{12}||x_1(t) - x_1^{\varepsilon}||x_2(t - \tau_2(t)) - x_2^{\varepsilon}|$$
$$-2r_*(x_2(t) - x_2^{\varepsilon})^2$$
$$+2r^*|\alpha_{21}||x_1(t - \tau_1(t)) - x_1^{\varepsilon}||x_2(t) - x_2^{\varepsilon}|, \; t \neq t_k, \; k = 1, 2, \dots.$$

Using the inequality $2|a||b| \leq a^2 + b^2$, we get

$$^{c}D_{+}^{\alpha}V(t, x_1(t), x_2(t))$$

$$\leq -2r_*(x_1(t) - x_1^{\varepsilon})^2 + r^*|\alpha_{12}|\left((x_1(t) - x_1^{\varepsilon})^2 + (x_2(t - \tau_2(t)) - x_2^{\varepsilon})^2\right)$$
$$-2r_*(x_2(t) - x_2^{\varepsilon})^2$$
$$+r^*|\alpha_{21}|\left((x_2(t) - x_2^{\varepsilon})^2 + (x_1(t - \tau_1(t)) - x_1^{\varepsilon})^2\right), \; t \neq t_k, \; k = 1, 2, \dots.$$

Then, we have

$$^{c}D_{+}^{\alpha}V(t, x_1(t), x_2(t))$$

$$\leq 2\left[-r_* + (\alpha_{12} + \alpha_{21})r^*\right]\left((x_1(t) - x_1^{\varepsilon})^2 + (x_2(t) - x_2^{\varepsilon})^2\right)$$
$$< -2cV(t, x_1(t), x_2(t)), \; t \neq t_k, \; k = 1, 2, \dots,$$

whenever $V(t + \theta, x_1(t + \theta), x_2(t + \theta))) \leq V(t, x_1(t), x_2(t)), \; \theta \in [-\tau_0, 0]$.
Also,

$$V(t_k^+, x_1(t_k^+), x_2(t_k^+))$$

$$= \left(x_1(t_k^+) - x_1^{\varepsilon}\right)^2 + \left(x_2(t_k^+) - x_2^{\varepsilon}\right)^2$$
$$= (1 + d_{1k})^2\left(x_1(t_k) - x_1^{\varepsilon}\right)^2 + (1 + d_{2k})^2\left(x_2(t_k) - x_2^{\varepsilon}\right)^2$$
$$\leq V(t_k, x_1(t_k), x_2(t_k)), \; k = 1, 2, \dots.$$

Since all conditions of Theorem 4.17 are satisfied, the system (4.108) is parametrically uniformly asymptotically stable for all permissible values of e_{12} and e_{21}. We can therefore conclude that the equilibrium x^{ε} is uniformly asymptotically stable at $e_{12} = e_{21} = 1$, i.e. for $e_{ij} \in [0, 1], \; i, j = 1, 2, \; i \neq j$ it remains stable.

4.3 Fractional Impulsive Models in Economics

In this section we shall discuss two fractional models in economics: a model of price fluctuations in a single commodity market, and a Solow-type model with endogenous delays. Both models are considered under impulsive perturbations at fixed moments of time and nonlinear impulsive operators. Qualitative properties of the solutions will be investigated.

4.3.1 Price Fluctuations Models

Trade cycles, business cycles, and fluctuations in the price and supply of various commodities have attracted the attention of researchers for a long time (see [Belair and Mackey 1989], [Weidenbaum and Vogt 1988]). Deterministic delay differential and integro-differential models for the dynamics of price adjustment in a single commodity market have also received increasing interest. Indeed, it is well known that periodic (almost periodic) fluctuations can be caused by time delayed influences. In the case of commodities such delays naturally arise from the necessary time to construct and eventually the time to transport the product to the market place. Many researchers investigated the effects of such delays on the price dynamics (see [Belair and Mackey 1989], [Liz and Röst 2013], [Mackey 1989], [Moreno 2002], [Rus and Iancu 1993]).

A general time-delay model for the market price $p(t)$ fluctuations in commodity markets may be written as

$$\begin{cases} \dot{p}(t) = f(p(t), p(t - \tau(t)))p(t), t \geq 0, \\ p(t) = \varphi_0(t), t \in [-\tau, 0], \end{cases}$$

where $\tau(t)$ corresponds to the transmission delay and satisfies $0 \leq \tau(t) \leq \tau$ ($\tau = const.$), $f : \mathbb{R}_+ \times \mathbb{R}_+ \to \mathbb{R}_+$ is a continuous function, and the initial function $\varphi_0 : [-\tau, 0]$ is continuous.

It is found in [He and Zheng 2010] that an increase in memory length not only can destabilize the market price, resulting in oscillatory market price characterized by a Hopf bifurcation, but also can stabilize an otherwise unstable market price, leading to stability switching as the memory length increases.

On the other hand, many processes in delayed price fluctuations models are characterized by the fact that at certain moments of time they experience a change of state abruptly. The economic shocks that shifts the price curve are examples of impulsive phenomena that can affect the transient behavior of the price fluctuation. Also, the technology shock (sometimes called the Solow residual or the total factor productivity shock) could reflect changes in prices of non-traded or non-measured inputs to production (e.g. raw materials), changes in the rules that govern the conduct of business, etc. (see [Dejong, Ingram and Whiteman 2000]). Among the most dramatic market shocks to input costs were the oil crises of the 1970s, see [Iacobucci, Trebilcock and Haider 2001]. Oil prices have been fluctuating dramatically again for last few years. There are, also, some perturbations in prices that can be caused by natural disasters and the governments. Different impulsive control approaches for financial models have been proposed in [Basin and Pinsky 1988], [Jeanblanc-Picqué 1993], [Korn 1999], [Oyelami and Ale 2013], [Sun, Qiao and Wu 2005]. Recently, several authors investigated the effects of some impulsive jumps on the price fluctuation (see [Stamov, Alzabut, Atanasov and Stamov 2011], [Stamov and Stamov 2013], [Stamova 2009]).

The magnitude of the financial variables such as foreign exchange rates, gross domestic product, interest rates, production, and stock market prices can have very long memory; the reason for describing financial systems using a fractional nonlinear model is that it simultaneously possesses memory and chaos (see [Chen, Chai and Wu 2011]). In the last decade, some progress in studying fractional financial models has been made (see [Dadras and Momeni 2010], [Danca, Garrappa, Tang and Chen 2013], [Hu and Chen 2013], [Laskin 2000], [Skovranek, Podlubny and Petráš 2012], [Wang, Huang and Shi 2011], [Zeng, Chen and Yang 2013]).

The periodic orbit theory of nonlinear economic dynamics can explain the onset of financial crisis in the international economy and can be used to implement financial stabilization policies (see [Chain 2000]). But, if we consider the effects of some financial, delay and impulsive factors, then the assumption of almost periodicity is more realistic, more important and more general. Though one can deliberately periodically fluctuate economic parameters in a theoretical study, price fluctuations in nature are hardly periodic. That is, almost periodicity is more likely to accurately describe natural price fluctuations in single commodity market.

In this section, we formulate a fractional-order impulsive time-delay model for price fluctuations in commodity markets. The existence of almost periodic solutions which are subject to short-term perturbations during their development will be investigated.

We shall consider the following fractional-order impulsive delay model for price fluctuations in commodity markets

$$\begin{cases} {}_{t_0}^c D_t^\alpha p(t) = F(p(t), p_t)p(t), \, t \neq t_k, \\ \Delta p(t_k) = p(t_k^+) - p(t_k) = I_k(p(t_k)), \, k = \pm 1, \pm 2, ..., \end{cases} \tag{4.112}$$

where $0 < \alpha < 1$, $\{t_k\} \in \mathcal{B}$, $B_v = \{p \in \mathbb{R}_+ : \, p < v\}$, $v > 0$ and $\Lambda \subset B_v$ where $\Lambda \neq \emptyset$, $F : \Lambda \times PC[[-\tau, 0], \Lambda] \to \mathbb{R}$, $I_k : \Lambda \to \mathbb{R}$, $k = \pm 1, \pm 2, ...,$ are functions which characterize the magnitude of the impulse effect at the times t_k, $p(t_k)$ and $p(t_k^+)$ are respectively the price levels before and after the impulse effects at t_k, and for $t \in \mathbb{R}$, $p_t \in PC[[-\tau, 0], \Lambda]$ is defined by $p_t(\theta) = p(t + \theta)$, $-\tau \leq \theta \leq 0$.

Let $t_0 \in \mathbb{R}_+$ and $\varphi_0 \in PC[\mathbb{R}, \Lambda]$. Denote by $p(t) = p(t; t_0, \varphi_0)$, $p \in \Lambda$, the solution of equation (4.112), satisfying the initial conditions

$$\begin{cases} p(t; t_0, \varphi_0) = \varphi_0(t - t_0), \, t_0 - \tau \leq t \leq t_0, \\ p(t_0^+; t_0, \varphi_0) = \varphi_0(0). \end{cases} \tag{4.113}$$

The solutions $p(t) = p(t; t_0, \varphi_0)$ of (4.112) are, in general, piecewise continuous functions with points of discontinuity of the first kind t_k at which they are left-continuous, that is, at the moments t_k, $k = \pm 1, \pm 2, ...,$ the following relations are satisfied:

$$p(t_k^-) = p(t_k) \text{ and } p(t_k^+) = p(t_k) + I_k(p(t_k)).$$

We assume that the functions F and I_k are smooth enough on $\Lambda \times PC[[-\tau,0],\Lambda]$ and Λ, respectively, to guarantee the existence, uniqueness and continuability of a solution $p(t) = p(t;t_0,\varphi_0)$ of the IVP (4.112) (4.113) on $[t_0 - \tau,\infty)$ for any initial data $(t_0,\varphi_0) \in \mathbb{R}_+ \times PC[\mathbb{R},\Lambda]$. We also suppose that the functions $p + I_k(p)$ are invertible in Λ, $k = \pm1,\pm2,....$ Since, we shall study almost periodic properties of the model (3.112), we shall also consider such solutions for which the continuability to the left of t_0 is guaranteed.

We introduce the following conditions:

H4.29 The sequence of functions $\{I_k(p)\}$, $k = \pm1,\pm2,...$ is almost periodic uniformly with respect to $p \in \Lambda$.

H4.30 The function $\varphi_0 \in PC[\mathbb{R},\Lambda]$ is almost periodic.

Let conditions H3.1, H4.29 and H4.30 hold and let $\{s'_m\}$ be an arbitrary sequence of real numbers. Then there exists a subsequence $\{s_n\}$, $s_n = s'_{m_n}$ such that the sequence $\varphi_0(t + s_n)$ converges uniformly to the function $\varphi_0^s(t)$ and the set of sequences $\{t_k - s_n\}$, $k = \pm1,\pm2,...$ is convergent to the sequence t_k^s uniformly with respect to $k = \pm1,\pm2,...$, as $n \to \infty$.

By $\{k_{n_i}\}$ we denote the sequence of integer numbers such that the subsequence $\{t_{k+n_i}\}$ converges uniformly to t_k^s with respect to k as $i \to \infty$. From H4.29 it follows that there exists a subsequence of the sequence $\{k_{n_i}\}$ such that the sequence $\{I_{k+k_{n_i}}(p)\}$ converges uniformly to a limit denoted by $I_k^s(p)$ as $i \to \infty$.

Then for an arbitrary sequence $\{s'_m\}$, the problem (4.112), (4.113) moves to the initial value problem

$$
\begin{cases}
{}_{t_0}^c D_t^\alpha p(t) = F^s(p(t),p_t)p(t), \ t \neq t_k^s, \\
p(t) = \varphi_0^s(t - t_0), \ t \in [t_0 - \tau,t_0], \\
p(t_0^+) = \varphi_0^s(0), \\
\Delta p(t) = I_k^s(p(t)), \ t = t_k^s, \ k = \pm1,\pm2,....
\end{cases}
\tag{4.114}
$$

The last problem we denote by $H(F,\varphi_0,I_k,t_k)$.

First, some stability results for the zero solution of the problem (4.114) will be stated. That is why we introduce the following condition

H4.31 $I_k(0) = 0$, $k = \pm1,\pm2,....$

Note that, if we need to study the stability of a non-zero equilibrium, then the standard approach is first to locate the equilibria, then select one that is of interest, translate it to the origin, and lastly determine its stability properties. The translation of the equilibrium is justied by the fact that a stability analysis can be developed without loss of generality for the equilibrium at the origin and the universally used for other equilibria of the model.

We shall use piecewise continuous Lyapunov functions $W : \mathbb{R} \times \Lambda \to \mathbb{R}_+$ which belong to the class W_0, and satisfy the following condition:

H4.32 $W(t,0) = 0, t \in \mathbb{R}$.

Theorem 4.19 Assume that conditions H3.1, H4.29–H4.31 hold, there exists a function $W \in W_0$ such that H4.32 is true,

$$w_1(|p|) \leq W(t,p) \leq w_2(|p|), \ w_1, w_2 \in K, \ (t,p) \in [t_0, \infty) \times \Lambda, \qquad (4.115)$$

$$W(t^+, \varphi(0) + I_k^s(\varphi)) \leq W(t, \varphi(0)), \ t = t_k^s, \ k = \pm 1, \pm 2, ..., \qquad (4.116)$$

and the inequality

$$^cD_+^\alpha W(t, \varphi(0)) \leq -w_3 W(t, \varphi(0)), t \neq t_k^s, k = \pm 1, \pm 2, ... \qquad (4.117)$$

is valid whenever $W(t + \theta, \varphi(\theta)) \leq W(t, \varphi(0))$ for $-\tau \leq \theta \leq 0, \ t \in [t_0, \infty)$, $w_3 = const > 0, 0 < \alpha < 1, \ \varphi \in PC[[t_0, \infty), \Lambda]$, then the zero solution of (4.114) is uniformly asymptotically stable for $t \geq t_0$.

The proof of Theorem 4.19 is essentially a repetition of the arguments used in the proof of Theorem 2.14, and we omit the details here.

Theorem 4.20 If in Theorem 4.19 condition (4.115) is replaced by the condition

$$|p| \leq W(t,p) \leq \gamma(v)|p|, \ (t,p) \in [t_0, \infty) \times \Lambda, \qquad (4.118)$$

where $\gamma(v) > 0$, then the zero solution of (4.114) is Mittag–Leffler stable for $t \geq t_0$.

Proof. Let $v = const > 0$ and $\Lambda \subset B_v$. Let $\varphi_0 \in PC[\mathbb{R}, \Lambda]$ and $p(t) = p(t; t_0, \varphi_0)$ be the solution of the problem (4.114). Let $W(t,p), \ p \in \Lambda$. From (4.116) and (4.117) it follows by Corollary 1.5, that for $t \geq t_0$ the following inequality is valid

$$W(t, p(t; t_0, \varphi_0)) \leq \sup_{-\tau \leq \theta \leq 0} W(t_0^+, \varphi_0(\theta)) E_\alpha(-w_3(t - t_0)^\alpha).$$

From the above inequality and (4.118), we obtain

$$|p(t; t_0, \varphi_0)| \leq \sup_{-\tau \leq \theta \leq 0} W(t_0^+, \varphi_0(\theta)) E_\alpha(-w_3(t - t_0)^\alpha)$$

$$\leq \gamma(v)\|\varphi_0\|_\tau E_\alpha(-w_3(t - t_0)^\alpha), \ t \geq t_0.$$

Let $m = \gamma(v)\|\varphi_0\|_\tau$. Then we have

$$|p(t; t_0, \varphi_0)| \leq m E_\alpha(-w_3(t - t_0)^\alpha), \ t \geq t_0,$$

where $m \geq 0$ and $m = 0$ holds only if $\varphi_0(\theta) = 0$ for $\theta \in [-\tau, 0]$, which implies that the zero solution of (4.114) is Mittag–Leffler stable for $t \geq t_0$.

Now, we shall prove our results about almost periodic solutions.

Theorem 4.21 Assume that:

1. Conditions H3.1, H4.29–H4.31 are met.

2. There exist functions $V \in V_2$ and $w_1, w_2 \in K$ such that

$$w_1(|p-q|) \leq V(t,p,q) \leq w_2(|p-q|), \ (t,p,q) \in [t_0,\infty) \times \Lambda \times \Lambda,$$
(4.119)

$$V(t^+, \varphi(0) + I_k(\varphi), \phi(0) + I_k(\phi))$$

$$\leq V(t, \varphi(0), \phi(0)), \ t = t_k, \ k = \pm 1, \pm 2, ...,$$
(4.120)

and the inequality

$${}^c D_+^\alpha V(t, \varphi(0), \phi(0)) \leq -w_3 V(t, \varphi(0), \phi(0)), \ t \neq t_k, \ k = \pm 1, \pm 2, ...$$
(4.121)

is valid whenever $V(t+\theta, \varphi(\theta), \phi(\theta)) \leq V(t, \varphi(0), \phi(0)), \ -\tau \leq \theta \leq 0,$ $\varphi, \phi \in PC[[t_0,\infty), \Lambda], \ w_3 = const > 0.$

3. There exists a solution $p(t; t_0, \varphi_0)$ of (4.112) such that

$$p(t; t_0, \varphi_0) < v, \text{ where } t \geq t_0, \ v > 0.$$

Then for the equation (4.112) there exists a unique almost periodic solution $\tilde{p}(t)$ such that:

1. $\tilde{p}(t) \leq v_1, \ 0 < v_1 < v.$

2. $H(\tilde{p}(t), t_k) \subset H(F, \varphi_0, I_k, t_k).$

3. $\tilde{p}(t)$ is uniformly asymptotically stable for $t \geq t_0.$

Proof. Let $\{s_m\}$ be an arbitrary sequence of real numbers such that $s_m \to \infty$ as $m \to \infty$ and $\{s_m\}$ moves the problem (4.112), (4.113) to (4.114).

For any real number ξ, let $m_0 = m_0(\xi)$ be the smallest value of m such that $s_{m_0} + \xi \geq t_0$. Since $p(t; t_0, \varphi_0) \leq v_1$ for all $t \geq t_0$ then $p(t + s_m; t_0, \varphi_0) \leq v_1$ for $t \geq \xi, \ m \geq m_0.$

Let $I \subset (\xi, \infty)$ be compact. Then for any $\varepsilon > 0$, choose an integer $n_0(\varepsilon, \xi) \geq m_0(\xi)$ so large that for $l \geq m \geq n_0(\varepsilon, \xi)$ and $t \in (\xi, \infty)$ it follows

$$w_2(2v_1) E_\alpha(-w_3(\xi + s_m - t_0)^\alpha) < w_1(\varepsilon),$$
(4.122)

where $w_3 = const > 0.$

Consider the function $V(\sigma, p(\sigma), p(\sigma + s_l - s_m)).$

For $\sigma > t_0,$ and $V(\sigma + \theta, p(\sigma + \theta), p(\sigma + \theta + s_l - s_m)) \leq V(\sigma, p(\sigma), p(\sigma + s_l - s_m)), \ \theta \in [-\tau, 0]$ from (4.120), (4.121), (4.122) and Corollary 3.2 it follows

$$V(t + s_m, p(t + s_m), p(t + s_l))$$

$$\leq \sup_{-\tau \leq \theta \leq 0} V(t_0^+, \varphi_0(\theta), \phi_0(\theta + s_l - s_m)) E_\alpha(-w_3(\xi + s_m - t_0)^\alpha) < w_1(\varepsilon).$$

Then, from (4.119) we have

$$|p(t + s_m) - p(t + s_l)| < \varepsilon,$$

for $l \geq m \geq n_0(\varepsilon, \xi)$, $t \in I$.

Consequently, there exists a function $\tilde{p}(t)$ such that $p(t + s_m) - \tilde{p}(t) \to 0$ for $m \to \infty$, which is bounded by v_1.

Since ξ is arbitrary it follows that $\tilde{p}(t)$ is defined uniformly on $t \in I$.

Next we shall show that $\tilde{p}(t)$ is a solution of (4.114).

As $p(t + s_j) \in B_{v_1}$ it follows that there exists $n_1(\varepsilon) > 0$ such that if $l \geq m \geq n_1(\varepsilon)$ then

$$|F(p(t + s_m), p_{t+s_m})p(t + s_m) - F(p(t + s_l), p_{t+s_l})p(t + s_l)| < \varepsilon,$$

and

$$|{}_{t_0}^c D_t^\alpha p(t + s_m) - {}_{t_0}^c D_t^\alpha p(t + s_l)| \leq \varepsilon, \ t + s_m, t + s_l \neq t_k^s$$

which show that $\lim_{m \to \infty} {}_{t_0}^c D_t^\alpha p(t + s_m)$ exists uniformly on all compact subsets of $\mathbb{R}$.

Then $\lim_{m \to \infty} {}_{t_0}^c D_t^\alpha p(t + s_m) = {}_{t_0}^c D_t^\alpha \tilde{p}(t)$ and

$$ {}_{t_0}^c D_t^\alpha \tilde{p}(t) = \lim_{m \to \infty} [F(p(t + s_m), p_{t+s_m})p(t + s_m)$$

$$-F(\tilde{p}(t), \tilde{p}_{t+s_m})\tilde{p}(t) + F(\tilde{p}(t), \tilde{p}_{t+s_m})\tilde{p}(t)]$$

$$= F^s(p(t), p_t)p(t), \ t \neq t_k^s. \tag{4.123}$$

On the other hand for $t + s_m = t_k^s$ it follows

$$\tilde{p}(t_k^{s+}) - \tilde{p}(t_k^s) = \lim_{m \to \infty} (p(t_k^s + s_m + 0) - p(t_k^s + s_m))$$

$$= \lim_{m \to \infty} I_k^s(p(t_k^s + s_m)) = I_k^s(\tilde{p}(t_k^s)). \tag{4.124}$$

From H4.30 we get that for the sequence $\{s_m\}$ there exists a subsequence $\{s_n\}$, $s_n = s_{m_n}$ such that the sequence $\{\varphi_0(t + s_n)\}$ converges uniformly to the function φ_0^s. From (4.123) and (4.124) it follows that $\tilde{p}(t)$ is a solution of (4.114).

We shall prove that $\tilde{p}(t)$ is an almost periodic function.

Let the sequence $\{s_m\}$ moves the problem (4.112), (4.113) to a problem at $H(F, \varphi_0, I_k, t_k)$. For any $\varepsilon > 0$ there exists $m_0(\varepsilon) > 0$ such that, if $l \geq m \geq m_0(\varepsilon)$, then

$$w_2(2v_1)E_\alpha(-w_3(s_m)^\alpha) < \frac{w_1(\varepsilon)}{2},$$

and

$$|F(p(\kappa + s_m), p_{\kappa+s_m})p(\kappa + s_m) - F(p(\kappa + s_l), p_{\kappa+s_l})p(\kappa + s_l)| < \frac{w_1(\varepsilon)w_3}{2L_V},$$

where $p \in PC[[t_0, \infty), \Lambda]$, $w_3 = const > 0$.

Consider the function $V(\tau_\varepsilon + \sigma, \tilde{p}(\sigma), \tilde{p}(\sigma + s_l - s_m))$, where $t \le \sigma \le t + s_m$.
Then

$$^cD_+^\alpha V(\tau_\varepsilon + \sigma, \tilde{p}(\sigma), \tilde{p}(\sigma + s_l - s_m))$$

$$\le -w_3 V(\tau_\varepsilon + \sigma, \tilde{p}(\sigma), \tilde{p}(\sigma + s_l - s_m)) + L_V |F^s(\tilde{p}(\sigma), \tilde{p}_\sigma)\tilde{p}(\sigma)$$

$$-F^s(\tilde{p}(\sigma), \tilde{p}_{\tau_\varepsilon + \sigma})\tilde{p}(\sigma)|$$

$$\le -w_3 V(\tau_\varepsilon + \sigma, \tilde{p}(\sigma), \tilde{p}(\sigma + s_l - s_m)) + \frac{w_1(\varepsilon)w_3}{2}. \qquad (4.125)$$

On the other hand,

$$V(\tau_\varepsilon + t_k^s, \tilde{p}(t_k^s) + I_k^s(\tilde{p}(t_k^s)), \tilde{p}(t_k^s + s_l - s_m) + I_k^s(\tilde{p}(t_k^s + s_l - s_m)))$$

$$\le V(\tau_\varepsilon + t_k^s, \tilde{p}(t_k^s), \tilde{p}(t_k^s + s_l - s_m)). \qquad (4.126)$$

From (4.125), (4.126) and Corollary 3.2 it follows

$$V(\tau_\varepsilon + t + s_m, \tilde{p}(t + s_m), \tilde{p}(t + s_l))$$

$$\le \sup_{-\tau \le \theta \le 0} V(t_0^+ + \tau_\varepsilon, \varphi_0(\theta + s_m), \phi_0(\theta + s_l - s_m))E_\alpha(-w_3(s_m)^\alpha) + \frac{w_1(\varepsilon)}{2} < w_1(\varepsilon).$$

Now from last inequality we get

$$|\tilde{p}(t + s_m) - \tilde{p}(t + s_l)| < \varepsilon, \ l \ge m \ge m_0(\varepsilon). \qquad (4.127)$$

From the definition of the sequence $\{s_m\}$ for $l \ge m \ge m_0(\varepsilon)$ it follows that

$$\rho(t_k + s_m, t_k + s_l) < \varepsilon,$$

where $\rho(\{t_k^{(1)}\}, \{t_k^{(2)}\})$ is the distance in $\mathscr{B}$.

From (4.127) and the last inequality we obtain that the sequence $\tilde{p}(t + s_m)$ converges uniformly to the function $\tilde{p}(t)$.

The assertion 1 and 2 of Theorem 4.21 follow immediately. We shall prove the assertion 3.

Let $\overline{\omega}(t)$ be an arbitrary solution of (4.114). Set

$$u(t) = \overline{\omega}(t) - \tilde{p}(t),$$

$$g^s(t, u(t)) = F^s(u(t) + \tilde{p}(t), u(t) + \tilde{p}_t)(u(t) + \tilde{p}(t))$$

$$-F^s(\tilde{p}(t), \tilde{p}_t)\tilde{p}(t),$$

$$\gamma_k^s(u) = I_k^\alpha(u + \tilde{p}) - I_k^s(\tilde{p}).$$

Now we consider the system

$$\begin{cases} ^c_{t_0}D_t^\alpha u(t) = g^s(t, u(t)), \ t \ne t_k^s, \\ \Delta u(t_k^s) = \gamma_k^s(u(t_k^s)), \ k = \pm 1, \pm 2, ..., \\ u(t_0^+) = u_0, \ t_0 \in \mathbb{R}_+. \end{cases} \qquad (4.128)$$

Let $W(t, u(t)) = V(t, \tilde{p}(t), \tilde{p}(t) + u(t))$.

Then from Theorem 4.19 it follows that the zero solution $u(t) = 0$ of system (4.128) is uniformly asymptotically stable for $t \geq t_0$ and $\tilde{p}(t)$ is uniformly asymptotically stable for $t \geq t_0$. The proof of Theorem 4.21 is complete.

Theorem 4.22 If in Theorem 4.21 condition (4.119) is replaced by the condition

$$|p - q| \leq V(t, p, q) \leq \gamma(v)|p - q|, \ (t, p, q) \in [t_0, \infty) \times \Lambda \times \Lambda,$$

where $\gamma(v) > 0$, then the almost periodic solution $\tilde{p}(t)$ of (4.112) is Mittag–Leffler stable.

The proof of Theorem 4.22 is analogous to that of Theorem 4.21, except Theorem 4.20 is applied for the function $W(t, u(t)) = V(t, \tilde{p}(t), \tilde{p}(t) + u(t))$.

4.3.2   Solow-type Models

In this section, we shall introduce fractional-order Solow-type models. Sufficient conditions for the Mittag–Leffler stability of their states will be derived.

In the basic Solow model, (see [Solow 1956]) it is supposed that the single composite commodity is produced by labor and capital under the standard neoclassical conditions. The Solow's model may be stated by the following ordinary differential equation

$$\dot{k} = sf(k) - nk, \tag{4.129}$$

where k is capital-labor ratio, $k = K/L$, K and L are the stocks of capital and labor, respectively, n is the population growth rate ($n = \dot{L}/L$), s is the (constant) saving rate, and $f(k)$ is the production function in intensive form satisfying the conditions:

$$f'(k) > 0 \text{ for all } k,$$
$$f''(k) < 0 \text{ for all } k,$$
$$f'(k) \to 0 \text{ as } k \to \infty,$$
$$f'(k) \to \infty \text{ as } k \to 0,$$
$$f(0) = 0,$$
$$f(k) \to \infty \text{ as } k \to \infty.$$

In the above equation $\dot{k} = dk/dt$ denotes the derivative with respect to time.

The Solow-type growth models have been studied extensively in the literature. Many remarkable results concerned with qualitative properties of the equilibrium states of such equations have been obtained (see [Accinelli and Brida 2007], [Acemoglu 2009], [Agénor 2004], [Barro and Sala-i-Martin 2004],

[Boucekkine, Licandro and Christopher 1997], [Deardorff 1970], [Dohtani 2010], [Fanti and Manfredi 2003], [Ferrara 2011], [Guerrini 2006], [Matsumoto and Szidarovszky 2013], [Stamova, Emmenegger and Stamov 2010], [Stamova and Stamov 2012], [Stamova and Stamov A 2013]).

Some of the main directions in which the classical growth model (4.129) has been widely and empirically generalized are the following:

Accounting for the Depreciation of Capital. A more realistic approach to the process is to consider the net increase in the stock of physical capital as equal to gross investment less depreciation. If δ denotes the depreciation rate, then we have the following dynamic equation

$$\dot{k} = sf(k) - (\delta + n)k, \qquad (4.130)$$

which is a generalization of the model (4.129).

The model (4.130) has been studied by many authors (see [Accinelli and Brida 2007], [Acemoglu 2009], [Agénor 2004] and the references therein). Guerrini also modified the Solow-type model (4.130) by considering a population growth rate $n(t)$ that is non-constant over time (see [Guerrini 2006]).

Accounting for the Level of Technology. When the technological development is accounted, then the corresponding Solow-type economic equation becomes

$$\dot{k} = sf(k) - (\delta + n_s + g)k, \qquad (4.131)$$

where g denotes the rate of the technological progress. It is often assumed in the theory of economic growth that technological knowledge of society grows exponentially. For some results on such Solow-type models see [Acemoglu 2009] and [Gandolfo 2009].

Accounting for Delay Factors. Motivated by the fact that the current rate of change of the supply of labor is related to past fertility, and thus to past levels of wage, following a prescribed pattern of delay, several authors generalized the classical models and incorporated endogenous delays. For example, Fanti and Manfredi (see [Fanti and Manfredi 2003]) show that persistent oscillations may occur in a Solow model when the rate of change of the labor supply is correctly assumed to depend (even in the simplest manner) on past demographic behaviors. They introduced a distributed delay in the population term in the original Solow growth model, and proposed the following integro-differential equation:

$$\dot{k}(t) = sf(k(t)) - \left[\int_{-\infty}^{t} n\Big(f(k(\tau))\Big) \mathscr{K}(t-\tau)d\tau \right] k(t), \qquad (4.132)$$

where the saving rate s is such that $0 < s < 1$. The term $n\Big(f(k(\tau))\Big)$, $\tau < t$, captures past (rather than current) income-related fertility, and $\mathscr{K}(t-\tau)$ is the corresponding delay kernel.

Accounting for Impulsive Effects. During the last couple of decades, several authors have investigated the effects of some impulsive jumps on the capital-labor ratio. These jumps may be driven by the dynamics of investments, technological changes, populations changes and some others. In a series of papers Stamova and her co-authors proposed some impulsive generalizations of the existing Solow-type models (see [Stamova, Emmenegger and Stamov 2010], [Stamova and Stamov 2012], [Stamova and Stamov A 2013]).

Let $t_0 \in \mathbb{R}_+$, $t_0 < t_1 < t_2 < ... < t_i < ...$ and $\lim_{i \to \infty} t_i = \infty$. An impulsive generalization of the model (4.132) is given by (see [Stamova 2009])

$$
\begin{cases}
\dot{k}(t) = sf(k(t)) - \left[\int_{-\infty}^{t} n\Big(f(k(\tau))\Big) \mathscr{K}(t-\tau)d\tau \right] k(t), \ t \neq t_i, \\
\Delta k(t_i) = k(t_i^+) - k(t_i) = P_i k(t_i), \ i = 1,2,...,
\end{cases}
\tag{4.133}
$$

where $k : [t_0, \infty) \to \mathbb{R}$, $f : \mathbb{R} \to \mathbb{R}$, $\mathscr{K} : \mathbb{R} \to \mathbb{R}_+$ is the delay kernel function, $0 < s < 1$, $n : \mathbb{R} \to \mathbb{R}$, $t_i < t_{i+1} < ...$, $i = 1,2,...$, are the moments of impulsive perturbations, due to which the capital-labor ratio k changes from a position $k(t_i)$ to the position $k(t_i^+)$, P_i are constants which represent the magnitudes of the impulse effects at the moments t_i.

A large number of empirical studies have investigated long-range dependence properties of some financial variables. The concepts of self-similarity, scaling, fractional processes and long-range dependence have been repeatedly used to describe properties of financial time series such as stock prices, foreign exchange rates, market indices, a gross domestic product, interest rates, and production. One way to incorporate long-term memory in the economic and fnancial models is to generalize the concept of differentiating to include fractional values. See for example, [Baleanu, Diethelm, Scalas and Trujillo 2012], [Chen, Chai and Wu 2011], [Cont 2005], [Cunado, Gil-Alana and Pérez de Gracia 2009], [Danca, Garrappa, Tang and Chen 2013], [Laskin 2000].

In this section we shall extend some of the above Solow-type growth models to the fractional-order case. These fractional-order models will propose explanations for some of the features of the long run path of the economies for which the neoclassical Solow's model was unable to account. Indeed, in the context of growth models, long memory arises naturally as the result of the inclusion of cross-sectional heterogeneity in a Solow-type growth model and thus, justifies the use of fractional techniques in analyzing the qualitative properties of the model. The evidence in favor of long memory may also be due to the effect of aggregation (see [Cunado, Gil-Alana and Pérez de Gracia 2009] and the references therein).

First, we shall consider the following fractional-order equation:

$$
\begin{cases}
{}_{t_0}^{c}D_t^\alpha k(t) = sf(k(t)) - \big(\delta + n_s(t) + g(t)\big)k(t), \ t \neq t_i, \\
\Delta k(t_i) = k(t_i^+) - k(t_i) = \Upsilon_i(k(t_i)), \ i = 1,2,...,
\end{cases}
\tag{4.134}
$$

where $0 < \alpha < 1$, $n_s, g : [t_0, \infty) \to \mathbb{R}$, Υ_i are continuous impulsive operators, $\Upsilon_i :$ $\mathbb{R} \to \mathbb{R}$, $i = 1, 2, \ldots$.

In the model (4.134) we consider both, the labor growth rate $n_s(t)$ and the rate of the technological progress $g(t)$ as nonconstant but variable over time.

Let $k_0 \in \mathbb{R}$. Denote by $k(t) = k(t; t_0, k_0)$, $k \in \mathbb{R}$ the solution of equation (4.134), satisfying the initial condition

$$k(t_0^+; t_0, k_0) = k_0. \tag{4.135}$$

The solution $k(t) = k(t; t_0, k_0)$ of the IVP (4.134), (4.135) is a piecewise continuous function with points of discontinuity of the first kind at the moments t_i, $i = 1, 2, \ldots$ at which it is continuous from the left, i.e. the following relations are valid:

$$k(t_i^-) = k(t_i), \ k(t_i^+) = k(t_i) + \Upsilon_i(k(t_i)), \ i = 1, 2, \ldots.$$

Due to the applicable point of view, we are interested in only those solutions of (4.134) corresponding to initial conditions of the form

$$k(t_0^+) = k_0 > 0.$$

We introduce the following conditions:

H4.33 $f(k) > 0$ for $k > 0$, $f(0) = 0$ and the function f is Lipschitz continuous with respect to $k \in \mathbb{R}$ on each of the sets (t_{i-1}, t_i), with a Lipschitz constant $L_f > 0$, i.e. for $k_1, k_2 \in \mathbb{R}$, and for $t \in \mathbb{R}$, $t \neq t_i$, i=1,2,..., it follows

$$|f(k_1(t)) - f(k_2(t))| \leq L_f |k_1(t) - k_2(t)|.$$

H4.34 The functions Υ_i are continuous on $\mathbb{R}$, and $\Upsilon_i(0) = 0$, $i = 1, 2, \ldots$.

H4.35 $t_i < t_{i+1}$, $i = 1, 2, \ldots$ and $t_i \to \infty$ as $i \to \infty$.

Further on we shall use piecewise continuous Lyapunov functions $V : [t_0, \infty) \times \mathbb{R} \to \mathbb{R}_+$ such that $V \in V_0$.

Let $\hat{k}_0 \in \mathbb{R}$. Denote by $\hat{k}(t) = k(t; t_0, \hat{k}_0)$, $k \in \mathbb{R}$ the solution of equation (4.134), satisfying the initial condition

$$\hat{k}(t_0^+; t_0, \hat{k}_0) = \hat{k}_0.$$

Theorem 4.23 Assume that:

1. Conditions H4.33–H4.35 hold.

2. The model parameters are such that

$$sL_f < \delta + n_s(t) + g(t), \ t \geq t_0, \ t \neq t_i, \ i = 1, 2, \ldots.$$

3. The functions Υ_i are such that

$$\Upsilon_i(k(t_i)) = -\sigma_i k(t_i), \quad 0 < \sigma_i < 2,$$

$i = 1, 2, \ldots$.

Then the solution $\hat{k}(t)$ of (4.134) is Mittag–Leffler stable.

Proof. We define a Lyapunov function

$$V(t, k, \hat{k}) = |k(t) - \hat{k}(t)|.$$

Then, for $t \geq t_0$ and $t = t_i$, from condition 3 of the theorem we obtain

$$V(t_i^+, k(t_i^+), \hat{k}(t_i^+)) = |k(t_i^+) - \hat{k}(t_i^+)|$$

$$= |k(t_i) - \hat{k}(t_i) - \sigma_i(k(t_i) - \hat{k}(t_i))| = |1 - \sigma_{ik}||k(t_i) - \hat{k}(t_i)|$$

$$< |k(t_i) - \hat{k}(t_i)| = V(t_i, k(t_i), \hat{k}(t_i)), \quad i = 1, 2, \ldots. \tag{4.136}$$

Let $t \geq t_0$ and $t \in [t_{i-1}, t_i)$. Then for the upper right-hand derivative ${}^cD_+^\alpha V(t, k(t), \hat{k}(t))$ along the solutions of (4.134), we get

$${}^cD_+^\alpha V(t, k(t), \hat{k}(t)) \leq sgn(k(t) - \hat{k}(t)) {}_{t_0}^c D_t^\alpha (k(t) - \hat{k}(t))$$

$$\leq s|f(k(t)) - f(\hat{k}(t))| - (\delta + n_s(t) + g(t)) |k(t) - \hat{k}(t)|.$$

Using H4.33, we get

$${}^cD_+^\alpha V(t, k(t), \hat{k}(t)) \leq \left(sL_f - (\delta + n_s(t) + g(t))\right) |k(t) - \hat{k}(t)|$$

$$= \left(sL_f - (\delta + n_s(t) + g(t))\right) V(t, k(t), \hat{k}(t)), \quad t \geq t_0, \, t \neq t_i, \, i = 1, 2, \ldots.$$

By virtue of condition 2 of Theorem 4.23, there exits a real number $w > 0$ such that

$${}^cD_+^\alpha V(t, k(t), \hat{k}(t)) \leq -wV(t, k(t), \hat{k}(t)), \tag{4.137}$$

$t \neq t_i, t > t_0$.

Then using (4.136), (4.137) and Corollary 1.3, we get

$$V(t, k(t), \hat{k}(t)) \leq V(t_0^+, k(t_0^+), \hat{k}(t_0^+)) E_\alpha(-w(t - t_0)^\alpha), t \in [t_0, \infty).$$

So,

$$|k(t) - \hat{k}(t)| \leq |k_0 - \hat{k}_0| E_\alpha(-w(t - t_0)^\alpha), \quad t \geq t_0.$$

Let $m = |k_0 - \hat{k}_0|$. Then

$$|k(t) - \hat{k}(t)| \leq m E_\alpha(-w(t - t_0)^\alpha), \quad t \geq t_0,$$

where $m \geq 0$ and $m = 0$ holds only if $k_0 = \hat{k}_0$, which implies that the solution $\hat{k}(t)$ of (4.134) is Mittag–Leffler stable. Since the Mittag–Leffler stability of a solution implies its asymptotic stability, the $\hat{k}(t)$ of (4.134) is asymptotically stable.

Theorem 4.23 can be applied to any solution of interest, including equilibrium states for $t \to \infty$. Note that, when $0 < \alpha < 1$, it follows that the Caputo fractional equation (4.134) has the same equilibrium points as the integer-order equation $\dot{k} = sf(k) - (\delta + n_s(t) + g(t))k$, $t \geq t_0$.

Next, we shall consider the impulsive fractional Solow-type model with endogenous delay

$$\begin{cases} {}_{t_0}^{c}D_t^{\alpha}k(t) = sf(k(t)) - \left[\delta + \int_{-\infty}^{t} n_s\Big(f(k(\tau))\Big)\mathscr{K}(t-\tau)d\tau\right]k(t), \, t \neq t_i, \\ \Delta k(t_i) = k(t_i^+) - k(t_i) = \Upsilon_i(k(t_i)), \, i = 1, 2, ..., \end{cases}$$

$$(4.138)$$

where $0 < \alpha < 1$, $t \geq t_0$, $k : [t_0, \infty) \to \mathbb{R}$, $f : \mathbb{R} \to \mathbb{R}$, $\mathscr{K} : \mathbb{R} \to \mathbb{R}_+$ is the delay kernel function, $0 < s < 1$, $n_s : \mathbb{R} \to \mathbb{R}$, δ denotes the depreciation rate, $t_i < t_{i+1} < ...$, $i = 1, 2, ...$, are the moments of impulsive perturbations and Υ_i are continuous impulsive operators, $\Upsilon_i : \mathbb{R} \to \mathbb{R}$.

Remark 4.9 The equation (4.138) is a fractional order generalization of the impulsive Solow-type equation with endogenous delay (4.133) accounting for the depreciation of capital. It can be used in economic studies of business cycles in situations when the capital-labor ratio $k(t)$ is subject to shock effects and we allow for a much richer flexibility in the dynamic behavior by using fractional derivatives.

In the model (4.138) we shall consider the labor growth rate n_s bounded over time.

We introduce the following conditions:

H4.36 The delay kernel $\mathscr{K} : \mathbb{R} \to \mathbb{R}_+$ is continuous, and there exists a positive number μ such that

$$\int_{-\infty}^{t} \mathscr{K}(t-\tau)d\tau \leq \mu < \infty$$

for all $t \in [t_0, \infty)$, $t \neq t_i$, $i = 1, 2, ...$

H4.37 The function n_s is continuous on $\mathbb{R}$, $0 < n_s(f(k)) < \mathscr{N}$ for $k > 0$, $n_s(f(0)) = 0$, and there exists a positive constant v such that

$$|n_s(f(k_1(t))) - n_s(f(k_2(t)))| \leq v|f(k_1(t)) - f(k_2(t))|$$

for all $k_1, k_2 \in \mathbb{R}$ and $t \in [t_0, \infty)$, $t \neq t_i$, $i = 1, 2, ...$

H4.38 There exist positive constants m and $M < \infty$ such that

$$m \le k(t) \le M, \ t \in [t_0, \infty),$$

where $k(t)$ is any solution of (4.138).

We also assume that all the functions $f, n_s, \mathscr{K}, \Upsilon_i, i = 1, 2, \ldots$ and the integral

$$\int_{-\infty}^{t} n_s\big(f(k(\tau))\big)g(t - \tau)d\tau$$

are smooth enough on their domains, so to guarantee the existence and uniqueness of a solution $k(t)$ of the equation (4.138) on $[t_0, \infty)$ for a suitable initial function.

Let $\tilde{k}_{10}, \tilde{k}_{20} \in CB[(-\infty, 0], \mathbb{R}]$, and let $\tilde{k}_1(t) = \tilde{k}_1(t; t_0, \tilde{k}_{10})$, $\tilde{k}_2(t) = \tilde{k}_2(t; t_0, \tilde{k}_{20})$ be two solutions of (4.138) for all $t \ge t_0$ with initial conditions

$$\tilde{k}_1(t; t_0, \tilde{k}_{10}) = \tilde{k}_{10}(t - t_0), t \in (-\infty, t_0]; \ \tilde{k}_1(t_0^+) = \tilde{k}_{10}(0),$$

$$\tilde{k}_2(t; t_0, \tilde{k}_{20}) = \tilde{k}_{20}(t - t_0), t \in (-\infty, t_0]; \ \tilde{k}_2(t_0^+) = \tilde{k}_{20}(0).$$

In the following, we shall suppose that

$$\tilde{k}_1(t) = \tilde{k}_{10}(t - t_0) \ge 0, \ \sup \tilde{k}_{10}(\theta) < \infty, \ \tilde{k}_{10}(0) > 0,$$

$$\tilde{k}_2(t) = \tilde{k}_{20}(t - t_0) \ge 0, \ \sup \tilde{k}_{20}(\theta) < \infty, \ \tilde{k}_{20}(0) > 0.$$

Theorem 4.24 Assume that:

1. Conditions H4.33–H4.38 and condition 3 of Theorem 4.23 hold.

2. There exists a non-negative constant w such that

$$sL_f + Mv\mu L_f + \mathscr{N}\mu \le \delta - w.$$

Then the solution $\tilde{k}_1(t)$ of (4.138) is Mittag–Leffler stable.

Proof. We define a Lyapunov function

$$V(t, \tilde{k}_1, \tilde{k}_2) = |\tilde{k}_1(t) - \tilde{k}_2(t)|.$$

Then, for $t \ge t_0$ and $t = t_i$, from condition 3 of the Theorem 4.23 we obtain

$$V(t_i^+, \tilde{k}_1(t_i^+), \tilde{k}_2(t_i^+)) = |\tilde{k}_1(t_i^+) - \tilde{k}_2(t_i^+)|$$

$$= |\tilde{k}_1(t_i) - \tilde{k}_2(t_i) - \sigma_i(\tilde{k}_1(t_i) - \tilde{k}_2(t_i))| = |1 - \sigma_{ik}||\tilde{k}_1(t_i) - \tilde{k}_2(t_i)|$$

$$< |\tilde{k}_1(t_i) - \tilde{k}_2(t_i)| = V(t_i, \tilde{k}_1(t_i), \tilde{k}_2(t_i)), \ i = 1, 2, \ldots. \tag{4.139}$$

Let $t \geq t_0$ and $t \in [t_{i-1}, t_i)$. Then, using H4.33, for the upper right-hand derivative $^cD_+^\alpha V(t, k(t), \dot{k}(t))$ along the solutions of (4.138), we get

$$^cD_+^\alpha V(t, \tilde{k}_1(t), \tilde{k}_2(t)) \leq sgn(\tilde{k}_1(t) - \tilde{k}_2(t))_{t_0}^c D_t^\alpha (\tilde{k}_1(t) - \tilde{k}_2(t))$$

$$\leq (sL_f - \delta)|\tilde{k}_1(t) - \tilde{k}_2(t)|$$

$$+ \left| \left(\int_{-\infty}^t n_s \Big(f(\tilde{k}_1(\tau)) \Big) \mathscr{K}(t-\tau)d\tau \right) \tilde{k}_1(t) - \left(\int_{-\infty}^t n_s \Big(f(\tilde{k}_2(\tau)) \Big) \mathscr{K}(t-\tau)d\tau \right) \tilde{k}_2(t) \right|$$

$$= (sL_f - \delta)|\tilde{k}_1(t) - \tilde{k}_2(t)|$$

$$+ \left| \left(\int_{-\infty}^t n_s \Big(f(\tilde{k}_1(\tau)) \Big) \mathscr{K}(t-\tau)d\tau \right) \Big(\tilde{k}_1(t) - \tilde{k}_2(t) + \tilde{k}_2(t) \Big) \right.$$

$$\left. - \left(\int_{-\infty}^t n_s \Big(f(\tilde{k}_2(\tau)) \Big) \mathscr{K}(t-\tau)d\tau \right) \tilde{k}_2(t) \right|$$

$$\leq (sL_f - \delta)|\tilde{k}_1(t) - \tilde{k}_2(t)|$$

$$+ \left| \int_{-\infty}^t n_s \Big(f(\tilde{k}_1(\tau)) \Big) \mathscr{K}(t-\tau)d\tau \right| |\tilde{k}_1(t) - \tilde{k}_2(t)|$$

$$+ \left(\int_{-\infty}^t \left| n_s \Big(f(\tilde{k}_1(\tau)) \Big) - n_s \Big(f(\tilde{k}_1(\tau)) \Big) \right| \mathscr{K}(t-\tau)d\tau \right) |\tilde{k}_2(t)|.$$

By H4.33, H4.36, H4.37 and H4.38, it follows that for $t \geq t_0$ and $t \neq t_i$, $i = 1, 2, ...$, we have

$$^cD_+^\alpha V(t, \tilde{k}_1(t), \tilde{k}_2(t)) \leq (sL_f - \delta)|\tilde{k}_1(t) - \tilde{k}_2(t)|$$

$$+ \mathscr{N}\mu|\tilde{k}_1(t) - \tilde{k}_2(t)| + Mv L_f \int_{-\infty}^t |\tilde{k}_1(\tau) - \tilde{k}_2(\tau)| \mathscr{K}(t-\tau)d\tau. \qquad (4.140)$$

Using the Razumikhin condition

$$V(\tau, \tilde{k}_1(\tau), \tilde{k}_2(\tau)) \leq V(t, \tilde{k}_1(t), \tilde{k}_2(t)), \ \tau \in (-\infty, t], \ t \geq t_0,$$

we have

$$|\tilde{k}_1(\tau) - \tilde{k}_2(\tau)| \leq |\tilde{k}_1(t) - \tilde{k}_2(t)|, \ \tau \in (-\infty, t], t \neq t_i, i = 1, 2, ... \qquad (4.141)$$

Then, from (4.140), (4.141) and condition 2 of Theorem 4.24, we obtain

$$^cD_+^\alpha V(t, \tilde{k}_1(t), \tilde{k}_2(t)) \leq wV(t, \tilde{k}_1(t), \tilde{k}_2(t)), \qquad (4.142)$$

for $t \geq t_0$ and $t \neq t_i$, $i = 1, 2, ...$

From (4.139) and (4.142), we obtain

$$V(t,\tilde{k}_1(t),\tilde{k}_2(t)) \leq \sup_{-\infty < \theta \leq 0} V(t_0^+,\tilde{k}_{10}(\theta),\tilde{k}_{20}(\theta))E_\alpha(-w(t-t_0)^\alpha), \quad (4.143)$$

for all $t \geq t_0$.

Then, from (4.143), we deduce the inequality

$$|\tilde{k}_1(t) - \tilde{k}_2(t)| \leq |\tilde{k}_{10} - \tilde{k}_{20}|_\infty E_\alpha(-w(t-t_0)^\alpha), \quad t \geq t_0.$$

This shows that the solution $\tilde{k}_1(t)$ of equation (4.138) is Mittag–Leffler stable.

Finally, we shall consider a fractional-order model with a neoclassical production function having the Cobb–Douglas form $f(k) = k^q$, where $0 < q < 1$ (see [Cobb and Douglas 1928]). When a Cobb–Douglas production function is applied, the model (4.132) becomes

$$\dot{k}(t) = sk^q(t) - \left[\int_{-\infty}^t n\left(k^q(\tau)\right)\mathscr{K}(t-\tau)d\tau\right]k(t). \quad (4.144)$$

The idea of using a Cobb-Douglas production function at the core of a Solow-type growth model is developed in many context in the literature. In this particular case, the researchers of the theory of economic growth are essentially interested in the effects of forces of a "fundamental" nature and they often assumed that the function n is linear and increasing, i.e. it is considered that $n(k^q) = n_s k^q$, where $n_s \geq 0$ is a constant parameter, tuning the reaction of the rate of change of the labor supply to changes in per-capita income. Therefore, the model (4.144) reduces to the model

$$\dot{k}(t) = sk^q(t) - n_s\left[\int_{-\infty}^t k^q(\tau)\mathscr{K}(t-\tau)d\tau\right]k(t). \quad (4.145)$$

Here, we consider the following impulsive fractional-order generalization of model (4.145)

$$\begin{cases} {}_{t_0}^C D_t^\alpha k(t) = sk^q(t) - n_s\left[\int_{-\infty}^t k^q(\tau)\mathscr{K}(t-\tau)d\tau\right]k(t), t \geq t_0, t \neq t_i, \\ \\ \Delta k(t_i) = k(t_i^+) - k(t_i) = Q_i(k(t_i)), i = 1,2,..., \end{cases} \quad (4.146)$$

where $0 < \alpha < 1$, $t_0 \in \mathbb{R}_+$, t_i, $i = 1,2,...$, are the moments of impulsive perturbations and satisfy $t_0 < t_1 < t_2 < ...$ and $\lim_{i\to\infty} t_i = \infty$, $Q_i(k(t_i))$ represents the abrupt change of the state $k(t)$ at the impulsive moment t_i.

We introduce the following condition:

H4.39 There exists a non-zero equilibrium k^* of the equation (4.146) such that

$$s = n_s k^* \int_{-\infty}^t \mathscr{K}(t-\tau)d\tau, \; Q_i(k^*) = 0, \; i = 1,2,....$$

Remark 4.10 The problems of existence and uniqueness of solutions of fractional-order impulsive integro-differential equations have been investigated by several authors. For some efficient sufficient conditions see, for example, [Anguraj and Maheswari 2012], [Gao, Yang and Liu 2013], [Suganya, Mallika Arjunan and Trujillo 2015], [Xie 2014] and the references therein.

Theorem 4.25 Assume that:

1. Conditions H4.35, H4.36, H4.38 and H4.39 hold.
2. There exists a positive continuous function $a(t)$ such that

$$\frac{k^{q-1}(t) - (k^*)^{q-1}}{k(t) - k^*} \leq -a(t)$$

for all $k \in \mathbb{R}$, $k \neq k^*$ and for all $t \in [t_0, \infty)$, $t \neq t_i$, $i = 1, 2, \ldots$.

3. The functions $Q_i : \mathbb{R} \to \mathbb{R}$ are such that

$$Q_i(k(t_i)) = -\sigma_i(k(t_i) - k^*), \quad 0 < \sigma_i < 2, \ i = 1, 2, \ldots.$$

4. There exists a non-negative constant w such that

$$w + q n_s m^{q-1} \mu \leq s a(t), \ t \geq t_0, \ t \neq t_i, \ i = 1, 2, \ldots.$$

Then the equilibrium k^* of (4.1469) is Mittag–Leffler stable.

Proof. Consider the Lyapunov function

$$V(t, k) = |k - k^*|.$$

For $t > t_0$ and $t = t_i$, from condition 3 of Theorem 4.25, we have

$$V(t_i^+, k(t_i^+)) = |k(t_i^+) - k^*| = |k(t_i) - \sigma_i(k(t_i) - k^*) - k^*|$$

$$= |1 - \sigma_i||k(t_i) - k^*| < |k(t_i) - k^*| = V(t_i, k(t_i)). \tag{4.147}$$

Consider the upper right-hand fractional derivative in the Caputo sense

$$^c D_+^\alpha V(t, k(t))$$

of the function $V(t, k(t))$ with respect to (4.146). For $t \geq t_0$ and $t \neq t_i$, $i = 1, 2, \ldots$, we have

$$^c_{t_0} D_t^\alpha V(t, k(t)) = \text{sgn}(k(t) - k^*) \, ^c_{t_0} D_t^\alpha k(t)$$

$$= \text{sgn}(k(t) - k^*) \left[s k^{q-1}(t) - n_s \int_{-\infty}^t \mathscr{K}(t - \tau) k^q(\tau) d\tau \right] k(t).$$

Since k^* is an equilibrium of (4.146), we obtain

$$^cD_+^\alpha V(t,k(t)) \leq k(t)\left[|k(t)-k^*|s\frac{k^{q-1}(t)-(k^*)^{q-1}}{k(t)-k^*}\right.$$

$$\left.+n_s\int_{-\infty}^t \mathcal{K}(t-\tau)|k^q(\tau)-(k^*)^q|d\tau\right], \ t\neq t_i, \ i=1,2,....$$

The function $k^q(t)$ is differentiable on any closed interval contained in $[t_0,t_1]\cup(t_i,t_{i+1})$, $i=1,2,...$, and the inequalities $m\leq k(t)\leq M$ hold for all $t\geq t_0, t\neq t_i, i=1,2,....$

Therefore,

$$|k_1^q(t)-k_2^q(t)|=|q||k^{q-1}(t)||k_1(t)-k_2(t)|\leq qm^{q-1}|k_1(t)-k_2(t)|$$

for $k_1(t)\leq k(t)\leq k_2(t)$, $k_1, k_2\in\mathbb{R}$ and for all $t\geq t_0, t\neq t_i, i=1,2,....$

From the last estimate for $t\neq t_i$, $i=1,2,...$, we obtain

$$^cD_+^\alpha V(t,k(t))$$

$$\leq k(t)\left[-sa(t)|k(t)-k^*|+n_s\int_{-\infty}^t(t-\tau)qm^{q-1}|k(\tau)-k^*|d\tau\right].$$

From condition 4 of Theorem 4.25, for any solution k of (4.146) such that $V(\tau,k(\tau))\leq V(t,k(t))$, $\tau\in(-\infty,t]$, $t\geq t_0$, we have

$$^cD_+^\alpha V(t,k(t))\leq -wk(t)|k(t)-E_1|\leq -wmV(t,k(t)), \tag{4.148}$$

$t\geq t_0$ and $t\neq t_i, i=1,2,....$

Then for $t\geq t_0$, from (4.147) and (4.148), we deduce the inequality

$$|k(t)-k^*|\leq |k_0-k^*|_\infty E_\alpha(-wm(t-t_0)^\alpha),$$

which shows that the equilibrium k^* of equation (4.146) is Mittag–Leffler (and hence, asymptotically) stable.

4.4   Notes and Comments

Theorems 4.1 and 4.2 are adapted from [Stamova 2014b]. Theorems 4.3–4.9 are new. Some interesting results for fractional order neural networks are reported in [Chen, Chai, Wu, Ma and Zhai 2013], [Chen, Qu, Chai, Wu and Qi 2013], [Huang, Zhao, Wang and Li 2012], [Kaslik and Sivasundaram 2012a], [Rakkiyappan, Cao and Velmurugan 2015], [Wang, Yu and Wen 2014], [Wu, Hei and Chen 2013], [Yu, Hu, Jiang and Fan 2014], [Zhang, Yu and Hu 2014], [Zhou, Li and Zhua 2008].

The results in Section 4.2.1 are new. Theorems 4.10, 4.11 and 4.12 for the fractional Lasota-Wazewska model are new. The Lotka–Volterra model (4.83) is an extension of the classical (integer order) Lotka–Volterra systems to the Caputo fractional order case. Theorem 4.13 is new. Theorem 4 .14 furnishes further evidence that the presence of impulsive perturbations as well as uncertain terms in regulatory mechanisms may introduce destabilizing effects. Similar models of integer order are investigated in [Stamov 2012] and [Stamov and Stamov 2013]. The parametric stability notion was introduced by Siljak in collaboration with Ikeda and Ohta in [Ikeda, Ohta and Siljak 1991]. The results on the parametric stability for impulsive Kolmogorov fractional-order models, listed in Section 4.2.4, are new.

The theorems for the price fluctuation model discussed in Section 4.3 are new. The proposed model is an extension of the existing impulsive time-varying models for the dynamics of price adjustment in a single commodity market (see [Stamov 2012], [Stamov and Stamov 2013], [Stamova 2009]). The results for the fractional order Solow-type models are new. For some more results on such models see [Cunado, Gil-Alana and Pérez de Gracia 2009].

References

[Abbas, Banerjee and Momani 2011] Abbas, S., Banerjee, M., and S. Momani. 2011. Dynamical analysis of fractional-order modified logistic model. *Comput. Math. Appl.* 62: 1098–1104.

[Abbas and Benchohra 2010] Abbas, S. and M. Benchohra. 2010. Impulsive partial hyperbolic functional differential equations of fractional order with state-dependent delay. *Fract. Calc. Appl. Anal.* 13: 225–44.

[Abbas, Benchohra and N'Guérékata 2012] Abbas, S., Benchohra, M. and G.M. N'Guérékata. 2012. *Topics in Fractional Differential Equations*. New York: Springer.

[Accinelli and Brida 2007] Accinelli, E. and J.G. Brida. 2007. Population growth and the Solow–Swan model. *Int. J. Ecol. Econ. Statistics* 8: 54–63.

[Acemoglu 2009] Acemoglu, D. 2009. *Introduction to Modern Economic Growth*. Princeton: Princeton University Press.

[Afonso, Bonotto, Federson and Gimenes 2012] Afonso, S.M., Bonotto, E.M., Federson, M. and L.P. Gimenes. 2012. Boundedness of solutions of retarded functional differential equations with variable impulses via generalized ordinary differential equations. *Math. Nachr.* 285: 545–61.

[Agarwal, Cuevas and Soto 2011] Agarwal, R.P., Cuevas, C., and H. Soto. 2011. Pseudo-almost periodic solutions of a class of semilinear fractional differential equations. *J. Appl. Math. Comput.* 37: 625–34.

[Agénor 2004] Agénor, P.-R. 2004. *The Economics of Adjustment and Growth*. Cambridge: Harvard University Press.

[Agrawal, Srivastava and Das 2012] Agrawal, S.K., Srivastava, M., and S. Das. 2012. Synchronization between fractional-order Ravinovich–Fabrikant and Lotka–Volterra systems. *Nonlinear Dynam.* 69: 2277–88.

[Aguila-Camacho, Duarte-Mermoud and Gallegos 2014] Aguila-Camacho, N., Duarte-Mermoud, M.A., and J.A. Gallegos. 2014. Lyapunov functions for fractional order systems. *Comm. Nonlinear Sci. Numer. Simul.* 19: 2951–57.

[Ahmad and Nieto 2011] Ahmad, B. and J.J. Nieto. 2011. Existence of solutions for impulsive anti-periodic boundary value problems of fractional order. *Taiwanese J. Math.* 15: 981–93.

[Ahmad and Sivasundaram 2010] Ahmad, B. and S. Sivasundaram. 2010. Existence of solutions for impulsive integral boundary value problems of fractional order. *Nonlinear Anal. Hybrid Syst.* 4: 134–41.

[Ahmad and Stamov 2009a] Ahmad, S. and G. Tr. Stamov. 2009. Almost periodic solutions of *N*-dimensional impulsive competitive systems. *Nonlinear Anal. Real World Appl.* 10: 1846–53.

[Ahmad and Stamov 2009b] Ahmad, S. and G. Tr. Stamov. 2009. On almost periodic processes in impulsive competitive systems with delay and impulsive perturbations. *Nonlinear Anal. Real World Appl.* 10: 2857–63.

[Ahmad and Stamova 2007a] Ahmad, S. and I.M. Stamova. 2007. Asymptotic stability of an *N*-dimensional impulsive competitive system. *Nonlinear Anal. Real World Appl.* 8: 654–63.

[Ahmad and Stamova 2007b] Ahmad, S. and I.M. Stamova. 2007. Asymptotic stability of competitive systems with delays and impulsive perturbations. *J. Math. Anal. Appl.* 334: 686–700.

[Ahmad and Stamova 2008] Ahmad, S., and I.M. Stamova. 2008. Global exponential stability for impulsive cellular neural networks with time-varying delays. *Nonlinear Anal.* 69: 786–95.

[Ahmad and Stamova 2012] Ahmad, S. and I.M. Stamova. 2012. Stability criteria for impulsive Kolmogorov-type systems of nonautonomous differential equations. *Rend. Istit. Mat. Univ. Trieste* 44: 19–32.

[Ahmad and Stamova 2013] Ahmad, S. and I.M. Stamova (Eds.) 2013. *Lotka-Volterra and Related Systems: Recent Developments in Population Dynamics*. Berlin: Walter de Gruyter.

[Ahmed, El-Sayed and El-Saka 2007] Ahmed, E., El-Sayed, A.M.A., and H.A.A. El-Saka. 2007. Equilibrium points, stability and numerical solutions of fractional-order predator-prey and rabies models. *J. Math. Anal. Appl.* 325: 542–53.

[Anguraj and Maheswari 2012] Anguraj, A. and M.L. Maheswari. 2012. Existence of solutions for fractional impulsive neutral functional infinite delay integrodifferential equations with nonlocal conditions. *J. Nonlinear Sci. Appl.* 5: 271–80.

[Arbib 1987] Arbib, M.A. 1987. *Brains, Machines and Mathematics*. New York: Springer-Verlag.

[Arik and Tavsanoglu 2000] Arik, S. and V. Tavsanoglu. 2000. On the global asymptotic stability of delayed cellular neural networks. *IEEE Trans. Circuits Syst.* I 47: 571–74.

[Babakhani, Baleanu and Khanbabaie 2012] Babakhani, A., Baleanu, D., and R. Khanbabaie. 2012. Hopf bifurcation for a class of fractional differential equations with delay. *Nonlinear Dynam.* 69: 721–29.

[Bacchelli and Vessella 2006] Bacchelli, V. and S. Vessella. 2006. Lipschitz stability for a stationary 2D inverse problem with unknown polygonal boundary. *Inverse Probl.* 22: 1627–58.

[Bagley 2007] Bagley, R. 2007. On the equivalence of the Riemann-Liouville and the Caputo fractional order derivatives in modeling of linear viscoelastic materials. *Fract. Calc. Appl. Anal.* 10: 123–26.

[Bai 2011] Bai, C. 2011. Impulsive periodic boundary value problems for fractional differential equation involving Riemann–Liouville sequential fractional derivative. *J. Math. Anal. Appl.* 384: 211–31.

[Bainov and Dishliev 1997] Bainov, D.D. and A.B. Dishliev. 1997. The phenomenon "beating" of the solutions of impulsive functional differential equations. *Commun. Appl. Anal.* 1: 435–41.

[Bainov, Kostadinov and Myshkis 1988] Bainov, D.D., Kostadinov, S.I., and A.D. Myshkis. 1988. Bounded and periodic solutions of differential equations with impulse effect in a Banach space. *Differential Integral Equations* 1: 223–30.

[Bainov and Simeonov 1989] Bainov, D.D. and P.S. Simeonov. 1989. *Systems with Impulsive Effect: Stability Theory and Applications*. Chichester: Ellis Horwood. [copublished: New York, Wiley (1993)]

[Bainov and Simeonov 1992] Bainov, D.D. and P.S. Simeonov 1992. *Integral Inequalities and Applications*. Dordrecht: Kluwer.

[Baleanu, Diethelm, Scalas and Trujillo 2012] Baleanu, D., Diethelm, K., Scalas, E., and J.J. Trujillo. 2012. *Fractional Calculus: Models and Numerical Methods*. Singapore: World Scientific.

[Baleanu, Sadati, Ghaderi, Ranjbar, Abdeljawad and Jarad 2010] Baleanu, D., Sadati, S.J., Ghaderi, R., Ranjbar, A., Abdeljawad, T., and F. Jarad. 2010. Razumikhin stability theorem for fractional systems with delay. *Abstr. Appl. Anal.* 2010: 1–9. http://dx.doi.org/10.1155/2010/124812

[Ballinger and Liu, 1997] Ballinger, G. and X. Liu. 1997. Permanence of population growth models with impulsive effects. *Math. Comput. Modelling* 26: 59–72.

[Banaś and Zaj̧ac 2011] Banaś, J., and T. Zaj̧ac. 2011. A new approach to the theory of functional integral equations of fractional order. *J. Math. Anal. Appl.* 375: 375-87.

[Barro and Sala-i-Martin 2004] Barro, R.J. and X. Sala-i-Martin. 2004. *Economic Growth*. New York: McGraw-Hill.

[Basin and Pinsky 1988] Basin, M.V. and M.A. Pinsky. 1988. Impulsive control in Kalman-like filtering problems. *J. Appl. Math. Stoch. Anal.* 11: 1–8.

[Belair and Mackey 1989] Belair, J. and M.C. Mackey. 1989. Consumer memory and price fluctuations in commodity markets: An integro-differential model. *J. Dynam. Differential Equations* 1: 299–525.

[Bellassoued and Yamamoto 2007] Bellassoued, M., and M. Yamamoto. 2007. Lipschitz stability in determining density and two Lamé coefficients. *J. Math. Anal. Appl.* 329: 1240–59.

[Benchohra, Henderson and Ntouyas 2006] Benchohra, M., Henderson, J., and S. Ntouyas. 2006. *Impulsive Differential Equations and Inclusions*. New York: Hindawi Publishing Corporation.

[Benchohra, Henderson, Ntouyas and Ouahab 2008]
Benchohra, M., Henderson, J., Ntouyas, S.K., and A. Ouahab. 2008. Existence results for fractional order functional differential equations with infinite delay. *J. Math. Anal. Appl.* 338: 1340–50.

[Benchohra and Slimani 2009] Benchohra, M. and B.A. Slimani. 2009. Existence and uniqueness of solutions to impulsive fractional differential equations. *Electron. J. Dierential Equations* 2009, no. 10: 1–11. http://ejde.math.txstate.edu/Volumes/2009/10/benchohra.pdf

[Bernfeld, Corduneanu and Ignatyev 2003] Bernfeld, S.R., Corduneanu, C., and A.O. Ignatyev. 2003. On the stability of invariant sets of functional differential equations. *Nonlinear Anal.* 55: 641–56.

[Bhalekar, Daftardar-Gejji, Baleanu and Magin 2011] Bhalekar, S., Daftardar-Gejji, V., Baleanu, D., and R. Magin. 2011. Fractional Bloch equation with delay. *Comput. Math. Appl.* 61: 1355–65.

[Bochner 1927] Bochner, S. 1927. Beitrage zur Theorie der fastperiodischen Funktionen, I: Funktionen einer Variaben. *Math. Ann.* 96: 119–47. (in German)

[Bochner 1933] Bochner, S. 1933. Homogeneous systems of differential equations with almost periodic coefficients. *J. London Math. Soc.* 8: 283–88.

[Bochner and Neumann 1935] Bochner, S. and J. von Neumann. 1935. Almost periodic functions of groups. II. *Trans. Amer. Math. Soc.* 37: 21–50.

[Bohr 1925] Bohr, H. 1925. Zur Theorie der Fastperiodischen Funktionen. II: Zusammenhang der fastperiodischen Funktionen mit Funktionen von unendlich vielen Variabeln; gleichmssige Approximation durch trigonometrische Summen. *Acta Math.* 46: 101–14. (in German)

[Bohr 1926] Bohr, H. and O. Neugebauer. 1926. Uber lineare Differential-gleichungen mit konstanten Koeffizienten und fastperiodischer rechter seite. *Nachr. Ges. Wiss. Geottingen. Math.-Phys. Klasse.* 8–22.

[Boucekkine, Licandro and Christopher 1997] Boucekkine, R., Licandro, O., and P. Christopher. 1997. Differential-difference equations in economics: on the numerical solutions of vintage capital growth model. *J. Econ. Dynam. Control* 21: 347–62.

[Burton 1985] Burton, T.A. 1985. *Stability and Periodic Solutions of Ordinary and Functional Differential Equations*. New York: Academic Press.

[Burton 2011] Burton, T.A. 2011. Fractional differential equations and Lyapunov functionals. *Nonlinear Anal.* 74: 5648–62.

[Cao and Bai 2014] Cao, Y. and Bai, C. 2014. Finite-time stability of fractional-order BAM neural networks with distributed delay. *Abstr. Appl. Anal.* 2014: 1–8. http://dx.doi.org/10.1155/2014/634803

[Cao and Chen 2012] Cao, J. and H. Chen. 2012. Impulsive fractional differential equations with nonlinear boundary conditions. *Math. Comput. Modelling* 55: 303–11.

[Cao and Wang 2005] Cao, J. and J. Wang. 2005. Global exponential stability and periodicity of recurrent neural networks with times delays. *IEEE Trans. Circuits Syst. I* 52: 920–31.

[Caponetto, Dongola, Fortuna and Petráš 2010] Caponetto, R., Dongola, G., Fortuna, L., and I. Petráš. 2010. *Fractional Order Systems: Modeling and Control Applications*. River Edge: World Scientific.

[Celentano 2012] Celentano, L. 2012. *New Results on Practical Stability for Linear and Nonlinear Uncertain Systems*. Rome: Nuova Cultura.

[Chain 2000] Chian, A.C.-L. 2000. Nonlinear dynamics and chaos in macroeconomics. *Int. J. Theor. Appl. Finan.* 3: 601.

[Chang and Nieto, 2009] Chang, Y.K. and J.J. Nieto. 2009. Existence of solutions for impulsive neutral integro-differential inclusions with nonlocal initial conditions via fractional operators. *Funct. Anal. Optim.* 30: 227–44.

[Chauhan and Dabas 2011] Chauhan, A. and J. Dabas. 2011. Existence of mild solutions for impulsive fractional-order semilinear evolution equations with nonlocal conditions. *Electron. J. Differential Equations* 2011, no. 107: 1–10. http://ejde.math.unt.edu/Volumes/2011/107/chauhan.pdf

[Chen 2002] Chen, Y. 2002. Global stability of neural networks with distributed delays. *Neural Netw.* 15: 867–71.

[Chen, Cao and Huang 2004] Chen, A., Cao, J., and L. Huang. 2004. Exponential stability of BAM neural networks with transmission delays. *Neurocomputing* 57: 435–54.

[Chen, Chai and Wu 2011] Chen, L., Chai, Y., and R. Wu. 2011. Control and synchronization of fractional-order financial system based on linear control. *Discrete Dyn. Nat. Soc.* 2011: 1–21. http://dx.doi.org/10.1155/2011/958393

[Chen, Chai, Wu, Ma and Zhai 2013] Chen, L., Chai, Y., Wu, R., Ma, T., and H. Zhai. 2013. Dynamic analysis of a class of fractional-order neural networks with delay. *Neurocomputing* 111: 190–94.

[Chen, Chen and Wang 2009] Chen, F., Chen, A., and X. Wang. 2009. On the solutions for impulsive fractional functional differential equations. *Differ. Equ. Dyn. Syst.* 17: 379–91.

[Chen, Qu, Chai, Wu and Qi 2013] Chen, L., Qu, J., Chai, Y., Wu, R., and G. Qi. 2013. Synchronization of a class of fractional-order chaotic neural networks. *Entropy* 15, no. 8 (August): 3265–76. http://www.mdpi.com/1099-4300/15/8/3265

[Choi and Koo 2011] Choi, S.K. and N. Koo. 2011. The monotonic property and stability of solutions of fractional differential equations. *Nonlinear Anal.* 74: 6530–36.

[Chua and Yang 1988a] Chua, L.O. and L. Yang. 1988. Cellular neural networks: theory. *IEEE Trans. Circuits Syst.* 35: 1257–72.

[Chua and Yang 1988b] Chua, L.O. and L. Yang. 1988. Cellular neural networks: applications. *IEEE Trans. Circuits Syst.* 35: 1273–90.

[Çiçek Yaker and Gücen 2014] Çiçek, M., Yaker, C., and M.B. Gücen. 2014. Practical stability in terms of two measures for fractional order systems in Caputo's sense with initial time difference. *J. Franklin Inst.* 351: 732–42.

[Cobb and Douglas 1928] Cobb, C.W. and P.H. Douglas. 1928. A theory of production. *Amer. Econ. Rev.* 18: 139–65.

[Cont 2005] Cont, R. 2005. Long range dependence in financial markets, In *Fractals in Engineering*, ed. J. Levy-Vehel, and E. Lutton, 159–79. London: Springer.

[Corduneanu and Ignatyev 2005] Corduneanu, C. and A.O. Ignatyev. 2005. Stability of invariant sets of functional differential equations with delay. *Nonlinear Funct. Anal. Appl.* 10: 11–24.

[Cunado, Gil-Alana and Pérez de Gracia 2009] Cunado, J., Gil-Alana, L.A., and F. Pérez de Gracia. 2009. AK growth models: new evidence based on fractional integration and breaking trends. *Rech. Econ. Louvain* 75: 131–49. http://www.cairn.info/revue-recherches-economiques-de-louvain-2009-2-page-131.htm

[Cunningham 1954] Cunningham, W.J. 1954. A nonlinear differential-difference equation of growth. *Proc. Natn. Acad. Sci. U.S.A.* 40: 708–13.

[Dadras and Momeni 2010] Dadras, S. and H.R. Momeni. 2010. Control of a fractional-order economical system via sliding mode. *Physica A* 389: 2434–42.

[Dalec'kii and Krein 1974] Dalec'kii, Ju. L., and M.G. Krein. 1974. *Stability of Solutions of Differential Equations in Banach Space*. Providence: American Mathematical Society.

[Danca, Garrappa, Tang and Chen 2013] Danca, M.-F., Garrappa, R., Tang, W.K.S., and G. Chen. 2013. Sustaining stable dynamics of a fractional-order chaotic financial system by parameter switching. *Comput. Math. Appl.* 66: 702–16.

[Dannan and Elaydi 1986] Dannan, F. and S. Elaydi. 1986. Lipschitz stability of nonlinear systems of differential equations. *J. Math. Anal. Appl.* 113: 562–77.

[Das 2011] Das, S. 2011. *Functional Fractional Calculus*. Berlin, Heidelberg: Springer.

[Das and Gupta 2011] Das, S., and P.K. Gupta. 2011. A mathematical model on fractional Lotka–Volterra equations. *J. Theoret. Biol.* 277: 1–6.

[Deardorff 1970] Deardorff, A. 1970. Growth paths in the Solow neoclassical growth model. *Q. J. Econ.* 84: 134–39.

[Debbouche and Baleanu 2011] Debbouche, A. and D. Baleanu. 2011. Controllability of fractional evolution nonlocal impulsive quasilinear delay integro-differential systems. *Comput. Math. Appl.* 62: 1442–50.

[Debbouche and El-Borai 2009] Debbouche, A. and M.M. El-Borai. 2009. Weak almost periodic and optimal mild solutions of fractional evolution equations. *Electron. J. Differential Equations* 2009, no. 46: 1–8. http://ejde.math.txstate.edu/Volumes/2009/46/debbouche.pdf

[Dejong, Ingram and Whiteman 2000] Dejong, D., Ingram, B., and C. Whiteman. 2000. Keynesian impulses versus Solow residuals: identifying sources of business cycle fluctuation. *J. Appl. Econom.* 15: 311–29.

[Diethelm 2010] Diethelm, K. 2010. *The Analysis of Fractional Differential Equations. An Application-oriented Exposition Using Differential Operators of Caputo Type*. Berlin: Springer.

[Ding and Nieto 2013] Ding, H.-S., and J.J. Nieto. 2013. A new approach for positive almost periodic solutions to a class of Nicholson's blowflies model. *J. Comput. Appl. Math.* 253: 249–54.

[Dohtani 2010] Dohtani, A. 2010. Growth-cycle model of Solow–Swan type. *Int. J. Econ. Behav. Organ.* 76: 428–44.

[Dong, Chen and Sun 2006] Dong, L., Chen, L., and L. Sun. 2006. Extinction and permanence of the predator-prey system with stocking of prey and harvesting of predator impulsively. *Math. Methods Appl. Sci.* 29: 415–25.

[D'Onofrio 2002] D'Onofrio, A. 2002. Stability properties of pulse vaccination strategy in SEIR epidemic model. *Math. Biosci.* 179: 57–72.

[Dou, Chen and Li 2004] Dou, J.W., Chen, L.S., and K.T. Li. A monotone-iterative method for finding periodic solutions of an impulsive competition system on tumor-normal cell interaction. *Discrete Contin. Dyn. Syst.* Ser. B 4: 555–62.

[Duarte-Mermoud, Aguila-Camacho, Gallegos and Castro-Linares 2015] Duarte-Mermoud, M.A., Aguila-Camacho, N., Gallegos, J.A., and R. Castro-Linares. 2015. Using general quadratic Lyapunov functions to prove Lyapunov uniform stability for fractional order systems. *Comm. Nonlinear Sci. Numer. Simul.* 22: 650–659.

[El-Borai 2004] El-Borai, M.M. 2014. The fundamental solutions for fractional evolution equations of parabolic type. *J. Appl. Math. Stoch. Anal.* 3: 197–211.

[El-Borai and Debbouche 2009] El-Borai, M.M., and A. Debbouche. 2009. Almost periodic solutions of some nonlinear fractional differential equations. *Int. J. Contemp. Math. Sci.* 4: 1373–1387. http://www.m-hikari.com/ijcms-password2009/25-28-2009/debboucheIJCMS25-28-2009-2.pdf

[El-Saka, Ahmed, Shehata and El-Sayed 2009] El-Saka, H.A., Ahmed, E., Shehata, M.I., and A.M.A. El-Sayed. 2009. On stability, persistence, and Hopf bifurcation in fractional-order dynamical systems, *Nonlinear Dynam.* 56: 121–26.

[El-Sayed, El-Mesiry and El-Saka 2007] El-Sayed, A.M.A., El-Mesiry, A.E.M., and H.A.A. El-Saka. 2007. On the fractional-order logistic equation. *Appl. Math. Lett.* 20: 817–23.

[El-Sayed, Gaaraf and Hamadalla 2010] El-Sayed, A.M.A., Gaaraf, F.M., and E.M.A. Hamadalla. 2010. Stability for a non-local non-autonomous system of fractional order differential equations with delays. *Electron. J. Differential Equations* 2010, no. 31: 1–10. http://ejde.math.txstate.edu/Volumes/2010/31/elsayed.pdf

[El-Sayed, Rida and Arafa 2009] El-Sayed, A.M.A., Rida, S.Z., and A.A.M. Arafa. 2009. Exact solutions of fractional-order biological population model. *Commun. Theor. Phys.* 52: 992–96.

[Fanti and Manfredi 2003] Fanti, L. and P. Manfredi. 2003. The Solow's model with endogenous population: a neoclassical growth cycle model. *J. Econ. Dev.* 28: 103–15.

[Fečkan, Zhou and Wang 2012] Fečkan, M., Zhou, Y. and J.R. Wang. 2012. On the concept and existence of solution for impulsive fractional differential equations. *Commun. Nonlinear Sci. Numer. Simulat.* 17: 3050–60.

[Ferrara 2011] Ferrara, M. 2011. A note on the Solow economic growth model with Richards population growth law. *Appl. Sci.* 13: 36–39.

[Fink 1974] Fink, A.M. 1974. *Almost Periodic Differential Equations*. Berlin: Springer.

[Fink and Seifert 1969] Fink, A.M. and G. Seifert. 1969. Lyapunov functions and almost periodic solutions for almost periodic systems. *J. Differential Equations* 5: 307–13.

[Gandolfo 2009] Gandolfo, G. 2009. *Economic Dynamics*. Berlin: Springer.

[Gao, Yang and Liu 2013] Gao, Z., Yang, L., and G. Liu. 2013. Existence and uniqueness of solutions to impulsive fractional integro-differential equations with nonlocal conditions. *Appl. Math.* 4 (June): 859–63. http://dx.doi.org/10.4236/am.2013.46118

[Gelfand and Shilov 1959] Gelfand, I.M. and G.E. Shilov. 1959. *Generalized Functions*. Moscow: Nauka. (in Russian).

[Georgescu and Zhang 2010] Georgescu, P. and H. Zhang. 2010. An impulsively controlled predator-pest model with disease in the pest. *Nonlinear Anal. Real World Appl.* 11: 270–87.

[Ghorbel and Spong 2000] Ghorbel, F. and M.W. Spong. 2000. Integral manifolds of singularly perturbed systems with application to rigid-link flexible-joint multibody systems. *Int. J. Nonlinear Mech.* 35: 133–55.

[Gopalsamy 1992] Gopalsamy, K. 1992. *Stability and Oscillation in Delay Differential Equations of Population Dynamics*. Dodrecht: Kluwer.

[Gopalsamy and Leung 1997] Gopalsamy, K. and I.K.C. Leung. 1997. Convergence under dynamical thresholds with delays. *IEEE Trans. Neural Netw.* 8: 341–48.

[Guerrini 2006] Guerrini, L. 2006. The Solow–Swan model with a bounded population growth rate. *J. Math. Econ.* 42: 14–21.

[Guo and Jiang 2012] Guo, T.L., and W. Jiang. 2012. Impulsive fractional functional differential equations. *Comput. Math. Appl.* 64: 3414–24.

[Gyori and Ladas 1991] Gyori, I., and G. Ladas. 1991. *Oscillation Theory of Delay Differential Equations. With Applications*. London: Oxford University Press.

[Halanay 1966] Halanay, A. 1966. An invariant surface for some linear singularly perturbed systems with time lag. *J. Differential Equations* 2: 33–46.

[Halanay and Wexler 1971] Halanay, A. and D. Wexler. 1971. *Qualitative Theory of Impulse Systems*. Moscow: Mir. (in Russian)

[Hale 1977] Hale, J.K. 1977. *Theory of Functional Differential Equations*. New York: Springer.

[Haykin 1998] Haykin, S. 1998. *Neural Networks: A Comprehensive Foundation*. Ehglewood Cliffs: Prentice-Hall.

[He, Chen and Li 2010] He, M., Chen, F., and Z. Li. 2010. Almost periodic solution of an impulsive differential equation model of plankton allelopathy. *Nonlinear Anal. Real World Appl.* 11: 2296–2301.

[He and Zheng 2010] He, X.Z., and M. Zheng. 2010. Dynamics of moving average rules in a continuous-time financial market model. *J. Econ. Behav. Organ.* 76: 615–34.

[Henderson and Ouahab 2009] Henderson, J. and A. Ouahab. 2009. Fractional functional differential inclusions with finite delay. *Nonlinear Anal.* 70: 2091–2105.

[Heymans and Podlubny 2005] Heymans, N. and I. Podlubny. 2005. Physical interpretation of initial conditions for fractional differential equations with Riemann–Liouville fractional derivatives. *Rheol. Acta* 37: 1–7.

[Hilfer 2000] Hilfer, R. 2000. *Applications of Fractional Calculus in Physics*. Singapore: World Scientific.

[Hilfer and Seybold 2006] Hilfer, R. and H.J. Seybold. 2006. Computation of the generalized Mittag–Leffler function and its inverse in the complex plane. *Integral Transforms Spec. Funct.* 17: 637–52.

[Hopfield 1984] Hopfield, J.J. 1984. Neurons with graded response have collective computational properties like those of two-stage neurons. *Proc. Natl. Acad. Sci. USA* 81: 3088–92.

[Hu 2013] Hu, H. 2013. Permanence for nonautonomous predator-prey Kolmogorov systems with impulses and its applications. *Appl. Math. Comput.* 223: 54–75.

[Hu and Chen 2013] Hu, Z. and W. Chen. 2013. Modeling of macroeconomics by a novel discrete nonlinear fractional dynamical system. *Discrete Dyn. Nat. Soc.* 2013: 1–9. http://dx.doi.org/10.1155/2013/275134

[Hu, Lu, Zhang and Zhao 2015] Hu, J.B., Lu, G.P., Zhang, S.B., and L.-D. Zhao. 2015. Lyapunov stability theorem about fractional system without and with delay. *Commun. Nonlinear Sci. Numer. Simulat.* 20: 905–13.

[Huang and Cao 2003] Huang, X., and J. Cao. 2003. Almost periodic solutions of shunting inhibitory cellular neural networks with time-varying delays. *Phys. Lett. A* 314: 222–31.

[Huang, Zhao, Wang and Li 2012] Huang, X., Zhao, Z., Wang, Z., and Y. Li. 2012. Chaos and hyperchaos in fractional-order cellular neural networks. *Neurocomputing* 94: 13–21.

[Hui and Chen 2005] Hui, J., and L. Chen. 2005. Periodicity in an impulsive logistic equation with a distributed delay. *IMA J. Appl. Math.* 70: 479–87.

[Hutchinson 1948] Hutchinson, G.F. 1948. Circular causal systems in ecology. *Ann. New York Acad. Sci.* 50: 221–46.

[Iacobucci, Trebilcock and Haider 2001] Iacobucci, E.M., Trebilcock, M.J., and H. Haider. 2001. *Economic Shocks: Defining a Role for Government*. Toronto: C. D. Howe Institute.

[Ikeda, Ohta and Siljak 1991] Ikeda, M., Ohta, Y., and D.D. Siljak. 1991. Parametric stability. In *New Trends in Systems Theory*, ed. G. Conte, A.M. Perdon, and B. Wyman, 7:1–20. Boston: Birkhuser.

[Imanuvilov and Yamamoto 2001] Imanuvilov, O. and M. Yamamoto 2001. Global Lipschitz stability in an inverse hyperbolic problem by interior observations. *Inverse Probl.* 17: 717–28.

[Jeanblanc-Picqué 1993] Jeanblanc-Picqué, M. 1993. Impulse control method and exchange rate. *Math. Finance* 3: 161–77.

[Jiang and Lu 2007] Jiang, G. and Q. Lu. Impulsive state feedback control of a predator-prey model. *J. Comput. Appl. Math.* 200: 193–207.

[Joelianto and Sutarto 2009] Joelianto, E. and H.Y. Sutarto. 2009. Controlled switching dynamical systems using linear impulsive differential equations. In *Intelligent Unmanned Systems: Theory and Applications*, ed. A. Budiyono, B. Riyanto, and E. Joelianto, 192: 227–44. Berlin: Springer.

[Kaslik and Sivasundaram 2012a] Kaslik, E. and S. Sivasundaram. 2012. Nonlinear dynamics and chaos in fractional order neural networks. *Neural Netw.* 32: 245–56.

[Kaslik and Sivasundaram 2012b] Kaslik, E. and S. Sivasundaram. 2012. Non-existence of periodic solutions in fractional-order dynamical systems and a remarkable difference between integer and fractional-order derivatives of periodic functions. *Nonlinear Anal. Real World Appl.* 13: 1489–97.

[Khadra, Liu and Shen 2003] Khadra, A., Liu, X., and X. Shen. 2003. Application of impulsive synchronization to communication security. *IEEE Trans. Circuits Systems I* 50: 341–51.

[Kilbas, Srivastava and Trujillo 2006] Kilbas, A., Srivastava, H., and J. Trujillo. 2006. *Theory and Applications of Fractional Differential Equations.* Amsterdam: Elsevier.

[Kiryakova 1994] Kiryakova, V. 1994. *Generalized Fractional Calculus and Applications.* Pitman Research Notes in Math. no 301, Harlow: Longman.

[Kolmanovskii and Myshkis 1999] Kolmanovskii, V.B. and A.D. Myshkis. 1999. *Introduction to the Theory and Applications of Functional Differential Equations.* Dordrecht : Kluwer Academic Publishers.

[Kolmanovskii and Nosov 1986] Kolmanovskii, V.B. and V.R. Nosov. 1986. *Stability of Functional-Differential Equations.* London: Academic Press.

[Korn 1999] Korn, R. 1999. Some applications of impulse control in mathematical finance. *Math. Oper. Res.* 50: 493–518.

[Kosko 1987] Kosko, B. 1987. Adaptive bidirectional associative memories. *Appl. Opt.* 26: 4947–60.

[Kosko 1988] Kosko, B. 1988. Bi-directional associative memories. *IEEE Trans. Syst. Man Cybern.* 18: 49–60.

[Kosko 1992] Kosko, B. 1992. *Neural Networks and Fuzzy Systems—A Dynamical Systems Approach to Machine Intelligence.* Englewood Cliffs: Prentice-Hall.

[Kosmatov 2013] Kosmatov, N. 2013. Initial value problems of fractional order with fractional impulsive conditions. *Results Math.* 63: 1289–1310.

[Kou, Adimy and Ducrot 2009] Kou, C., Adimy, M., and A. Ducrot. 2009. On the dynamics of an impulsive model of hematopoiesis. *Math. Model. Nat. Phenom.* 4: 89–112.

[Krasovskii 1963] Krasovskii, N.N. 1963. *Stability of Motion.* Stanford: Stanford University Press.

[Kuang 1993] Kuang, Y. 1993. *Delay Differential Equations with Applications in Population Dynamics*. Boston: Academic Press.

[Kulenovic, Ladas and Sficas 1989] Kulenovic, M.R.S., Ladas, G., and Y.G. Sficas. 1989. Global attractivity in population dynamics. *Comput. Math. Appl.* 18: 925–28.

[Kulev and Bainov 1991] Kulev, G.K. and D.D. Bainov. 1991. Lipschitz stability of impulsive systems of differential equations. *Internat. J. Theoret. Phys.* 30: 737–56.

[Lakshmikantham 2008] Lakshmikantham, V. 2008. Theory of fractional functional differential equations. *Nonlinear Anal.* 69: 3337–43.

[Lakshmikantham, Bainov and Simeonov 1989] Lakshmikantham, V., Bainov, D.D., and P.S. Simeonov. 1989. *Theory of Impulsive Differential Equations*. Teaneck: World Scientific.

[Lakshmikantham and Leela 1969] Lakshmikantham, V. and S. Leela. 1969. *Differential and Integral Inequalities: Theory and Applications*. New York: Academic Press.

[Lakshmikantham, Leela and Martynyuk 1989] Lakshmikantham, V., Leela, S., and A.A. Martynyuk. *Stability Analysis of Nonlinear Systems*. New York: Marcel Dekker.

[Lakshmikantham, Leela and Martynyuk 1990] Lakshmikantham, V., Leela, S., and A.A. Martynyuk. 1990. *Practical Stability Analysis of Nonlinear Systems*. Singapore: World Scientific.

[Lakshmikantham, Leela and Sambandham 2008] Lakshmikantham, V., Leela, S., and M. Sambandham. 2008. Lyapunov theory for fractional differential equations. *Commun. Appl. Anal.* 12: 365–76.

[Lakshmikantham, Leela and Vasundhara Devi 2009] Lakshmikantham, V., Leela, S., and J. Vasundhara Devi. 2009. *Theory of Fractional Dynamic Systems*. Cambridge: Cambridge Scientific Publishers.

[Lakshmikantham, Matrosov and Sivasundaram 1991] Lakshmikantham, V., Matrosov, V.M., and S. Sivasundaram. 1991. *Vector Lyapunov Functions and Stability Analysis of Nonlinear Systems*. Dordrecht: Kluwer.

[Lakshmikantham and Vatsala 2008] Lakshmikantham, V., and A.S. Vatsala. 2008. Basic theory of fractional differential equations. *Nonlinear Anal.* 69: 2677–82.

[LaSalle 1960] LaSalle, J.P. 1960. Some extensions of Liapunov's second method. *IRE Trans.* CT-7: 520–27.

[LaSalle and Lefschetz 1961] LaSalle, J.P., and S. Lefschetz. 1961. *Stability by Lyapunov Direct Method and Application*. New York: Academic Press.

[Laskin 2000] Laskin, N. 2000. Fractional market dynamics. *Physica A* 287: 482–92.

[Lazarević and Spasić 2009] Lazarević, M. P., and A. M. Spasić. 2009. Finite-time stability analysis of fractional order time-delay systems: Gronwall's approach. *Math. Comput. Modelling* 49: 475–81.

[Leibniz 1695] Leibniz, G.W. 1695. Letter from Hanover, Germany to G.F.A. L'Hospital, September 30, 1695, in *Mathematische Schriften* 1849, reprinted 1962, Hildesheim, Germany (Olms Verlag) 2: 301–02.

[Levitan and Zhikov 1983] Levitan, B.M. and V.V. Zhikov. 1983. *Almost Periodic Functions and Differential Equations*. Cambridge: Cambridge University Press.

[Li, Chen and Li 2013] Li, X., Chen, F., and X. Li. 2013. Generalized anti-periodic boundary value problems of impulsive fractional dierential equations. *Commun. Nonlinear Sci. Numer. Simul.* 18: 28–41.

[Li, Chen and Podlubny 2010] Li, Y., Chen, Y., and I. Podlubny. 2010. Stability of fractional-order nonlinear dynamic systems: Lyapunov direct method and generalized Mittag-Leffler stability. *Comput. Math. Appl.* 59: 1810–21.

[Li and Fan 2007] Li, W.T., and Y.H. Fan. 2007. Existence and global attractivity of positive periodic solutions for the impulsive delay Nicholson's blowflies model. *J. Comput. Appl. Math.* 201: 55–68.

[Li, Liao, Yang and Huang 2005] Li, C., Liao, X., Yang, X., and T. Huang. 2005. Impulsive stabilization and synchronization of a class of chaotic delay systems. *Chaos* 15: 043103.

[Li, Qian and Chen 2011] Li, C., Qian, D., and Y.Q. Chen. 2011. On Riemann–Liouville and Caputo derivatives. *Discrete Dyn. Nat. Soc.* 2011: 1–15. http://dx.doi.org/10.1155/2011/562494

[Li, Wang, Zhang and Yang 2009] Li, D., Wang, S., Zhang, X., and D. Yang. 2009. Impulsive control of uncertain Lotka–Volterra predator-prey system. *Chaos Solitons Fractals* 41: 1572–77.

[Li and Ye 2013] Li, Y. and Y. Ye. 2013. Multiple positive almost periodic solutions to an impulsive non-autonomous Lotka–Volterra predator-prey system with harvesting terms. *Commun. Nonlinear Sci. Numer. Simulat.* 18: 3190–3201.

[Li and Zhang 2011] Li, C.P., and F.R. Zhang. 2011. A survey on the stability of fractional differential equations. *Eur. Phys. J.* 193: 27–47.

[Liu 1995] Liu, J. 1995. Uniform asymptotic stability via Liapunov–Razumikhin technique. *Proc. Amer. Math. Soc.* 123: 2465–71.

[Liu 2007] Liu, B. 2007. *Uncertainty Theory*. 2nd Edition, Berlin: Springer.

[Liu 2012] Liu, Y. 2012. An analytic method for solving uncertain differential equations. *J. Uncert. Syst.* 6: 244–49.

[Liu 2013] Liu, Y. 2013. Existence of solutions for impulsive differential models on half lines involving Caputo fractional derivatives. *Commun. Nonlinear Sci. Numer. Simul.* 18: 2604–25.

[Liu and Ballinger 2002] Liu, X., and G. Ballinger. 2002. Existence and continuability of solutions for differential equations with delays and state-dependent impulses. *Nonlinear Anal.* 51: 633–47.

[Liu, Huang and Chen 2012] Liu, Y., Huang, Z., and L. Chen. 2012. Almost periodic solution of impulsive Hopfield neural networks with finite distributed delays. *Neural Com. Appl.* 21: 821–31.

[Liu and Jiang 2014] Liu, K.W., and W. Jiang. 2014. Stability of fractional neutral systems. *Adv. Differ. Equ.* 78: 1–9. http://www.advancesindifferenceequations.com/content/pdf/1687-1847-2014-78.pdf

[Liu, Liu and Liao 2004] Liu, B., Liu, X., and X. Liao. 2004. Robust stability of uncertain dynamical systems. *J. Math. Anal. Appl.* 290: 519–33.

[Liu and Rohlf 1998] Liu, X. and K. Rohlf. 1998. Impulsive control of a Lotka–Volterra system. *IMA J. Math. Control Inform.* 15: 269–84.

[Liu and Takeuchi 2007] Liu, X. and Y. Takeuchi. 2007. Periodicity and global dynamics of an impulsive delay Lasota–Wazewska model. *J. Math. Anal. Appl.* 327: 326–41.

[Liu, Teo and Hu 2005] Liu, X., Teo, K.L., and B. Hu. 2005. Exponential stability of impulsive high-order Hopfield-type neural networks with time-varying delays. *IEEE Trans. Neural Netw.* 16: 1329–39.

[Liu and Wang 2007] Liu, X., and Q. Wang. 2007. The method of Lyapunov functionals and exponential stability of impulsive systems with time delay. *Nonlinear Anal.* 66: 1465–84.

[Liu, Yu and Zhu 2008] Liu, H., Yu, J., and G. Zhu. 2008. Global behaviour of an age-infection-structured HIV model with impulsive drug-treatment strategy. *J Theor. Biol.* 253: 749–54.

[Liu and Zhao 2012] Liu, Y. and S. Zhao. 2012. Controllability analysis of linear time-varying systems with multiple time delays and impulsive effects. *Nonlinear Anal. Real World Appl.* 13: 558–68.

[Liz and Röst 2013] Liz, E. and G. Röst. 2013. Global dynamics in a commodity market model. *J. Math. Anal. Appl.* 398: 707–14.

[Long and Xu 2008] Long, S. and D. Xu. 2008. Delay-dependent stability analysis for impulsive neural networks with time varying delays. *Neurocomputing* 71: 1705–13.

[Luo and Shen 2001] Luo, Z. and J. Shen. 2001. Stability and boundedness for impulsive functional differential equations with infinite delays. *Nonlinear Anal.* 46: 475–93.

[Lyapunov 1950] Lyapunov, A.M. 1950. *General Problem on Stability of Motion*. Moscow: Grostechizdat. (in Russian)

[Mackey 1989] Mackey, M. 1989. Commodity price fluctuations: price dependent delays and nonlinearities as explanatory factors. *J. Econ. Theory* 48: 495–509.

[Magin 2006] Magin, R.L. 2006. *Fractional Calculus in Bioengineering*. Redding: Begell House.

[Mahto and Abbas 2013] Mahto, L., and S. Abbas. 2013. Approximate controllability and optimal control of impulsive fractional functional differential equations. *J. Abstr. Differ. Equ. Appl.* 4: 44–59.

[Mahto, Abbas and Favini 2013] Mahto, L., Abbas, S., and A. Favini. 2013. Analysis of Caputo impulsive fractional order differential equations with applications. *Int. J. Differ. Equ.* 2013: 1–25. http://arxiv.org/pdf/1205.3619v1.pdf

[Markoff 1933] Markoff, A. 1933. Stabilitt im Liapounoffschen Sinne und Fastperiodizitt. *Math. Z.* 36: 708–38. (in German)

[Matsumoto and Szidarovszky 2013] Matsumoto, A. and F. Szidarovszky. 2013. Asymptotic behavior of a delay differential neoclassical growth model. *Sustainability* 5: 440–55.

[Miller and Ross 1993] Miller, K.S. and B. Ross. 1993. *An Introduction to the Fractional Calculus and Differential Equations*. New York: John Wiley.

[Mil'man and Myshkis 1960] Mil'man, V.D. and A.D. Myshkis. 1960. On the stability of motion in the presence of impulses. *Siberian Math. J.* 1: 233–37. (in Russian)

[Mitropol'skii, Fodchuk and Klevchuk 1986] Mitropol'skii, Y.A., Fodchuk, V.I., and I.I. Klevchuk. 1986. Integral manifolds, stability and bifurcation of solutions of singularly perturbed functional-differential equations. *Ukrain. Math. J.* 38: 335–40.

[Mophou 2010] Mophou, M.G. 2010. Existence and uniqueness of mild solutions to impulsive fractional differential equations. *Nonlinear Anal.* 72: 1604–15.

[Moreno 2002] Moreno, D. 2002. Prices, delay and the dynamics of trade. *J. Econ. Theory* 104: 304–39.

[Naito 1970] Naito, T. 1970. Integral manifolds for linear functional differential equations on some Banach space. *Funkcial. Ekvac.* 13: 199–212.

[Neugebauer 1957] Neugebauer, O. 1957. *The Exact Sciences in Antiquity*. Providence: Braun University Press.

[Ortigueira and Coito 2010] Ortigueira, M.D., and F.J. Coito. 2010. System initial conditions vs derivative initial conditions. *Comput. Math. Appl.* 59: 1782–89.

[Oyelami and Ale 2013] Oyelami, B.O. and S.O. Ale. 2013. Application of E^p-stability to impulsive financial model. *Int. J. Anal. Appl.* 2: 38–53.

[Pachpatte 2005] Pachpatte, B. 2005. On generalizations of Bihari's inequality. *Soochow J. Math.* 31: 261–71.

[Palmer 1975] Palmer, K.J. 1975. Linearization near an integral manifold. *J. Math. Anal. Appl.* 51: 243–55.

[Pandit and Deo 1982] Pandit, S.G. and S.G. Deo. 1982. *Differential Systems Involving Impulses*. Berlin: Springer-Verlag.

[Papaschinopoulos 1997] Papaschinopoulos, G. 1997. Linearization near the integral manifold for a system of differential equations with piecewise constant argument. *J. Math. Anal. Appl.* 215: 317–33.

[Pazy 1983] Pazy, A. 1983. *Semigroups of Linear Operators and Applications to Partial Differential Equations*. New York: Springer-Verlag.

[Peng and Yao 2011] Peng, J. and K. Yao. 2011. A new option pricing model for stocks in uncertainty markets. *Int. J. Oper. Res.* 8: 18–26.

[Perestyuk and Cherevko 2002] Perestyuk, N.A., and I.M. Cherevko. 2002. Investigation of the integral manifolds of singularly perturbed functional differential equations. *Math. Notes* (Miskolc) 3: 47–58.

[Podlubny 1999] Podlubny, I. 1999. *Fractional Differential Equations*. San Diego: Academic Press.

[Qiao, Liu and Forys 2013] Qiao, M., Liu, A., and U. Forys. 2013. Qualitative analysis of the SICR epidemic model with impulsive vaccinations. *Math. Methods Appl. Sci.* 36: 695–706.

[Qiuxiang and Rong 2006] Qiuxiang, F., and Y. Rong. 2006. On the Lasota–Wazewska model with piecewise constant argument. *Acta Math. Sci.* (English) 26: 371–78.

[Rakkiyappan, Cao and Velmurugan 2015] Rakkiyappan, R., Cao, J., and G. Velmurugan. 2015. Existence and uniform stability analysis of fractional-order complex-valued neural networks with time delays. *IEEE Trans. Neural Netw. Learn. Syst.* 26: 84–97.

[Razumikhin 1988] Razumikhin, B.S. 1988. *Stability of Hereditary Systems*. Moscow: Nauka. (in Russian)

[Rehman and Eloe 2013] Rehman, M. and P. Eloe. 2013. Existence and uniqueness of solutions for impulsive fractional differential equations. *Appl. Math. Comput.* 224: 422–31.

[Rouche, Habets and Laloy 1977] Rouche, H., Habets, P., and M. Laloy. 1977. *Stability Theory by Lyapunov's Direct Method*. New York: Springer-Verlag.

[Rus and Iancu 1993] Rus, A.T. and C. Iancu. 1993. A functional-differential model for price fluctuations in a single commodity market. *Studia Univ. Babe-Bolyai Math.* 2: 9–14.

[Sadati, Baleanu, Ranjbar, Ghaderi and Abdeljawad 2010] Sadati, S.J., Baleanu, D., Ranjbar, A., Ghaderi, R., and T. Abdeljawad. 2010. Mittag–Leffler stability theorem for fractional nonlinear systems with delay. *Abstr. Appl. Anal.* 2010: 1–7. http://dx.doi.org/10.1155/2010/108651

[Sakamoto 1990] Sakamoto, K. 1990. Invariant manifolds in singular perturbation problems for ordinary dierential equations. *Proc. Roy. Soc. Edinburgh* 116 A: 45–78.

[Samoilenko and Perestyuk 1995] Samoilenko, A.M. and N.A. Perestyuk. 1995. *Differential Equations with Impulse Effect*. Singapore: World Scientific.

[Seifert 1966] Seifert, G. 1966. Almost periodic solutions for almost periodic systems of ordinary differential equations. *J. Differential Equations* 2: 305–19.

[Shao 2014] Shao, J. 2014. New integral inequalities with weakly singular kernel for discontinuous functions and their applications to impulsive fractional differential systems. *J. Appl. Math.* 2014: 1–5. http://dx.doi.org/10.1155/2014/252946

[Shen 1999] Shen, J. 1999. Razumikhin techniques in impulsive functional differential equations. *Nonlinear Anal.* 36: 119–30.

[Skovranek, Podlubny and Petráš 2012] Skovranek, T., Podlubny, I., and I. Petráš. 2012. Modeling of the national economics instate-space: a fractional calculus approach. *Econ. Model.* 29: 1322–27.

[Smith and Wahl 2004] Smith, R.J. and L.M. Wahl. 2004. Distinct effects of protease and reverse transcriptase inhibition in an immunological model of HIV-1 infection with impulsive drug effects. *Bull. Math. Biol.* 66: 1259–83.

[Solow 1956] Solow, R. 1956. A contribution to the theory of economic growth. *Q.J. Econ.* 70: 65–94.

[Song and Cao 2007] Song, Q.K. and J.D. Cao. 2007. Global exponential stability of bidirectional associative memory neural networks with distributed delays. *J. Comp. Appl. Math.* 202: 266–79.

[Song, Xin and Huang 2012] Song, X., Xin, X., and W. Huang. 2012. Exponential stability of delayed and impulsive cellular neural networks with partially Lipschitz continuous activation functions. *Neural Netw.* 29–30: 80–90.

[Stamov 1996] Stamov, G.T. 1996. Affinity integral manifolds of linear singularly perturbed systems of impulsive differential equations. *SUT J. Math.* 32: 121–31.

[Stamov 1997] Stamov, G.T. 1997. Integral manifolds of perturbed linear impulsive differential equations. *Appl. Anal.* 64: 39–56.

[Stamov 2004] Stamov, G. Tr. 2004. Impulsive cellular neural networks and almost periodicity. *Proc. Japan Acad. Ser. A Math. Sci.* 80: 198–203.

[Stamov 2008] Stamov, G.T. 2008. Almost periodic models in impulsive ecological systems with variable diffusion. *J. Appl. Math. Comput.* 27: 243–55.

[Stamov 2009a] Stamov, G. Tr. 2009. On the existence of almost periodic solutions for impulsive Lasota–Wazewska model. *Appl. Math. Lett.* 22: 516–20.

[Stamov 2009b] Stamov, G.T. 2009. Lyapunov's functions and existence of integral manifolds for impulsive differential systems with time-varying delay. *Methods Appl. Anal.* 16: 291–98.

[Stamov 2009c] Stamov, G.T. 2009. Uncertain impulsive differential-difference equations and stability of moving invariant manifolds. *J. Math. Sci.* 161: 320–26.

[Stamov 2009d] Stamov, G. Tr. 2009. Almost periodic models of impulsive Hopfield neural networks. *J. Math. Kyoto Univ.* 49: 57–67.

[Stamov 2010a] Stamov, G. Tr. 2010. Almost periodic solutions in impulsive competitive systems with infinite delays. *Publ. Math. Debrecen* 76: 89–100.

[Stamov 2010b] Stamov, G. Tr. 2010. Almost periodicity and Lyapunov's functions for impulsive functional differential equations with infinite delays. *Canad. Math. Bull.* 53: 367–77.

[Stamov 2012] Stamov, G.T. 2012. *Almost Periodic Solutions of Impulsive Differential Equations*. Berlin: Springer.

[Stamov 2015] Stamov, G.T. 2015. Impulsive fractional integrodifferential equations and Lyapunov method for existence of almost periodic solutions. *Math. Probl. Eng.* 2015: 1–9. http://www.hindawi.com/journals/mpe/2015/861039/

[Stamov and Alzabut 2011] Stamov, G. Tr. and J.O. Alzabut. 2011. Almost periodic solutions in the PC-space for uncertain impulsive dynamical systems. *Nonlinear Anal.* 74: 4653–59.

[Stamov, Alzabut, Atanasov and Stamov 2011] Stamov, G. Tr., Alzabut, J.O., Atanasov, P., and A.G. Stamov. 2011. Almost periodic solutions for an impulsive delay model of price fluctuations in commodity markets. *Nonlinear Anal. Real World Appl.* 12: 3170–76.

[Stamov and Stamov 2013] Stamov, G. Tr., and A.G. Stamov. 2013. On almost periodic processes in uncertain impulsive delay models of price fluctuations in commodity markets. *Appl. Math. Comput.* 219: 5376–83.

[Stamov and Stamova 2001] Stamov, G.T. and I.M. Stamova. 2001. Second method of Lyapunov and existence of integral manifolds for impulsive differential-difference equations. *J. Math. Anal. Appl.* 258: 371–79.

[Stamov and Stamova 2007] Stamov, G. Tr., and I.M. Stamova. 2007. Almost periodic solutions for impulsive neural networks with delay. *Appl. Math. Model.* 31: 1263–70.

[Stamov and Stamova 2014a] Stamov, G.T. and I.M. Stamova. 2014. Almost periodic solutions for impulsive fractional differential equations. *Dyn. Syst.* 29: 119–32.

[Stamov and Stamova 2014b] Stamov, G. Tr. and I.M. Stamova. 2014. Integral manifolds of impulsive fractional functional differential systems. *Appl. Math. Lett.* 35: 63–66.

[Stamov and Stamova 2015a] Stamov, G.T. and I.M. Stamova. 2015. Impulsive fractional functional differential systems and Lyapunov method for existence of almost periodic solutions. *Rep. Math. Phys.* 75: 73–84.

[Stamov and Stamova 2015b] Stamov, G.T. and I.M. Stamova. 2015. Second method of Lyapunov and almost periodic solutions for impulsive differential systems of fractional order. *IMA J Appl. Math.* 80: 1619–33.

[Stamova 2007] Stamova, I.M. 2007. Vector Lyapunov functions for practical stability of nonlinear impulsive functional differential equations. *J. Math. Anal. Appl.* 325: 612–23.

[Stamova 2008] Stamova, I.M. 2008. Boundedness of impulsive functional differential equations with variable impulsive perturbations. *Bull. Aust. Math. Soc.* 77: 331–45.

[Stamova 2009] Stamova, I.M. 2009. *Stability Analysis of Impulsive Functional Differential Equations*. Berlin: Walter de Gruyter.

[Stamova 2010] Stamova, I.M. 2010. Impulsive control for stability of n-species Lotka–Volterra cooperation models with finite delays. *Appl. Math. Lett.* 23: 1003–07.

[Stamova 2011a] Stamova, I.M. 2011. Lyapunov–Razumikhin method for impulsive differential equations with "supremum". *IMA J. Appl. Math.* 76: 573–81.

[Stamova 2011b] Stamova, I.M. 2011. Existence and global asymptotic stability of positive periodic solutions of n-species delay impulsive Lotka–Volterra type systems. *J. Biol. Dyn.* 5: 619–35.

[Stamova 2014a] Stamova, I.M. 2014. Global stability of impulsive fractional differential equations. *Appl. Math. Comput.* 237: 605–12.

[Stamova 2014b] Stamova, I.M. 2014. Global Mittag–Leffler stability and synchronization of impulsive fractional-order neural networks with time-varying delays. *Nonlinear Dynam.* 77: 1251–60.

[Stamova 2015] Stamova, I.M. 2015. Mittag–Leffler stability of impulsive differential equations of fractional order. *Quart. Appl Math.* LXXIII: 525–35.

[Stamova 2016] Stamova, I.M. On the Lyapunov theory for functional differential equations of fractional order. *Proc. Amer. Math. Soc.* 144: 1581–93.

[Stamova, Emmenegger and Stamov 2010] Stamova, I.M., Emmenegger, J.F., and A.G. Stamov. 2010. Stability analysis of an impulsive Solow–Swan model with endogenous population. *Int. J. Pure Appl. Math.* 65: 243–55.

[Stamova and Henderson 2016] Stamova, I.M., and J. Henderson. 2016. Practical stability analysis of fractional-order impulsive control systems. *ISA Trans.* (to appear).

[Stamova, Ilarionov and Vaneva 2010] Stamova, I.M., Ilarionov, R., and R. Vaneva. 2010. Impulsive control for a class of neural networks with bounded and unbounded delays. *Appl. Math. Comput.* 216: 285–90.

[Stamova and Stamov 2011] Stamova, I.M., and G. Tr. Stamov. 2011. Impulsive control on global asymptotic stability for a class of impulsive bidirectional associative memory neural networks with distributed delays. *Math. Comput. Modelling* 53: 824–31.

[Stamova and Stamov 2012] Stamova, I.M. and A.G. Stamov. 2012. Impulsive control on the asymptotic stability of the solutions of a Solow model with endogenous labor growth. *J. Franklin Inst.* 349: 2704–16.

[Stamova and Stamov A 2013] Stamova, I.M. and A.G. Stamov. 2013. On the stability of the solutions of an impulsive Solow model with endogenous population. *Econ. Change Restruct.* 46: 203–17.

[Stamova and Stamov G 2013] Stamova, I.M. and G.T. Stamov. 2013. Lipschitz stability criteria for functional differential systems of fractional order. *J. Math. Phys.* 54: 043502 11 pp.

[Stamova and Stamov G 2014a] Stamova, I.M. and G.T. Stamov. 2014. On the stability of sets for delayed Kolmogorov-type systems. *Proc. Amer. Math. Soc.* 142: 591–601.

[Stamova and Stamov G 2014b] Stamova, I.M. and G.T. Stamov. 2014. Stability analysis of impulsive functional systems of fractional order. *Commun. Nonlinear Sci. Numer. Simulat.* 19: 702–09.

[Stamova and Stamov T 2014a] Stamova, I.M. and T. Stamov. 2014. Asymptotic stability of impulsive control neutral-type systems. *Internat. J. Contr.* 87: 25–31.

[Stamova and Stamov T 2014b] Stamova, I.M. and T. Stamov. 2014. Impulsive effects on global stability of models based on impulsive differential equations with "supremum" and variable impulsive perturbations. *Appl. Math. Mech.* (Engl. Ed.) 35: 85–96.

[Stamova, Stamov and Li 2014] Stamova, I.M., Stamov, T., and X. Li. 2014. Global exponential stability of a class of impulsive cellular neural networks with supremums. *Int. J. Adapt. Control* 28: 1227–39.

[Stamova, Stamov and Simeonova 2013] Stamova, I.M., Stamov, T., and N. Simeonova. 2013. Impulsive control on global exponential stability for cellular neural networks with supremums. *J. Vib. Control* 19: 483–90.

[Stamova, Stamov and Simeonova 2014] Stamova, I.M., Stamov, T., and N. Simeonova. 2014. Impulsive effects on the global exponential stability of neural network models with supremums. *Eur. J. Control* 20: 199–206.

[Sternberg 1969] Sternberg, S. 1969. *Celestial Mechanics*. Part I. New York: W. A. Benjamin.

[Strygin and Fridman 1984] Strygin, V.V. and E.M. Fridman. 1984. Asymptotic of integral manifolds of singularly perturbed systems of dierential equations with time lag. *Math. Nachr.* 117: 83–109.

[Su and Feng 2012] Su, Y. and Z. Feng. 2012. Existence theory for an arbitrary order fractional differential equation with deviating argument. *Acta Appl. Math.* 118: 81–105.

[Suganya, Mallika Arjunan and Trujillo 2015] Suganya, S., Mallika Arjunan, M. and J.J. Trujillo. 2015. Existence results for an impulsive fractional integro-differential equation with state-dependent delay. *Appl. Math. Comput.* 266: 54–69.

[Sun and Chen 2007] Sun, S.T. and L.S. Chen. 2007. Dynamic behaviors of Monod type chemostat model with impulsive perturbation on the nutrient concentration. *J. Math. Chem.* 42: 837–48.

[Sun, Qiao and Wu 2005] Sun, J., Qiao, F., and Q. Wu. 2005. Impulsive control of a financial model. *Physics Lett. A* 335: 282–88.

[Takeuchi 1996] Takeuchi, Y. 1966. *Global Dynamical Properties of Lotka-Volterra Systems*. Singapore: World Scientific.

[Tariboon, Ntouyas and Agarwal 2015] Tariboon, J., Ntouyas, S.K., and P. Agarwal. 2015. New concepts of fractional quantum calculus and applications to impulsive fractional q-difference equations. *Adv. Difference Equ.* 18: 1–19.
http://www.advancesindifferenceequations.com/content/pdf/s13662-014-0348-8.pdf

[Teng, Nie and Fang 2011] Teng, Z., Nie, L., and X. Fang. 2011. The periodic solutions for general periodic impulsive population systems of functional differential equations and its applications. *Comput. Math. Appl.* 61: 2690–2703.

[Vasundhara Devi, Mc Rae and Drici 2012] Vasundhara Devi, J., Mc Rae, F.A., and Z. Drici. 2012. Variational Lyapunov method for fractional differential equations. *Comput. Math. Appl.* 64: 2982–89.

[Wang 2012] Wang, H. 2012. Existence results for fractional functional differential equations with impulses. *J. Appl. Math. Comput.* 38: 85–101.

[Wang, Ahmad and Zhang 2012] Wang, G., Ahmad, B., and L. Zhang. 2012. On impulsive boundary value problems of fractional differential equations with irregular boundary conditions. *Abstr. Appl. Anal.* 2012: 1–18.
http://dx.doi.org/10.1155/2012/356132

[Wang, Chen and Nieto 2010] Wang, L., Chen, L., and J.J. Nieto. 2010. The dynamics of an epidemic model for pest control with impulsive effect. *Nonlinear Anal. Real World Appl.* 11: 1374–86.

[Wang, Ding and Qi 2015] Wang, Q., Ding, D.-S., and D.-L. Qi. 2015. Mittag–Leffler synchronization of fractional-order uncertain chaotic systems. *Chinese Phys. B* 24: 060508.

[Wang, Fečkan and Zhou 2011] Wang, J.R., Fečkan, M., and Y. Zhou. 2011. On the new concept of solutions and existence results for impulsive fractional evolution equations. *Dyn. Partial Differ. Equ.* 8: 345–61.

[Wang and Fu 2007] Wang, L. and X. Fu. 2007. A new comparison principle for impulsive differential systems with variable impulsive perturbations and stability theory. *Comput. Math. Appl.* 54: 730–36.

[Wang, Huang and Shi 2011] Wang, Z., Huang, X., and G. Shi. 2011. Analysis of nonlinear dynamics and chaos in a fractional order financial system with time delay. *Comput. Math. Appl.* 62: 1531–39.

[Wang and Lin 2014] Wang, J.R. and Z. Lin. On the impulsive fractional anti-periodic BVP modelling with constant coefcients. *J. Appl. Math. Comput.* 46: 107–21.

[Wang and Liu 2007a] Wang, Q. and X. Liu. 2007. Exponential stability of impulsive cellular neural networks with time delay via Lyapunov functionals. *Appl. Math. Comput.* 194: 186–98.

[Wang and Liu 2007b] Wang, Q. and X. Liu. 2007. Impulsive stabilization of delay differential systems via the Lyapunov-Razumikhin method. *Appl. Math. Lett.* 20: 839–45.

[Wang, Yang, Ma and Sun 2014] Wang, Z., Yang, D., Ma, T., and N. Sun. 2014. Stability analysis for nonlinear fractional-order systems based on comparison principle. *Nonlinear Dynam.* 75: 387–402.

[Wang, Yu and Niu 2012] Wang, L., Yu, M., and P. Niu. 2012. Periodic solution and almost periodic solution of impulsive Lasota–Wazewska model with multiple time-varying delays. *Comput. Math. Appl.* 64: 2383–94.

[Wang, Yu and Wen 2014] Wang, H., Yu, Y., and G. Wen. 2014. Stability analysis of fractional-order Hopfield neural networks with time delays. *Neural Netw.* 55: 98–109.

[Wang and Zhou 2011] Wang, J.R. and Y. Zhou. 2011. Analysis of nonlinear fractional control systems in Banach spaces. *Nonlinear Anal. Real World Appl.* 74: 5929–42.

[Wang, Zhou and Fečkan 2012] Wang, J. R., Zhou, Y., and M. Fečkan. 2012. On recent developments in the theory of boundary value problems for impulsive fractional differential equations. *Comput. Math. Appl.* 64: 3008–20.

[Wazewska–Czyzewska and Lasota 1976] Wazewska–Czyzewska, M., and A. Lasota. 1976. Mathematical problems of the dynamics of a system of red blood cells. *Mat. Stos.* 6: 23–40.

[Weidenbaum and Vogt 1988] Weidenbaum, M.L. and S.C. Vogt. 1988. Are economic forecasts any good? *Math. Comput. Modelling* 11: 1–5.

[Widjaja and Bottema 2005] Widjaja, J. and M.J. Bottema. 2005. Existence of solutions of diffusive logistic equations with impulses and time delay and stability of the steady-states. *Dyn. Contin. Discrete Impuls. Syst.* Ser. A 12: 563–78.

[Wu, Hei and Chen 2013] Wu, R., Hei, X., and L. Chen. 2013. Finite-time stability of fractional-order neural networks with delay. *Commun. Theor. Phys.* 60: 189–93.

[Xia 2007] Xia, Y. 2007. Positive periodic solutions for a neutral impulsive delayed Lotka-Volterra competition system with the effect of toxic substance. *Nonlinear Anal. Real World Appl.* 8: 204–21.

[Xie 2014] Xie, S. 2014. Existence results of mild solutions for impulsive fractional integro-differential evolution equations with infinite delay. *Fract. Calc. Appl. Anal.* 17: 1158–74.

[Xue, Wang and Jin 2007] Xue, Y., Wang, J., and Z. Jin. 2007. The persistent threshold of single population under pulse input of environmental toxin. *WSEAS Trans. Math.* 6: 22–29.

[Yakar, Çiçek and Gücen 2011] Yakar, C., Çiçek M., and M.B. Gücen. 2011. Boundedness and Lagrange stability of fractional order perturbed system related to unperturbed systems with initial time difference in Caputo's sense. *Adv. Difference Equ.* 54: 1–14. http://www.advancesindifferenceequations.com/content/pdf/1687-1847-2011-54.pdf

[Yan 2003] Yan, J. 2003. Existence and global attractivity of positive periodic solution for an impulsive Lasota–Wazewska model. *J. Math. Anal. Appl.* 279: 111–20.

[Yan 2010] Yan, P. 2010. Impulsive SUI epidemic model for HIV/AIDS with chronological age and infection age. *J. Theor. Biol.* 265: 177–84.

[Yan and Shen 1999] Yan, J., and J. Shen. 1999. Impulsive stabilization of impulsive functional differential equations by Lyapunov–Razumikhin functions. *Nonlinear Anal.* 37: 245–55.

[Yang 2001] Yang, T. 2001. *Impulsive Control Theory*. Berlin: Springer.

[Yang, Song, Liu and Zhao 2014] Yang, X., Song, Q., Liu, Y., and Z. Zhao. 2014. Uniform stability analysis of fractional-order BAM neural networks with delays in the leakage terms. *Abstr. Appl. Anal.* 2014: 1–16. http://dx.doi.org/10.1155/2014/261930

[Ye, Gao and Ding 2007] Ye, H., Gao, J., and Y. Ding. 2007. A generalized Gronwall inequality and its application to a fractional differential equation. *J. Math. Anal. Appl.* 328: 1075–81.

[Yoshizawa 1966] Yoshizawa, T. 1966. *Stability Theory by Lyapunov's Second Method*. Tokyo: The Mathematical Society of Japan.

[Yu, Hu, Jiang and Fan 2014] Yu, J., Hu, C., Jiang, H., and X. Fan. 2014. Projective synchronization for fractional neural networks. *Neural Netw.* 49: 87–95.

[Yukunthorn, Ntouyas and Tariboon 2015] Yukunthorn, W., Ntouyas, S., and J. Tariboon. 2015. Impulsive multiorders Riemann–Liouville fractional differential equations. *Discrete Dyn. Nat. Soc.* 2015: 1–9. http://dx.doi.org/10.1155/2015/603893

[Zecevic and Siljak 2010] Zecevic, A.I. and D.D. Siljak. 2010. *Control of Complex Systems: Structural Constraints and Uncertainty.* New York: Springer.

[Zeng, Chen and Yang 2013] Zeng, C., Chen, Y.-Q., and Q. Yang. 2013. Almost sure and moment stability properties of fractional order Black–Scholes model. *Fract. Calc. Appl. Anal.* 16: 317–31.

[Zhang 2003] Zhang, J. 2003. Globally exponential stability of neural networks with variable delays. *IEEE Trans. Circuits Syst.* I 50: 288–91.

[Zhang, Suda and Iwasa 2004] Zhang, J., Suda, Y., and T. Iwasa. 2004. Absolutely exponential stability of a class of neural networks with unbounded delay. *Neural Netw.* 17: 391–97.

[Zhang and Teng 2011] Zhang, L. and T. Teng. 2011. N-species non-autonomous Lotka–Volterra competitive systems with delays and impulsive perturbations. *Nonlinear Anal. Real World Appl.* 12: 3152–69.

[Zhang, Yu and Hu 2014] Zhang, S., Yu, Y., and W. Hu. 2014. Robust stability analysis of fractional-order Hopfield neural networks with parameter uncertainties. *Math. Probl. Eng.* 2014: 1–14. http://dx.doi.org/10.1155/2014/302702

[Zhang, Zhang and Zhang 2014] Zhang, X., Zhang, X., and M. Zhang. 2014. On the concept of general solution for impulsive differential equations of fractional order $q \in (0, 1)$. *Appl. Math. Comput.* 247: 72–89.

[Zhao 2002] Zhao, H. 2002. Global stability of bidirectional associative memory neural networks with distributed delays. *Phys. Lett. A* 30: 519–46.

[Zhao, Xia and Ding 2008] Zhao, Y., Xia, Y., and W. Ding. 2008. Periodic oscillation for BAM neural networks with impulses. *J. Appl. Math. Comput.* 28: 405–23.

[Zhong, Lin and Jiong 2001] Zhong, W., Lin, W., and R. Jiong. 2001. The stability in neural networks with delay and dynamical threshold effects. *Ann. Differential Equations* 17: 93–101.

[Zhou, Li and Zhua 2008] Zhou, S., Li, H., and Z. Zhua. 2008. Chaos control and synchronization in a fractional neuron network system. *Chaos Solitons Fractals* 36: 973–84.

[Zhou, Liu, Zhang and Jiang 2013] Zhou, X.F., Liu, S., Zhang, Z., and W. Jiang. 2013. Monotonicity, concavity, and convexity of fractional derivative of functions. *Sci. World J.* 2013: 1–6. http://dx.doi.org/10.1155/2013/605412

[Zhou and Wan 2009] Zhou, Q. and L. Wan. 2009. Impulsive effects on stability of Cohen–Grossberg-type bidirectional associative memory neural networks with delays. *Nonlinear Anal. Real World Appl.* 10: 2531–40.

[Zhou, Wang and Zhou 2013] Zhou, H., Wang, J., and Z. Zhou. 2013. Positive almost periodic solution for impulsive Nicholson's blowflies model with multiple linear harvesting terms. *Math. Methods Appl. Sci.* 36: 456–61.

[Zhou, Zhou and Wang 2011] Zhou, H., Zhou, Z., and Q. Wang. 2011. Positive almost periodic solution for a class of Lasota–Wazewska model with infinite delays. *Appl. Math. Comput.* 218: 4501–06.

Index